# A HISTORY OF EMBRYOLOGY

THE EIGHTH SYMPOSIUM OF
THE BRITISH SOCIETY
FOR DEVELOPMENTAL BIOLOGY

# A history of embryology

*EDITED BY*

## T.J.HORDER

*Department of Human Anatomy, University of Oxford*

## J.A.WITKOWSKI

*Royal Postgraduate Medical School, London*

## C.C.WYLIE

*St George's Hospital Medical School, London*

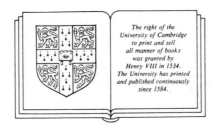

The right of the
University of Cambridge
to print and sell
all manner of books
was granted by
Henry VIII in 1534.
The University has printed
and published continuously
since 1584.

# CAMBRIDGE UNIVERSITY PRESS

CAMBRIDGE

LONDON    NEW YORK    NEW ROCHELLE

MELBOURNE    SYDNEY

Published by the Press Syndicate of the University of Cambridge
The Pitt Building, Trumpington Street, Cambridge CB2 1RP
32 East 57th Street, New York, NY 10022, USA
10 Stamford Road, Oakleigh, Melbourne 3166, Australia

First published 1986

Printed in Great Britain at The Bath Press, Avon

*British Library cataloguing in publication data*

British Society for Developmental Biology.
  *Symposium;* 8th
  A history of embryology: the Eighth Symposium
  of the British Society for Developmental Biology.
  1. Embryology – History
  I. Title   II. Horder, T.J.   III. Witkowski, J.A.
  IV. Wylie, C.C.
  591.3'3   QL953

*Library of Congress cataloguing in publication data*

British Society for Developmental Biology. Symposium
  (8th : 1983 : Nottingham, Nottinghamshire)
  A history of embryology.

  Includes index.
  1. Embryology – History – Congresses. I. Horder,
  T. J. II. Witkowski, J. (Jan), 1947–
  III. Wylie, C.C. (Christopher Craig), 1945–
  IV. Title. [DNLM: 1. Embryology – history – congresses.
  W3 BR459B 8th 1983h/QS 611.1 B862 1983h]
  QL953.B75   1983     574.3'3     85-16583

ISBN 0 521 25953 3

# Contents

Preface  *J. Needham*, F.R.S.                                                    vii

Editor's introduction                                                           ix

Chronological outline                                                           xviii

PART I                                                                          I

*An outline of concepts of generation and heredity up to the end of the nineteenth century*                                                                3

Weismann, Hydromedusae, and the biogenetic imperative: a reconsideration
*Frederick B. Churchill*                                                         7

Embryology and classical zoology in Great Britain
*M. Ridley*                                                                     35

*Was there a characteristic tradition in Britain determining the response to embryological issues?*                                                           69

Preformation or new formation – or neither or both
*Jane Maienschein*                                                             73

*Origins of the embryological tradition in the United States*                   109

T. H. Morgan and the split between embryology and genetics, 1910–35
*Garland E. Allen*                                                             113

*Experimental technique in the rise of American embryology*                     147

Ross Harrison and the experimental analysis of nerve growth: the origins of tissue culture
*J. A. Witkowski*                                                             149

*The emergence of experimental embryology in Germany*                           179

Hans Spemann and the organiser
*T. J. Horder and P. J. Weindling*                                             183

PART 2                                                    243
Early interactions between embryology and biochemistry
*J. Brachet*                                              245

Primary embryonic induction in retrospect
*Lauri Saxén and S. Toivonen*                            261

Structural and dynamical explanations in the world of neglected
   dimensions
*Robert Olby*                                            275

Form and strategy in biology: reflections on the career of
   C. H. Waddington
*Edward Yoxen*                                           309

Regeneration
*H. Wallace*                                             331

Gradients, position and pattern: a history
*L. Wolpert*                                             347

The role of genes in ontogenesis – evolving concepts from 1883 to
   1983 as perceived by an insect embryologist
*Klaus Sander*                                           363

Genome function in sea-urchin embryos: fundamental insights of
   Th. Boveri reflected in recent molecular discoveries
*Eric H. Davidson*                                       397

Reductionism and holism in biology
*Neil W. Tennant*                                        407

*Selected references to resources relating to the history of embryology*   435

*Index*                                                  463

# Preface

J.NEEDHAM, F.R.S.

It is a great privilege for me to have been asked to write the preface for this book; and it has also been a great pleasure to read in full the fascinating contributions which it contains. The symposium itself took place in April 1983, and I would like to recall that I travelled over to Nottingham for it with my old friend, Alfred Glücksmann of the Strangeways Institute, who was so well-known for his work on cell death in normal ontogenesis, as also for the relations between differentiation and malignancy. Unfortunately, I have also to recall that he never lived to see this book, because he died in July 1985.

I must have been, no doubt, the oldest inhabitant at the symposium, because I first started research in the Cambridge Biochemical Laboratory 64 years ago. About 35 years before that, Frederick Gowland Hopkins had been invited to Cambridge, and he stayed the night with the Professor of Physiology, Sir Michael Foster. In the morning, over their breakfast egg, Foster said to him, 'Now, Hopkins, this is something for you to do. Tell us how the haemoglobin is synthesised during the three weeks of development from the pure white and yolk of the egg at the beginning of development.' Hopkins never did in fact work on this, but the mantle of Elijah descended on me, and realising what a marvellous chemical factory the hen's egg was, I produced in 1931 a three-volume work, entitled '*Chemical Embryology*'.

But that dealt with metabolism *during* development, not with metabolism in relation to morphogenesis. It was in the very year following that publication that Bautzmann, Holtfreter, Spemann & Mangold announced the stability of the amphibian organiser to boiling and other forms of protein denaturation. And again in the following year, Waddington did the same thing for the chick embryo.

This gave the green light to all biochemists, and throughout the thirties, my wife and I and our collaborators worked with Conrad Waddington to try to elucidate the chemical nature of the primary inductor. But as everyone knows, the problem turned out to be much more subtle than had at first been supposed, and with our discovery of the 'masked inductor', there came a natural pause.

My views on the whole story are, I must say, much as I stated them in 1966 in the third reprint of *'Biochemistry and Morphogenesis'*. This book had originally been published in 1942, the very same year in which I left for China to become Scientific Counsellor at the British Embassy in Chungking. Unfortunately, the Cambridge University Press forgot to draw attention on the dust-cover to the presence of this historical review – 'Organiser Phenomena after Four Decades; Retrospect and Prospect'. However, the same historical 'reminiscences' were published soon afterwards in a book entitled *'Haldane and Modern Biology'*, edited by K. R. Dronamraju (1968). In the light of this, one can see with what excitement I have followed the exposés of T. J. Horder & P. J. Weindling, Jean Brachet (one of my old collaborators), Lauri Saxén & S. Toivonen and Edward Yoxen in this very volume. But in general the essays are of the highest interest. I like them all the more because my own *'History of Embryology'* (1934) ended with Karl Ernst von Baer at the beginning of the 19th century, and a second volume which would have carried the story down to the present time has never appeared. That was one of the reasons why I was so grateful for the splendid exposition of Jane Oppenheimer at the symposium, which unfortunately it was not possible to include in the present book.

In later times, Conrad Waddington always used to say that the concentration on the 'genetic code' and genetic engineering would never reveal all the secrets of life, including embryogenesis; and he used to predict that people would have to come back to classical, experimental and biochemical embryology in the end. I think that is now happening, and there can be no doubt of the enduring fascination that people find in the mysteries of embryonic development – the process which brings out in a relatively short time the fully-developed individual animal from the food-materials provided in the egg. Many of its subtleties have yet to be revealed. Hence what little Dr Harvey said in 1651 was absolutely right: 'For more, and abler, operations, are required for the Fabrick and Erection of living beings, than for their dissolution and the plucking of them down; for those things that easily and nimbly perish, are slow and difficult in their Rise and Complement'.

Sept. 85                                                                                    J.N.

# Editor's Introduction

'To put aside for a moment the matter-of-factness of an exact scientist, I will confess that I frequently have the feeling in my experimental work of holding a dialogue with someone who is considerably brighter than me.'
Spemann[1]

Unattainable though it must be, the historian's highest aspiration is to objectivity. We therefore begin by trying to define the background of this book so that the reader is in a position to make any necessary allowances. The planning of the volume was a response to a sense of puzzlement concerning the present state of the discipline of embryology, where, despite all our massive knowledge about embryos at the descriptive level and their basis in molecular and cell biology, the nature of embryological events is generally viewed as mysterious and unsolved. The volume is founded on the conviction that embryology represents a distinct and significant domain among biological phenomena, as open to satisfying explanation as any other, and that it is a subject which, since embryogenesis has been a precondition for the very existence of living forms throughout evolution, ought to occupy a key position in biology, many areas of which stand to benefit if it were better understood.

If, in attempting to pinpoint the nature of the difficulty, one turns to any current textbook of developmental biology one is struck, on the one hand, by the wide range of discrete biological topics which are brought in, and on the other by the absence of any sense of priorities or centre. If, like Tennant, one examines the words and concepts used by embryologists they are overwhelmingly descriptive, and even those used in some explanatory/mechanistic role frequently turn out to be mere analogies, or entities (e.g. gemmules, determinants, cytoplasmic factors, organisers, growth factors, morphogens, etc., etc.) invented as causal explanations for a given phenomenon which do little more than rephrase the problem and assume a simple one-to-one causal relationship which hardly does justice to the complexity of biological systems. Finally, and most notable of all, the words that seem to refer to the *most* characteristic of embryological phenomena (e.g. regeneration, regulation, multipotency, fields) are not only difficult to relate to specific, known mechanisms or means of measurement, but usually date back a century or more. It is here that the problem seems to lie; the lack of integration in any coherent explanatory framework and an apparent block to conceptual advance at the very centre. A number of examples are described and examined in this book; i.e. the closeness of the relationship of the work of Wolpert and Davidson to that of Driesch and Boveri, the continuing artificial separation of regenerative from embryonic phenomena discussed by Wallace. Our working hypothesis is that the current, primitive state of embryology as compared to other divisions of science can, at least in part, be explained in historical terms.

All the available historical surveys covering the field of embryology stop at about 1900, which was just the time when the subject was entering its most distinctively influential period. Although embryological matters have often been touched on in histories of genetics or general biology, and have merited an assortment of often distinguished specialist studies, the absence of any coherent treatment of this area of biology is becoming increasingly conspicuous. The structure of the book, and the way in which it seeks solutions, need some explanation. Multiple authorship was the only way in which we could hope to achieve any degree of expert coverage of the range of complex issues involved. Our contributors originally presented papers at a meeting of the British Society for Developmental Biology at Nottingham in April 1983, but the material has, in many respects, undergone extensive modification subsequently. The papers are arranged in two parts. The first group of papers documents what we regard as major historical aspects of the subject during the period 1880–1940, the period of 'classical' experimental embryology. Any attempt at a complete and systematic listing of events or personalities would have been impossibly voluminous and of spurious 'objectivity'; a *Cronological Outline* is provided in the hope that it will show the range and sequence of historical events that form the background of this period. Another way to get an immediate feel for the development of the subject is to refer to Willier and Oppenheimer's[2] collection of milestone papers, which, to a remarkable degree, captures the preoccupations and peaks of achievement characterising each successive decade of our period. We have arranged papers by country because, to an extent that the Editor's linking review sections seek to show, differing research traditions affected approaches to the subject in ways that had identifiable implications, and as part of our overall objective for these papers of re-creating the historical contexts of important episodes in order to illustrate how different were the circumstances under which earlier embryologists worked. The 'other worldliness' that we experience on reading older works and difficulties associated with reconstructing the meaning they had at the time are measures of how obligatory such reconstruction is and how misled one can otherwise be: they remind us to avoid the temptation of dismissing past aberrations as due to lesser mental powers and to expect that, when seen in context, even the most bizarre event will have a rational explanation.

World War II can usefully be taken as a watershed in the development of biological priorities affecting embryology: around that time there developed that puzzling lack of a centre, of any 'sense of progression'[3], that is still so much a feature of the subject. The second set of papers, reflecting this situation, abandons any attempt at historical sequence as such and addresses the present situation of the subject as much as the post-war period overall. The papers are in the form of studies, often by practising embryologists, which we hope, at the very least, provide documentation as well as insights into the use of history relevant to the analysis of historical forces operating on developmental biology today. Mindful of the fact that history is increasingly distorting the nearer we are to it, we have merely aimed at illustrative case studies of as wide a variety as possible. Perhaps the greatest service that a book such as this could fulfil is as a source of stimulus and guidance for further studies by historians and embryologists themselves. We have appended a *Selected Bibliography* arranged in categories which, it is hoped, will facilitate access to the wide and often poorly documented range of sources of evidence relevant to the history of the subject and will assist working embryologists in finding their way through the older data.

How seriously should today's scientist take a claim that his work is significantly determined by historical factors? Entertaining though

the exposing of past errors might be in themselves, if we want
to solve the problems of embryology would time not be better spent
in pursuing such things as the chemistry of induction? With the
half-life of the typical biological paper being some seven years – as
measured by the period over which it is cited[4] – it is easy to regard
science as a process of uniform linear advance, continually replacing
and improving what went before, published papers as ephemeral,
and concern with the origins of words or concepts as an irrelevance.
This perspective corresponds well enough with scientific practice in
the case of subjects involving relatively few and well defined vari-
ables or targets, especially those of a deductive, practical or applied
nature. But what about a subject like embryology, a 'model'[3] among
biological sciences? How does this image of science as step-wise
discovery describe and suit the particular nature of biology?

What Needham wrote with reference to epigenesis and preformation in 1934, 'an antithesis ... the subsequent history of which is
almost synonymous with the history of embryology'[5], remains valid
today. Though nowadays claimed as the starting point for modern
embryology, Roux's famous experiment (in which he showed
autonomous (mosaic') development of embryonic parts) was just as
much part of the history of genetics, as its relevance, through Weismann, as a test of the germ-line and chromosome theories demonstrates. It was the counter-example, so soon afterwards announced
by Driesch, which should perhaps more appropriately occupy the
place of honour because it introduced and defined a clear question
about embryonic events specifically. Whereas mosaic phenomena in
themselves reveal little about actual mechanisms and, like earlier
preformationist theories, effectively sidestep the problem by
referring everything to previous occurrences, Driesch's finding
(that parts of embryos can 'regulate' their fates according to their
surroundings) demonstrates unambiguously what the visible
emergence of organisation out of the relative homogeneity of the egg
had long implied; embryonic parts are initially capable of forming
many adult structures and therefore the origin of the diversity and
commitment of parts must be sought in the events and processes that
characterise embryogenesis. The way in which Driesch invoked the
entelechy as an explanatory agency and retreated into an almost
mystic vitalism is one of the better known cautionary tales of
biology. Spemann, in the course of a detailed confirmation of the
initial multipotency of embryonic parts, demonstrated, and estab-

lished experimental means of defining, sequential epigenetic events which showed that it was possible to analyse the emergence of orderly anatomical structure while avoiding arbitrary Drieschian solutions and preformation. Spemann's interpretations of the underlying mechanisms, and in particular the idea of the organiser, were put to the experimental tests at the time considered appropriate. The verification and identification of the organiser were sought by the attempted isolation and purification of an underlying chemical agent and it was the considerable sense of disillusionment generated by the resulting confused findings that signalled a comprehensive abandonment of not only the hypothetical mechanisms of the organiser but with them the epigenetic experimental data. The middle path between Roux and Driesch and, for the second time, an approach founded on a clear understanding of the central problem of embryology, had become blocked under somewhat traumatic circumstances.

It remains standard practice for textbooks of developmental biology to repeat the account of the experiments of Roux and Driesch, but this is more than a brief concession to history: it symbolises mental categories that are employed in our perceptions of possible explanations of embryology, and is itself a continuation of a process which has contributed to the failure of the subject to advance conceptually. This is the stuff of scientific myth making, several examples of which we shall come across in the book. Carried along under the impetus of novelty or fashion[6], and often assisted by misrepresentation or misunderstanding of alternative positions[7], what were, in their initial contexts, necessary and rational advances can later become fossilised into primitive assumptions that continue as preconditions for the future development of a subject. Given such conveniently distinct alternatives as mosaic and regulative forms of development it is not surprising that embryologists have tended to build on one or the other approach according to the category that applied to the organism they happened to study. There has been little incentive to escape from the implication that organisms themselves fall into two distinct categories. As with the split between genetics and embryology, the divide becomes increasingly difficult to bridge because subspecialisations arise and all the more difficult to re-think.

The fate of Spemann's contributions illustrates other effects of historical sequence. The eclipse of Spemann's theoretical position

meant that the experimental data underpinning it also had less interest. With the passage of time alone the data itself has become first
unfamiliar and eventually simply unknown. Again it had become
increasingly difficult with the passage of time to re-think the original
arguments, even though, as we suggest later, the way in which the
organiser had been found wanting reflected consideration of only a
narrow aspect of the experimental data and testing by methods
which were inappropriate to the operational and whole organism
nature of the dynamic processes under consideration. We have here
the potential for an element of positive feedback whereby the failure
of one biological methodology in the face of a currently more
successful one leads to yet further confirmation of the predominance
of the successful methodology and the possibility that desirable
alternatives are eventually driven out entirely. In such a process
historians of biology may have had an exacerbating, rather than
moderating, influence. In so far as they are inevitably dependent for
their examples on material supplied by scientists, they could hardly
have avoided concentrating on the 'success stories' of biology such as
the modern evolutionary synthesis, genetics and molecular biology.
We have had few precedents forewarning of the possibility of
reversals of fortune of the kind we have discussed in embryology
here: neo-Darwinism perhaps came nearest to it.[8] Philosophers of
science have similarly been swept along by scientific positivism[9],
and it was perhaps their belief that biological method could be
reduced to formal, logical categories and procedures that limited the
important efforts of theoretical biologists like Woodger, Weiss or
Waddington to increase awareness of biological methodology.

Increased technical power allowing observation of events in ever-
greater detail and terminological changes in the mechanisms
hypothesised in line with advances elsewhere in biology may serve to
disguise the fact that the central issues of embryology remain
unanswered: indeed such modern developments may make it less
easy to recognise and re-think those central issues. Today's concerns
in developmental biology (such issues as the identification of cytoplasmic factors in the egg, of genes controlling development, or the
clonal origin of cell types) have become so distanced from their
original conceptual and observational roots that the underlying preformationist implications (i.e. implicit reference of adult organisation directly back to preconditions already fully established at the
earliest stages of development; failure to explain regulation or why

the complexities of embryogenesis occur at all) of many of them are hard to uncover. The history of embryology thus raises the question of the adequacy of current methodology for the solution of the more complex of biological problems. These methods seem to have failed to prevent the continuation of mistaken approaches – still in innumerable demonstrations of mosaic behaviour in embryos, we see repeated Roux's failure to consider whether the experimental intervention created conditions that allowed regulation to show itself and his overinterpretation of the preconditions needed to explain the observed degree of mosaic behaviour. These methods appear to have allowed the increasing legitimisation of use of abstract and mathematical models as explanations of selected developmental phenomena, which, in the hands of some developmental biologists, seem almost to have become substitutes for the embryo itself. For no sustainable reason the phenomena of regeneration, which provide potentially the most direct evidence that – even in organisms with mosaic embryos – cells are open to epigenetic influences, are still not generally taken into consideration, as Wallace describes.

To raise general issues of biological method in the context of a book on embryology needs no apology because that subject has for a long time occupied a prominent place in discussions of this kind, particularly through its close association with theoretical biology. Indeed embryogenesis has frequently been taken as a test case (alongside other typically biological problem areas such as behaviour) with the added advantage that the potential for an acceptable, experimentally based explanatory framework seemed within reach. In biology, as embryology demonstrates so well, we are faced above all with a problem of complexity in terms of quantity and variety of potentially relevant information, where concepts, ordering principles or laws are relatively few and, as Glass has pointed out in relation to significant advances in genetics[10], may be increasingly rare as data accumulate. Whatever mechanisms science generates in the future to regulate its long-term aims – and these may well benefit increasingly from the more direct application to the actual scientific decision-making processes of individual scientists of the so-far largely independent and external methods and conclusions of sociologists and philosophers of science – historical considerations cannot be avoided. Historical awareness will remain the best insurance against unsuspected presuppositions and assumptions and a constant reminder that even scientific method itself is not inevitable.

History will continue to provide us with our models of good and bad science and their consequences. The more one examines the historical details of the work of figures such as Morgan, Weismann or Spemann, the more apparent it becomes that the emergence of important biological principles cannot be identified with any single chance observation, experiment or discovery, but that they are the result of a willingness to return repeatedly to old questions, and to re-arrange inexplicable bodies of data in new ways. Not so long ago there was no more burning issue for biologists than the question 'What is Life?'. The meaninglessness of this question to us now makes one wonder what equivalent pseudo-questions may be influencing our priorities today. It may well be that the very idea of 'the unsolved problem of embryology' is one such; the phrase itself almost presupposes a particular form of solution and invites particular forms of research, tending to invoke images of a single, all-revealing experiment or discovery. In fact it may be the case that we already have all the data we need to arrive at an understanding of embryology, and the approach appropriate to such a subject lies in the direction of integration and rearrangement of a complex of existing concepts. Certainly we can hardly expect the diverse branches of the biological sciences to have advanced independently at comparable rates such that evidence relevant to an eventual explanation becomes available in the appropriate sequence. The tendency to judge the validity of a published paper by its recency not only exposes us to the risk of needless repetition and rediscovery (as illustrated by recent demonstrations of the organiser effect described in Saxén & Toivonen's paper) but also means that, by ignoring older literature, we are effectively discarding information the potential uses of which, particularly in the case of descriptive data, have in no way changed. In so far as such considerations have made us ignorant of the literature of classical experimental embryology the consequences may prove to have been very damaging indeed.

Of some of the deficiencies of this volume the editors are painfully aware. Inevitably and regretfully we were obliged to exclude many obviously relevant areas of concern: the important contributions of many countries are not mentioned; topics obviously under-represented include botany, medical embryology, theoretical biology, neuroembryology, studies relating to growth and reproduction, differentiation, evolution and comparative embryology.

We wish, finally, to acknowledge with gratitude the generous financial assistance we have received from the Wellcome Trust, the Royal Society and the British Council and the privilege of access to the rich resources of the Radcliffe Science Library in Oxford. We are particularly grateful to Jane Ballinger, Drs Jane Green, D. Stocker, P. J. Windling for their major contributions during editing, to Professor Needham for his Preface and to Dr R. M. Clayton, the late Dr A. Glücksmann, Drs S. Løvtrup, D. R. Newth, J. M. Oppenheimer, S. Smith, D.E.S. Truman and E. N. Willmer for their substantial involvement at early stages in preparation of this book.

T.J.H.

# References

1. H. Spemann. (1928). *Zur Theorie der tierischen Entwicklung*, Rektoratsräde. Freiburg: Speyer u Kaerner.
2. B. H. Willier & J. M. Oppenheimer. (1974). *Foundations of Experimental Embryology*, 2nd edn. New York: Hafner.
3. P. B. Medawar. (1965). *Nature*, **207**, 1327–30.
4. A. J. Meadows. (1974). *Communication of Science*. London: Butterworths.
5. J. Needham. (1934) *A History of Embryology*, p. 22. Cambridge: Cambridge University Press.
6. H. B. Fell. (1960). *Science*, **132**, 1625–7. J. A. Witkowski. (1979). *Medical History*, **23**, 279–96.
7. L. Picken. (1956). *Nature*, **178**, 1162–5. B. Wallace. (1985). *Q. Rev. Biol.* **60**, 31–42.
8. P. J. Bowler. (1983). *The Eclipse of Darwin; anti-Darwinian Evolution Theory in the Decades around 1900*. Baltimore: Johns Hopkins University Press.
9. G. Gale. (1984). *Nature*, **312**, 491–5.
10. B. Glass. (1980). *Q. Rev. Biol.* **54**, 31–53.

# Chronological outline

1818    Meckel classifies human abnormalities

1821    Magendie founds *Journal de physiologie expérimentale*

1823    Chevallier brothers introduce achromatic microscope. Amici observes germination of pollen grains

1824    Prévost and Dumas describe cleavage of frog egg: sperm needed for fertilisation

Carus first sees polar bodies

1825    Rathke describes gill slits in pig embryo

Purkinje discovers germinal vesicle in hen's egg

1827    von Baer identifies mammalian ovum

1828    Wöhler achieves artificial synthesis of urea. von Baer: *Ueber Entwicklungsgeschichte der Tiere. Beobachtung und Reflexion*

1831    Brown describes cell nuclei

1834    von Baer associates cleavage furrow with cell division

Coste describes germinal vesicle in mammalian ovum

1835    Schwann discovers pepsin

First textbook on human development – Valentin: *Manual of the Development of the Foetus*

1837    Reichert describes homologies of middle-ear ossicles

Siebold and Sars describe division of invertebrate eggs

1838    Endlicher: *Grundzüge einer neuen Theorie der Pflanzenzeugung*

1839    Schwann introduces cell theory and identifies eggs as cells

1841    Kölliker suggests spermatozoa are of cellular origin

First histology textbook – Henle: *Allgemeine Anatomie*

1842    Chromosomes visualised in plants by Naegeli; dissolution of nucleus during mitosis described

First textbook of comparative embryology – Bischoff: *Entwicklungs-geschichte der Saugerthiere und des Menschen*

1843    Spermatozoa seen in ovum

J. Müller: *Elements of Physiology*

1844    Kölliker argues that the ovum is a cell

Heidelberg University creates separate departments of anatomy and physiology

1845 Steenstrup: *On the Alternation of Generations*
1847 Reichert observes asters
Liebig: *Chemistry and Its Application to Agriculture and Physiology*
1848 Leydig connects nuclear division with segmentation of cells
1849 Wagner and Leuckart show that spermatozoa are not parasites and are essential for fertilisation
Owen: *On Parthenogenesis*
Hofmeister identifies both gametes as cells in plants
1850 Schacht: *Entwicklungsgeschichte des Pflanzen – Embryon*
1851 Carmine introduced
Leuckart: *Über den Polmorphismus der Individuen oder die Erscheinungen der Arbeitstheilung in der Natur*
Nelson describes fertilisation in *Ascaris*
1852 Kölliker: *Textbook on Histology*
Remak argues that cleavage is comparable to cell division in general; that this is the origin of new cells and is initiated by nuclear division
1854 Newport performs the first experiment on embryos: point of sperm entry determines axis of developing embryo
Büchner: *Kraft und Stoff*
Graham studies electrochemical equilibria across semipermeable membranes
1856 Perkin introduces first aniline dyes
1858 Virchow: *Cellular Pathology* pronounces *Omnis Cellula e Cellula*
1859 Darwin: *The Origin of Species*
First zoological station founded – at Concarneau
T. H. Huxley compares embryonic germ-layers with coelenterate structure
1861 Brücke's theory of protoplasm
Gegenbaur concludes that all eggs in vertebrates are cells
Pasteur attacks theory of spontaneous generation
Kölliker: *Entwicklung des Mensches und der Höheren Tiere*
1862 Haematoxylin introduced
1863 Schultze promotes protoplasm as opposed to cell wall as defining feature of cells
1864 Fromman introduces first silver stain for neurons
Spencer discusses heredity by 'physiological units'
Schultze founds *Archiv für mikroscopische Anatomie*
1865 Schweigger-Seidel and La Valette St George argue that spermatozoa consist of nucleus and cytoplasm
Mendel first formulates laws of hybridisation
Schultze introduces osmic acid fixation
First textbook of plant physiology – Hofmeister: *Handbuch der physiologischen Botanik*
Early microtome built by His
'Culture' of blood cells by Schultze
1866 Haeckel's *Generelle Morphologie der Organismen* in which he suggests the nucleus as carrier of inheritance
1867 Pflüger starts *Archiv für die gesamte Physiologie*

1868    Darwin's provisional theory of pangenesis proposed in *The Variation of Animals and Plants under Domestication*
        His describes method for reconstruction of sectioned embryo material
        Rivet describes the simple microtome
1869    Strasburger studies behaviour of spermatozoa within ovum in fern
        Paraffin embedding introduced by Flemming
1871    Miescher describes a procedure for separating nuclei from other cell components
1872    Naples Zoological Station founded
        Abbé introduces oil-immersion lenses
        *Archive de Zoologie expérimentale et générale* founded
1873    Flemming gives modern interpretation of nuclear division
        Chromosome behaviour in nuclear division described by Schneider, Bütschli and Fol
        Fol describes asters as centres of attraction
1874    Haeckel introduces his *Gastraea-Theorie*
        Auerbach observes fusion of two pronuclei at fertilisation
        His: *Unsere Körperform und das physiologische Problem ihrer Entstehung*
        Bastian: *Evolution and the Origin of Life*
        Burdon-Sanderson created first Professor of Physiology in UK at University College
1875    O. Hertwig first uses sea-urchin to observe division of the aster of sperm cells and identifies fertilisation as fusion of sperm with egg nuclei, in parallel with Fol, Bütschli, van Beneden and Strasburger
        Balfour equates blastopore and primitive streak
1876    Bütschli sees fertilisation as trigger for polar body formation
        Centrosome first seen by van Beneden
        Balbiani suggests that elementary units line up to make chromosomes
        Methylene blue introduced by Caro
        Born describes wax reconstruction method
1877    O. Hertwig identifies polar body as cellular product of cell division
1878    Whitman's first paper on cell-lineage
        Shenk attempts *in vitro* fertilisation of mammalian ovum
1879    Fol shows that only *one* sperm is needed for fertilisation
        Strasburger, Flemming and Schleicher discuss nuclear division as cause of cell division
        Colloidin embedding introduced by Schieferdecker
1880    Splitting of chromosomes described by Strasburger in plant cells
1881    Roux: *Der Kampf der Theile im Organismus*
        Balfour: *A Treatise on Comparative Embryology*
1882    Flemming observes longitudinal splitting of chromosomes in animal cells
        Continuous ribbon wax sectioning introduced
        Lyon first applies centrifugation to eggs
1883    Golgi introduces his method of silver impregnation. van Beneden, introducing *Ascaris*, observes doubling of chromosome numbers after fertilisation; notes equivalence of maternal and paternal contributions and equiva-

lence of chromosomes in daughter cells in mitosis; infers persistence of chromosomes in cell cycle

Weismann's *Über Vererbung* introduces germ-line theory

Brooks: *The Law of Heredity*

Roux speculates on the equivalence of homologous chromosomes and qualitative differences within them

Pflüger studies effect of gravity on frog cleavage: theory of isotropy of egg cytoplasm

1884 Strasburger identifies male contribution to zygote as nuclear only

Multiple statements of nucleus as mediator of inheritance by O. Hertwig, Strasburger, Kölliker and Weismann

Naegeli introduces the notion of 'idioplasm'

O. Hertwig: *The Problem of Fertilisation and Theory of Heredity*

1885 Rabl notes the constancy of chromosome number in different species and infers continuity of individual chromosomes in cell cycle

Vital dyes introduced by Ehrlich

1886 First 'apochromatic' lens

1887 Boveri analyses meiosis as seen in polar body chromosome rearrangements in *Ascaris*

Boveri and van Beneden report role of sperm centrosome in control of cell division

*Journal of Morphology* founded by Whitman

Weismann introduces ideas of reduction division

O. Hertwig separates sea urchin embryo cells by shaking

Hertwig brothers develop technique of fertilising nucleus-free egg fragments in sea-urchin

1888 Roux tests Weismann theory in amphibia by killing one blastomere with a hot needle

Boveri shows that splitting of chromosomes is not due to pulling by spindle

Woods Hole Marine Biological Station founded

Roux establishes the *Institüt für Entwicklungsgeschichte und Entwicklungsmechanik* in Breslau

1889 de Vries: *Intracellular Pangenesis*

1890 O. Hertwig suggests reduction division as equivalent in egg and sperm

Loeb: theory of tropisms

1891 Waldeyer and Cajal promote neuron theory

Driesch shows regulation in fragmented sea-urchin embryos

Henking observes sex chromosomes and homologous pairing in meiosis

Driesch: *Die mathematisch-mechanische Betrachtung morphologischer Probleme der Biologie*

1892 Weismann: *Das Keimplasma: eine Theorie der Vererbung*

Herbst develops $Ca^{2+}$-free dissociation of embryo cells in sea-urchin and introduces lithium salts as experimental tool

First binocular dissecting microscope; Greenough

1893 Wolff describes regeneration of amphibian lens

O. Hertwig first divides an egg with a silk loop

Whitman writes on *The Inadequacy of the Cell Theory of Development*

1894    Rückert considers reduction division as halving the number of chromosomes
        Foundation of first embryological journal: W. *Roux Archiv für Entwicklungs-
        mechanik*
        O. Hertwig: *The Biological Problem of Today: Preformation or Epigenesis?*
        Driesch: *Analytische Theorie der organischen Entwicklung*
        O. Schultze produces double embryo by inverting an amphibian ovum
        Adrenalin extracted by Oliver and Schäfer
1895    First standardisation of anatomical nomenclature by His
        Bateson: *Materials for the Study of Variation*
1896    Wilson: *The Cell in Development and Inheritance*
        Boveri tests whether nucleus carries heredity qualities
        Ovarian hormones demonstrated by Krauer and Halban
        Born describes heteroplastic recombination of amphibian embryo fragments
1897    Roux publishes his *Entwicklungsmechanik* manifesto
        Morgan: *The Development of the Frog's Egg*
1898    Benda names mitochondria
1899    Loeb demonstrates artificial fertilisation
        Beijerinck isolates filtrable virus of tobacco mosaic disease
1900    Multiple rediscovery of Mendel by Correns, de Vries and Tschermak
1901    De Vries: *Mutations-Theorie*
        Spemann's first publication on lens induction
        Herbst: *Formative Reize in der tierischen Ontogenese*
        Rockefeller Institute founded
1902    Chromosome theory of heredity introduced by Wilson, Sutton, Cannon and
            Boveri
        Bayliss and Starling investigate hormones
        Hofmeister and Fischer hypothesise that all proteins are formed by the union
            of amino acids by means of a specific peptide bond
        Carnegie Institute founded
        Accessory chromosomes identified by McClung
1903    Boveri tests individuality of action of chromosomes in double fertilised sea-
            urchin monsters
        Dennarb: *At the Deathbed of Darwinism*
1904    *Journal of Experimental Zoology* founded
1905    Linkage groupings described by Bateson, Saunders and Punnett
1907    Harrison first uses tissue culture to investigate nerve fibre outgrowth
        Hopkins investigates vitamins
        Eugenics Laboratory founded at University College, London
1908    Driesch: *The Science and Philosophy of the Organism*
        Keibel: *Normentafeln der Entwicklungsgeschichte des Menschen*
        Hardy and Weinberg introduce their law linking Mendelian hypothesis with
            population studies
        M. Lewis first cultures mammalian cells *in vitro*
1909    Johannsen makes 'genotype'/'phenotype' distinction
        Jenkinson: *Experimental Embryology*
        Garrod: *The Inborn Errors of Metabolism* – first study in biochemical genetics
            demonstrating that a single gene mutation blocks a single metabolic step

Janssens hypothesises that exchanges between chromatids produce chiasmata

Correns and Baur study the non-Mendelian inheritance of chloroplast defects in variegated flowering plants

von Uexküll: *Umwelt Theorie*

1910  Morgan discovers the sex-linked white eye colour mutant in *Drosophila*

1911  Morgen uses chromosomal crossing-over for gene mapping

Hubrecht founds '*Institute Internationale d'Embryologie*'

Goodale uses vital dye to map amphibian embryo morphogenesis

Goldschmidt: *Einführung in die Vererbungswissenschaft*

*1912  First chair in Biochemistry in UK – Hopkins in Cambridge*

*First use of chorio-allantoic grafting in chicken egg by Murphy and Rous*

*Loeb: The Mechanistic Conception of Life*

1913  Henderson: *Fitness of the Environment*

Carnegie Department of Embryology established under Mall

1914  Bridges discovers nondisjunction in *Drosophila*

Kaiser-Wilhelm Institute for Biology founded in Berlin

1915  Child: *Individuality in Organisms*

Morgan: *The Mechanism of Mendelian Heredity*

1916  First establishment of timing of ovulation in humans

Lillie demonstrates role of hormones in sexual differentiation in freemartins

First purely genetics journal in USA founded: *Genetics*

1917  Thompson: *On Growth and Form*

Bridges discovers the first chromosome deficiency

Goldschmidt develops ideas on physiological action of genes and studies industrial melanism

1919  Lillie: *Problems of Fertilization*

Bridges describes chromosomal duplications in *Drosophila*

1921  Spemann and Mangold discover the organiser effect

Banting and Best isolate insulin

1922  German Committee for Racial Hygiene and Population established

1923  Svedberg introduces the first ultracentrifuge

Boycott and Diver describe delayed Mendelian inheritance of the direction of coiling in the shell of the snail *Limnaea*

Warburg defines anaerobic metabolism

Oestrin isolated by Allen and Doisy

1924  J. S. Huxley describes allometric growth equation

1925  Sturtevant analyses the Bar effect in *Drosophila*; discovery of position effect

1926  Sumner crystallises the first enzyme

UK Birth Control Investigation Committee founded

Whitehead: *Science and the Modern World*

Kammerer's work on inheritance of acquired characters discredited

1927  Belling introduces acetocarmine technique for staining chromosome squashes

Muller describes artificial induction of mutations in *Drosophila* by X-rays

Morgan: *Experimental Embryology*

1928  Griffith shows transformation of pneumococci from avirulence to virulence by extracts of killed virulent organisms

1929 Darlington hypothesises that chiasmata hold homologous chromosomes together at meiotic metaphase and thereby ensure their segregation at anaphase, so completing understanding of meiosis

   Corner identifies progesterone

   Fisher: *The Genetical Theory of Natural Selection*

1931 Needham: *Chemical Embryology*

   Sewall Wright discusses evolution in Mendelian populations; foundation of the synthetic theory of evolution

   Holtfreter describes his culture medium

1932 Knoll and Ruska describe prototype of the electron microscope

   Induction by a 'dead' organiser first demonstrated by Spemann and colleagues

   Theoretical Biology Club first meets

   J. S. Huxley: *Problems of Relative Growth*

1933 Painter identifies the giant salivary gland chromosomes in Diptera

   Bertalanffy: *Modern Theories of Development*

   Tiselius introduces procedure to separate charged molecules by electrophoresis

1934 First X-ray crystallography of protein by Bernal. Huxley and de Beer: *The Elements of Experimental Embryology*

   Schlesinger shows that bacteriophages are composed of DNA and protein

   Hybrid androgenic merogon technique introduced by Hadorn

   Morgan: *Embryology and Genetics*

1935 Beadle, Ephrussi, Kuhn and Butenandt study the biochemical genetics of eye pigment synthesis in *Drosophila* and *Ephestia*

   Stanley crystallises the tobacco mosaic virus

   Spemann awarded Nobel Prize

   Zenicke introduces first phase microscope

1936 Bittner shows that mammary carcinomas in mice are caused by a virus

   First non-organic agent (methylene blue) shown to mimic organiser by Needham and colleagues

1939 Ellis and Delbruck study the genetics of bacteriophages

   First 'Growth' Symposium in USA

   Holtfreter first shows selective reaggregation of cultured cells

1941 Beadle and Tatum study the biochemical genetics of *Neurospora*; the 'one- gene–one-enzyme' hypothesis

1942 The gene will be 'as dead as the dodo'; Goldschmidt

   Needham: *Biochemistry and Morphogenesis*

1943 Production of radioisotopes begun at Oak Ridge

Part 1

# An outline of concepts of generation and heredity up to the end of the nineteenth century

As a separate discipline embryology effectively did not exist until some time in this century; how distinct a discipline it is even today is a question that this book may help to resolve. What is certain is that prior to the attainment of the distinction between genotype and phenotype, heredity and development were lumped together, alongside phenomena like regeneration and asexual budding, and attributed, by way of explanation, to the manifestation of the *'germ'*.

In one way or other the origin of living forms must frequently have figured in philosophical and theological systems of the past. Aristotle made it the basis of his classification of animals, which he divided according to whether they originated by live birth, by *spontaneous generation* (meaning an origin independently of preexisting organisms), or from an egg. This was still the position only three hundred years ago, when Harvey began to question the distinctions. Harvey's advocacy of another Aristotelian concept, that of *epigenesis* (formation by gradual elaboration) appears to have contributed to the first expression of precisely the opposite notion in 1674; Descartes' failure to achieve a convincing mechanistic explanation of Harvey's observations of the chick embryo is thought to have led to Malebranche's formulation of the concept of *preformation*. At that time this stood for the idea that each successive generation of organism was preordained in the first created germ.

In the eighteenth century this subject aroused famous debates. By 1745 Maupertuis was defining difficulties associated with preformation. He argued that extinctions, malformations and the blending of parental contributions were incompatible with it. In 1745 Bonnet demonstrated parthenogenesis in aphids and an important issue of the day was the extent to which reproduction in different species was sexual. Sex in plants gradually became accepted and thus opened the way to the growth of hybridisation studies. Meanwhile embryology as such was just one piece in a still unsorted jigsaw of the *generation* of organisms, not yet clearly associated with sexual reproduction or the role of egg or sperm, and not distinguished from issues of variation and descent. A new way of considering embryos was introduced, first by Kielmeyer in 1793 and then by Oken and Meckel. This was the principle of *recapitulation* which sought to draw a parallelism between the hierarchies of increasing complexity seen in the animal kingdom and the increase of organisation seen in embryonic development. In the same period von Baer's descriptive studies of embryos – in terms of germ layers rather than cells, which had yet to be recognised – provided some solid grounding in fact.

The dominant forces in biology in the nineteenth century were cell and evolution theory. As Wilson put it (1925, *The Cell in Development and Heredity*, 3rd edn, p. 2).

The cell-theory and the evolution-theory are now closely affiliated; but the historian of biology is struck by the fact that for a long time they did not come within hailing distance of each other. The theory of evolution originally grew out of the study of natural history and took definite shape long before the finer structure of living bodies was made known.

The defining of transmission genetics would only be possible through the coming together of the two disciplines. Cell theory, introduced by Schwann in 1839, stood for the whole spectrum of approaches that includes reductionism, mechanistic philosophy, monism which associates inorganic and organic phenomena; and along with its openness to new technologies of science, cell theory was a direct spur to the rise of physiology, an association that would only be lost gradually during the next century. At the opposite end of the biological spectrum was the traditional approach of the morphologist/naturalist whose prime concern was the whole organism, its species variations and underlying principles of organisation and classification.

Darwin himself gave an influential start to the process of connecting the theory of descent with the material entities being described by cell theorists, when, in 1868, he reintroduced the ancient concept of *pangenesis*. According to Darwin's theory each adult organ, having acquired its characteristics partly from its parent and partly from its environment, releases particulate representations of itself (*gemmules*) which are incorporated, via the bloodstream, into the gametes and so form the basis of the next generation.

Embryos played an increasing but ambiguous role in the events of the second half of the nineteenth century. In the first place, increasingly, embryos were found to be favourable material on which to study the important issues of cell theory: for example, whether the ovum and spermatozoon were cells, the nature of fertilisation and the events of mitosis as studied in cleavage, the behaviour of nuclei and later chromosomes during fertilisation, mitosis and meiosis. Secondly the centrality of embryos was obscured by uncertainties about the universality of sex itself. In mid-century increasing knowledge about complex life histories, particularly in parasitic and metamorphosing species, led to the concept of *alternation of generation* (the alternation of sexually and asexually reproducing generations in a life cycle) and an emphasis on alternative forms of reproduction such as by budding, regeneration and by spore formation. 'The first three decades of the nineteenth century witnessed the triumph of spontaneous generation. In these years most of the foremost natural historians came to believe in spontaneous generation of infusorians and parasitic worms' (Farley, 1974, *The Spontaneous Generation Controversy*, p. 31). Such an interlinking of half-formed ideas, largely foreign to us now, explains the complexity of the organ systems and species chosen for study at this time. Thus, Weismann could move from studies of colouration patterns in caterpillars to studies of the origins of sex cells in hydroids: or Morgan, in 1903, could write a book in which he reviewed the topic of sexual reproduction, which was shortly to prove a vital route towards *Drosophila* genetics, having previously studied regeneration in annelids; the link was that they were all considered in equal terms to be examples of adaptive variation, whose Darwinian implications were still open to debate.

In Germany Darwin's theory became a matter of passionate controversy and the focus of the debate was to a considerable extent determined by Haeckel's presentation of Darwinism. With background allegiances to both cell theory and *Naturphilosophie*, Haeckel's search for evidence in support of Darwinism led him to resurrect, virtually without any change in thinking, the already discredited parallelism of stages of embryonic development and phylogenetic series (*biogenesis*). Only later, alongside a number of British biologists (Lankester, Huxley, etc.) did he focus attention on the germ layers as the most elementary and tangible proof of his position. As Russell put it 'The gastrulae theory gave point and substance to the biogenetic law' (*Form and Function*, 1916, p. 291).

These then are some of the perspectives of the period covered by the first two chapters of this book. The first concerns the derivation of Weismann's concept of *germ plasm*, a crucial step towards the concept of genotype. Just how 'primitive' understanding of hereditary mechanisms was in the very year (1884) of his discovery is vividly suggested by the following quotation from Naegeli describing his conception of hereditary mechanisms:

occasionally a definite colony of micellae, or a combination of such colonies, become active that is 'are thrown into definite conditions of tension or motion'. This local irritation, by means of dynamic influence, and the transmission of peculiar conditions of oscillation acting at a microscopical distance, governs the chemical and plastic processes. . . . Which micella group of the idioplasm becomes active during development depends upon its shape, upon the stimulation it has previously received, and finally, upon the position in the individual organism in which the idioplasm is placed.

Five years later de Vries expresses a recognisably more particulate viewpoint, and one that influenced Weismann himself.

In place of this dynamic hypothesis, de Vries assumes that the character of the cell is affected in a more material fashion. He is of opinion that, whilst the majority of the idioblasts or 'pangenae' (de Vries) remain inactive, others become active, and grow and multiply. Some of these then migrate from the nucleus into the protoplasm, in order to continue here their growth (O. Hertwig, 1895, *The Cell*, pp. 358–9).

Weismann's arrival at the concept of the germ plasm represents one of the most decisive and clearly definable steps in the linking of evidence from evolutionists and cell theorists, which would culminate in the acceptance of Mendel's laws and the chromosome theory of inheritance. The assumption of Lamarckian inheritance which had been universal – Darwin, Haeckel and Weismann himself until 1885 had made this assumption – meant that variation and adaptation required no further explanation. Weismann's achievement consisted in rearranging the familiar, diverse facts (embracing constancy of species type, parental roles in sexual reproduction, variation, form regeneration, etc., without recourse to Lamarckian inheritance) into a new pattern with new emphasis on the regularities behind Darwinian adaptation. He was able to relate such phenomena to the emerging understanding of cell, nuclear, and even chromosomes, behaviour in the cycle of sexual propagation. For the first time, in the concept of germ plasm, it was clearly appreciated how the somatic manifestations of successive generations of organisms did not themselves determine what was inherited by their progeny, and how they, and their gametes, were the product of a quite distinct germ-line passed down and rearranged at fertilisation. Contemporaneously many others were arriving at similar conceptual frameworks in quite different ways: the particular route taken by Weismann happened to rely on using the distinctions between germ layers in embryos to establish the continuity between the fertilised ovum and the gametes of the new generation. The relevance to the later, crucial genotype/phenotype distinction of the geneticists is obvious: the implications for embryology are less clear.

# Weismann, Hydromedusae, and the biogenetic imperative: a reconsideration

FREDERICK B. CHURCHILL

Department of History and Philosophy of Science, University of Indiana,
Bloomington, Indiana 47405, USA

## Introduction

When the historian of biology turns to nineteenth century embryology, he conjures forth an imposing structure. At one end of the
century exist the exemplary observations of Pander and von Baer and
at the other the dramatic experiments of Roux and Driesch. Firmly
settled between these two opposing buttresses rises the towering
edifice of classical descriptive embryology, solid in its discoveries,
magnificent in its tracery and fine details, and as defiant of and
removed from modern biology as a gothic cathedral is from today's
secular world. Few can doubt the real achievements of those artisans
who, in constructing this temple, eternally glorified the per-
severance and perspicacity of descriptive biologists. From von
Baer's discovery of the mammalian ovum, on through Rathke's
analysis of the brachial arches, Müller's, Reichert's and Huxley's
examination of the development of the vertebrate skeletal system and
the exquisite descriptions of invertebrate development by Kowalevsky, Metchnikoff and Kleinenberg, and terminating with the
monumental studies on the development of single organisms or
organ systems by Götte, Balfour, Semper, His and scores of others,
the spires of this cathedral rest on the securest of pillars.

Yet it would be a mistake to notice only the carefully chiselled
details. General laws of morphology motivated the overall architec-
ture. During the first half of the century, concerns about an
archetypal structure and a general parallelism between development
and the hierarchy of life promoted the embryo as the morphological
court of last appeal. By mid-century the Virchowian form of the
cell theory and Darwin's theory of descent added two important

7

CONSTANT PEIGNÉ PHOT.

Fig. 1. A characteristically 'staged' collecting party, which was highly prized by the naturalists of the last half of the nineteenth century. On his way to collect hydroids in the Loire estuary Weismann stands in the centre, presumably with van Rees on his left and a student assistant, Retzer, peering into a vial on his right. Of this picture Weismann wrote the amusing account in his Notizbuch for 1880/30/8: 'About 12 to St. Nazaire in 3. Class with buckets, nets; guides etc... photography in great heat in bathing costumes, 10 exposures! Boulevard de l'Océan! 2 pharmacies, both of them did not really know what absolute alcohol was! One felt that we could not use it at all, for it was too volatile, we would open the bottle and in 5 minutes nothing more would remain. 6 in Croisic, by foot for a pretty penny to the cliffs in order to look for larvae and Pelobates. But in vain. 9 o'clock grog with Mr. Méau.' The author wishes to thank Professor Helmut Risler for permission to reproduce this photograph in his family archives and for bringing to my attention the associated passage from the Notizbuch.

theoretical dimensions for understanding the cleaving egg and differentiation of structure. After the late sixties and early seventies Haeckel's biogenetic law and the doctrine of the germ-layers, initiated by Kowalevsky and examined at length by the Hertwig brothers, Lankester, and Balfour, created for many, but not all, a reliable framework for the orientation of further descriptive work.

The paper that follows is an attempt to transport our mind's eye back to this period when first-rate biologists passionately pursued descriptive embryology. We will examine in detail a specific work by the outstanding German zoologist August Weismann, who today is largely remembered for his heroic effort to bring embryology, heredity theory, cytology and evolution into a grand neo-Darwinian synthesis. The endeavour was a hundred years too early and understandably failed. But Weismann's major statement, *Das Keimplasma* of 1892, remains one of the great classics of fin de siècle biology. Besides pushing for an uncompromising neo-Darwinian explanation of evolution, *Das Keimplasma* incorporated some novel embryological and cytological concepts that soon became very much a part of biological discourse. Three of them continue to be closely identified with Weismann's name. The first, the concept of the continuity of the germ-plasm, was spelled out in detail by Weismann in 1885 and became the core of Weismann's neo-Darwinian crusade. It also formed the basis for the other two concepts. Weismann worked out the second, that is, the notion of a qualitative reduction division of the germ-plasm during gamete maturation, between 1887 and 1891 in collaboration with his assistant Chiyomatsu Ishikawa. This reduction division, later known as meiosis, became the essential explanatory core of classical genetics. Weismann made his third conceptual contribution in an effort to understand differentiation. Borrowing a suggestion made by Wilhelm Roux in 1883 and expanding it into an elaborate theory about Biophors, Determinants, and Ids, Weismann claimed that a qualitative nuclear division took place during embryogenesis. This third idea was severely criticised by many and accepted by almost no one except Weismann and his close followers. The experimental embryological work of Boveri, Driesch, and the early Spemann quickly put to rest any possibility of a qualitative mitotic division.

This paper deals only with the first of these three concepts, that is, with the belief in a continuity of the germ-plasm. What makes this discovery particularly appropriate for this volume on the history of modern embryology is that Weismann arrived at his discovery through the traditions of classical embryology. Weismann had received his early training as an histologist and descriptive embryologist. His *Habilitationsarbeit* completed in 1863 provided a revolutionary study on the development of *Diptera*. Deteriorating eyesight, however, soon forced Weismann to turn to problems in natural history and evolution theory. After 1874 with some respite in his eye illness Weismann again took up microscopical studies of development. This time he turned to the formation of summer and winter eggs in *Daphnia* and then to the origin of the germ cells in *Hydrozoa* and *Siphonophora*. All of these embryological studies by Weismann were outstanding exemplars of the reigning traditions of descriptive embryology; they were meticulous in detail and masterful in design. It is the last of these studies which concerns us here, for this led Weismann directly to the conception of the continuity of the germ-plasm.

But Weismann had incorporated more into his study than simply the art of making exacting observations. As we will see, his work on the rise of the germ cells in hydroids was informed by many of the theoretical constructs of the day: evolution theory, the biogenetic law, the germ-layer doctrine, and the significance of the processes of the division of labour and of the alternation of generations. These concepts formed the theoretical backdrop for Weismann's descriptive studies and out of a combination of theory and observation emerged Weismann's continuity concept. It is with the idea of examining exactly how theory and observations interact in biology that we turn to Weismann's study of hydroids. Above all, it is in this way we hope better to understand how the biogenetic imperative and the germ-layer doctrine promoted as much as hindered descriptive embryology in its advance toward the era of classical genetics.

## The biogenetic law and its historians

For the most part the biogenetic law has a notorious reputation. E. S. Russell, who wrote the best known historical study of its rise and fall, ascribed the late nineteenth century enthusiasm for the biogenetic law to its rendering of earlier transcendental ideas into an evolutionary and mechanistic framework. 'Homologies', he explained, 'were considered more interesting than analogies, vestigial organs more interesting than foetal and larval adaptations.

Convergence was anathema. The dead-weight of the past', he continued in a moralising tone, 'was appreciated at its full and more
than its full value.' The reader quickly discovers what Russell preferred: a functionally oriented embryology that included not only
the methods of Entwicklungsmechanik but a return to an
Aristotelian teleology and 'the fruitful Lamarckian conception of the
transforming power of function.'[1]

A generation later Gavin de Beer incorporated an historical survey
of the biogenetic law into his efforts to classify various ontological
modes. He hoped to free morphology from 'the trammels and fetters
of the theory of recapitulation . . .'[2] and homed in on the source of the
problem by showing that Haeckel and other nineteenth century
biologists wrongly claimed that there existed a close causal bond
between phylogeny and ontogeny. De Beer's intent for the most part
was to rid embryology of any such causal association. That was a role
for genetics. 'A living organism', he insisted in 1938, 'must be
studied from two distinct aspects', that is, ontogeny and phylogeny.
'Each of these aspects', he added, 'may make suggestions concerning the possible significance of events seen under the other, but does
not explain or translate them into simpler terms.'[3]

In his impressive volume on *Ontogeny and Phylogeny* Stephen
Gould has recently surveyed the same historical terrain covered by
Russell and de Beer. He spells out in more explicit terms certain
elements of the recapitulation process: in particular, the principles
of terminal addition of new traits and of their condensation. Just as
had Russell, Gould found Haeckel's materialism, his concept of the
individual, and his belief in the inheritance of acquired characters as
contributing to the overall mechanics of the biogenetic law. He also
shows how Edward Drinker Cope supplemented these principles
with laws about the differential growth of organs and how Alpheus
Hyatt developed a theory about the juvenile senescence of the species
in order to account for paedomorphosis.

As he turns to the downfall of the biogenetic law, Gould recites the
same countervailing forces identified by Russell, and de Beer. He
insists, however, that the coup de grace came with the rediscovery of
Mendel's laws. Classical genetics, Gould incisively and perceptively
asserts, 'Ultimately disproved the two "laws" of evolution that recapitulation required for its general occurrence – terminal addition
and condensation.'[4] Gould, as readers of his works will know, does
not share de Beer's compulsion to keep separate the 'causal–analytic

aspect' of embryology from the 'historical descriptive aspect' of systematics.

My intentions are not to revise the details or the theses of these three historical accounts. I find each in its own way valuable. To a certain extent each also explains why it was that certain biologists could be captivated by conceptions that in retrospect seem so foreign from the modern perspective. Take for example Haeckel's assertion that phylogeny causes ontogeny, or to put it more bluntly, that the past causes the present. Dressed out in a materialistic rendering of earlier transcendental investments; laced with late nineteenth century assumptions about evolution, individuality, and transmission inheritance; and embellished with the principles of terminal addition and condensation, Haeckel's causal assertions are perfectly rational. Yet we still have a problem, which only Gould has recognised. As he rightly notes we must distinguish between the 'rhetoric' of recapitulation used by popularisers, such as Haeckel, and 'the discussion of recapitulation as it occurred among practising scientists in the course of their normal professional work.'[5] Now, moreover, we need to ask why it was that a whole generation of professionals, biologists of the rank of Weismann, Balfour, Lankester, and Hyatt, subscribed to and invoked the biogenetic law. Their diverse backgrounds and assertions belie that they all were seduced by the same idealistic philosophy as Haeckel. Even more to the point, how were their scientific achievements really possible if the ground beneath their castles consisted of the quicksand of recapitulation?

What I would like to do is to forget Haeckel and the 'rhetoric' of recapitulation. Let us follow Gould's example and examine more closely the 'normal professional work' of the period. Even though we might find that the biogenetic law was fundamental and comprehensive in its purported causal explanation, I believe the historian must recognise that a far richer assemblage of methods and principles informed the biology of the day.

## Weismann and hydromedusae

I propose to examine some outstanding microscopical work that Weismann performed on Hydrozoa between 1878 and 1883.[6] This work led directly to Weismann's distinction between the germ-plasm and soma and to his major efforts to understand evolution in neo-Darwinian terms. I choose this study for a practical reason: Gould has already described accurately Weismann's 1876 recourses to the biogenetic law in the latter's examination of the markings of caterpillars.[7] My study, I hope, will give perspective to that antecedent work. There is a tactical reason, too. In a penetrating paper in 1948, an article which deserves to be better known to historians of biology, the eminent embryologist N. John Berrill and his Chinese student C. K. Liu examined in detail Weismann's *Hydromedusen* and condemned the theory of the continuity of the germ-plasm, which arose out of it, on the grounds that it was founded upon the notion of recapitulation. If this is indeed the case, Berrill and Liu have presented a much more impressive example than any found by de Beer or Gould to demonstrate the perniciousness of the biogenetic law.[8] After all, the notion of continuity of the germ-plasm, so closely identified with Weismann's name, presented the most important frame of reference for discussions of heredity and evolution at the turn of the century. I will return to Berrill and Liu's argument at the end of this paper.

Weismann began studying the production of the sex cells among hydroids shortly after he completed a four-year examination of the alternation of summer and winter generations in *Daphnia*. There was, of course, a connection between the two studies of these organisms that lay at opposite ends of the invertebrate spectrum. Both concerned the biology of reproduction. Above all else, Weismann focused upon such questions as: Whence the germ cells? What were the differences between fertilised and parthenogenetic eggs? What were the bonds between sexual and asexual generations? It was characteristic of Weismann and the period to cover a range of types while investigating such fundamental issues. To turn from *Daphnia* to hydromedusae was a natural methodological ploy.

The questions that Weismann was asking encompassed some of the most compelling biological problems of the day. They stemmed from the pre-Darwinian period when in 1842 Johannes Japetus Steenstrup had written his epochal account *On the Alternation of Generations* and when seven years later Richard Owen had first coined the term parthenogenesis.[9] In the 1850s Weismann's own mentor, Rudolf Leuckart and one of his most admired contemporaries, Theodore von Seibold, had worked intensively on parthenogenesis in insects. From Robert Olby's account of the origins of Mendelism and John Farley's recent study, focusing attention on

Wilhelm Hofmeister and Frederick Orpen Bower, we have learned that the botanists in the second half of the century were deeply involved in similar concerns about sexual and asexual reproduction.[10] Darwin, Spencer, Haeckel, and Gegenbaur were among the generalists who picked up the implications of such discoveries in the 1860s. For obvious reasons that we will come to, the same problems maintained their momentum into the next decade. Waldeyer, Kleinenberg, the Hertwig brothers, Balbiani, Balfour, Claus, and Van Beneden were among the zoologists who wrote major monographs on the embryology of the reproductive organs and products in the 1870s and who employed the new germ-layer doctrine as a framework for their quest. Weismann then was following a common pursuit when he became immersed in sex cell production in hydroids.

Between 1878 and 1882 Weismann doggedly examined dozens of species of hydroids or to use his term, hydromedusae. His collecting trips started on the French riviera, and they took him over the four-year period to Naples, to the estuary of the Loire, and to Roskoff on the northern coast of Brittany. He experimented with a number of methods of killing and preserving his specimens, of embedding and staining, which had been recommended to him by the staff at the Stazione Zoologica, by Arnold Lang, and in the recent monographs of the brothers Hertwig.[11] He also alternated between several methods of free-hand cutting and tissue maceration. Many of his illustrations were done from optical 'cuts' through fresh specimens. During these four years Weismann also wrote a number of short articles on his findings, but he reserved the full picture of his objectives and the details of his argument for the magnificent three hundred page monograph that celebrated its centennial in May 1983 [1883–1983].

*Die Entstehung der Sexualzellen bei den Hydromedusen* deserves to be restudied. It stands as an exemplar of the best of late nineteenth century *Wissenschaft*. It was handsomely printed and meticulously illustrated by the author. Its format included the traditional historical overview that gave perspective to that learned century. The body of the text presented a detailed description of Weismann's microscopical examinations. By the time he had completed his study, Weismann had examined forty-three species of marine hydroids. These were distributed among three major groups: 20 tubularians (that is, the modern *Anthomedusae*), 15 campanularians (that is, the modern *Leptomedusae*), and 8 siphonophors. He also incorporated into his discussions the freshwater hydra, a species of which had been thoroughly studied by Nicholaus Kleinenberg in the early years of the decade.[12] Weismann's collecting had been confined to the coast; so he did not examine any pelagic Trachomedusae (the modern *Trachylina*) that were to be so effectively employed three years later by William Keith Brooks for a complete phylogenetic revision of the entire coelenterate phylum.[13] In any case it is doubtful that Brooks' discoveries would have altered significantly the structure of Weismann's own argument.

Weismann, however, reserved his theoretical message for a seventy-page concluding analysis of the data. It was here that Weismann led his readers through a series of complex arguments that wound their way to conclusions far beyond the modest lives of the hydromedusae. It is here we must follow Weismann in order to understand the intellectual forces that dominated the biological thought of the day.

## Dislocation of the germ site

Weismann tells us that his initial intent was to render a complete histological study of the entire class of *Hydrozoa*. This soon turned out to be impractical, and since during the first year of his investigations he stumbled upon certain anomalies having to do with the origin of the sex cells, it probably took little prompting for him to concentrate on these. The most important of his findings concerned the tubularian genus *Eudendrium*, first found on the riviera but later examined from the bay of Naples, and elsewhere.[14] In brief, Weismann discovered that the germ cells did not arise consistently in either the ectoderm or the endoderm but that the germ-site could be found in both germ-layers. Furthermore, the germ site was far removed from the gonads or germ-maturation site and in *Eudendrium* appeared in the coenosarc of the polyp colony. This proximal location near the colony's point of attachment brought considerable confusion to the traditional notion of a biological individual, that is, of the zooid among the colonial polyps. This was because the germ cell in its migration from the germ site or point of origin to the maturation site must obviously migrate across the boundaries between two, perhaps three, generations of polyp individuals. What did this imply for the received view about the alternation of generations?[15]

Weismann was not the first to spot such irregularities with the hydromedusae,[16] but he was the first who tried systematically to explain the phenomena across a wide range of species. His program entailed a comparative histology of germ-cell production. The conceptions he used and problems he broached were embryological. With just a whiff of Weismann's empirical data in our nostrils, let us turn to the general conclusions he offered at the end of his monograph. The separation of the germ site from the maturation site was the phenomenon that first galvanised his attention.

Both the maturation sites and the germ sites differed widely in their anatomical location among the forty odd species that Weismann examined. The various locations of the two sorts of sites, however, stood in marked contrast. Weismann identified four distinct locations for the maturation sites: I, the manubrium of the free-swimming medusa; II, the ectoderm of the sessile medusa; III, the ectoderm of the sporosac; and IV, the epidermal wall of the polyp stalk.[17] These maturation sites, no matter where they were found, resided in the ectoderm. It was also evident that these locations corresponded with and were constrained by the evolution of the gonozooid, itself. Clearly, the first of these locations could apply only to those hydroids that actually produced free-swimming medusae. At the opposite end of the spectrum, monomorphic polyps, such as the hydra, had no opportunity to produce gametes anywhere but in the epidermis of the main body. Not surprisingly then, the maturation sites appeared to be closely associated with the gross morphological features of each species.

The appearance of the germ sites, that is the location where the germ cells first arose from undifferentiated embryological cells, was less predictable. As he had found with *Eudendrium*, these germ sites might occur either in the ectoderm or endoderm. Nevertheless, Weismann was again able to identify a set number of specific locations, here six in number: (1) those closely associated with the gonads in the ectoderm of the manubrium of the free-swimming or sessile medusa; (2) those situated in the entocodon, or to use the more poetic German term, the 'Glockenkern'; (3) those found in the endoderm of the gonophore; (4) those located in the endoderm of the blastostyle; (5) those residing in the endoderm of the side and main hydranths, that is, in the coenosarc of the hydroid colony; and finally (6) those situated in the ectoderm of the main hydranth.[18] This wide distribution of the germ sites bore little relation to the accepted taxonomic order. Closely related species displayed germ-site locations as diversely separated as the manubrium of the free-swimming medusa and the endoderm of the coenosarc. Distantly related species, on the other hand, might have germ sites in similar locations. Consistency could be guaranteed only within a given sex of a given species.

Weismann recognised, however, that when he compared the two sets of tabulations they revealed an increasing separation of the two different sites.[19] From a phylogenetic point of view the germ site became ever more dislocated from the maturation site. This dislocation, in fact, was a directional one, for the germ site appeared ever closer to the attachment point of the hydroid colony as Weismann's comparisons moved from the first to the fourth position of the maturation site. In his terminology there occurred an accelerated centripetal dislocation of the germ site. This dislocation was most obvious with the twenty species of tubularian polyps. Although the series was not as complete, Weismann observed a similar dislocation pattern among the fifteen species of campanularians, and he even claimed to find a similar pattern among the six species of siphonophors. Yet what had he accomplished? What problems had he solved through these correlations? And what new problems had he introduced?

## Evolution of the sporosac

To answer these questions we must recognise that Weismann had touched upon a controversial issue in hydroid morphology: the relationship between the polyp, the medusa, and the gonophore. The standard view presented by Allman, Huxley, Van Beneden, and Gegenbaur held that through a process of the division of labour the single polyp evolved into a hydroid colony containing morphologically diverse zooids, which performed different functions in the life of the colony. Thus there arose gastrozooids, gonozooids, and other specialised individuals for defence. As evolution progressed, this specialisation and segregation of tasks continued. The reproductive individuals became more elaborate; the sporosac evolved into medusoid gonophores, and these in turn evolved into free-swimming medusae. Carl Claus, Professor of Zoology and Comparative Anatomy at the University of Vienna and an authority on coelenterates, pictured this progressive sequence from hydra to medusa in his popular zoology textbook:

The sexual bodies appear at many different levels of morphological development, for they at first form simple swellings of the body wall filled with sexual materials (Hydra), at a more advanced stage they take over as a prominent extension of the body cavity or radial canal, in whose surroundings the sexual materials accumulate (*Clava squamata*), in an even more advanced stage there exists at the perifery of the bud a mantle-like covering with more or less radial vessels (*Tubularia, Coronata, Eudendrium ramosum* Van Ben.), and finally it comes to the formation of a small releasable medusa with mouth, floating bladder, tentacles, and periferal bodies (*Campanularia gelatinosa* Van Ben., *Sarsia Tubulosa*).[20]

There was, however, a contrary view that had first been suggested by Gottlieb von Koch in 1873, was endorsed by one of Haeckel's students, Otto Hamann, and had been halfheartedly sanctioned by the brothers Hertwig.[21] Their conception was contrary to a progressionist vision. They maintained that the gonophores and sporosac were degenerated medusae that had lost their locomotive capacity. Weismann had predicated the ordering of his four groups of maturation site locations on just this counter-intuitive regression, and by the time he had completed his analysis, there could be little doubt that degeneration of the medusae was the true origin of the sporosac.[22] His analysis was intricate and involved a detailed comparison of sporosac and medusa ontogeny. I will refer to only two of Weismann's demonstrations.

The first involved the entocodon or 'Glockenkern'. At that time this structure was considered to be a delaminated knot of ectodermal tissue that soon after its embryological appearance became closely surrounded by the endodermal lamella.[23] In hydroid species with free-swimming medusae, the entocodon develops into the ectodermal portions of the manubrium and the subumbrella. In species with sporosacs the entocodon remains a negligible ectodermal tissue situated next to the gastrodermal stem of the blastostyle (Fig. 2). A progressive evolution of the entocodon from sporosac to free medusa implied a teleological prescience on the part of the evolving phylum. A regressive evolution, that is, a degeneration from medusa to sporosac, simply consigned the remains of the entocodon to a nonfunctioning vestige. Weismann eschewed teleology and opted for the second explanation.[24]

The second argument had essentially the same thrust. It involved the endodermal lamella, which in the free medusae develops into the lining of the gastric cavity and the radial and ring canals. In

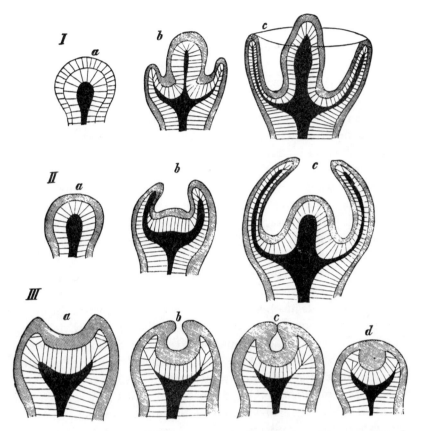

Fig. 2. Diagram of sequential phyletic steps in the evolution of the entocodon. In reference to the three horizontal rows the original caption reads:

I. Three stages in the ontogeny of the medusa bud; the oldest phyletic stage.

II. The same from a later phyletic stage.

III. The same ontogenetic stage (corresponding to stage b in rows I and II of four sequential phyletic stages. (Weismann, *Hydromedusen*, p. 260)

In this diagram the entocodon is clearly identifiable as the ectodermal knot in figure (d) of row III. Weismann interpreted the entocodon as a vestige of the ectodermal portion of the manubrium and subumbrella of the medusa, e.g., row I. In the final phylogenetic stage the endodermal lamella surrounding the entocodon is also the vestige of an endodermal tissue that in previous phylogenetic stages, i.e., figure (c), row I, formed the endodermal portion of the manubrium and subumbrella. On the same page in the text, Weismann explains: 'One sees that the reconstructed phylogenetic development of the entocodon is not merely a simple reversal of today's ontogenetic stages.' In other words, the regression of the medusa and the phylogeny of the entocodon (read in the sequence Ib, IIb, IIIa, b, c, and d) was not recapitulated in the ontogeny of IIId. In this particular example, germ-layer analysis rather than the biogenetic law served as an interpretive guide.

sporosacs one finds the 'vestigial' remains of this lamella perfectly intact, but serving no obvious function. Its 'presence', Weismann pondered, 'makes no sense at all if one assumes progressive development.'[25]

In both of these ontogenetic examples Weismann was arguing that evolution was not a symmetrical process. Forms could not come down the tree exactly as they had climbed it. Moreover, as he reconstructed the evolutionary stages that the degeneration of the medusae must have entailed, he found himself describing a sequence that was unlike the observed ontogeny of the medusae. 'One sees', he asserted in a key sentence, 'that this reconstructed phylogenetic development of the entocodon is not merely a simple reversal of today's ontogenetic stages.' In short, the ontogeny of degenerate structures DID NOT RECAPITULATE THE PHYLOGENETIC PAST.

Furthermore, Weismann insisted that each evolutionary stage bore some functional relationship to the life of the organism. This becomes clearer as he offered a neo-Darwinian scenario of the evolution and later degeneration of the medusa. Weismann's phylogenetic claims and neo-Darwinian scenarios have important implications for our perception of his use of the biogenetic law.[26] More on this later.

The upshot of Weismann's re-evaluation of the evolution of the sporosac was momentous. It led him to conclude that a regression of the medusa forced a dislocation of the maturation site from the ectoderm of the manubrium to the entocodon, that is, to the ontogenetic precursor of the ectoderm of the manubrium and subumbrella. A similar dislocation must have occurred with the germ site, which, you will remember, in the first position coincided with the maturation site. Weismann's histological examinations and the comparative tables of the location of the two sites told the rest of the story. As the medusa phylogenetically degenerated into a medusoid gonophore and then into a sporosac, the germ site became centripetally removed: first into the entocodon, then into the endoderm of the gonophore bud, further into the endoderm of the blastostyle, until in the most advanced evolutionary species the germ sites appeared in the endoderm and ectoderm of the stem hydranth.[27] (See Fig. 3.)

This phylogenetic separation of the germ site from the maturation site necessitated an ontogenetic migration of the germ cells. Since the germ cells were distinctly different from the other cells, it was a migration that the histologist could observe. Moreover, since he never found germ cells wandering into gastrozooids and since there always seemed to be the correct number of germ cells congregated at the point where a blastostyle with a set number of gonophores was about to develop, Weismann argued that there must exist a pre-established migration route, or as he called it, a 'Marsch-route' from germ site to gonads. The 'marching route of the single cells', he wrote, 'must be imprinted in a rather special way.'[28]

## Specificity of the germ-layers

Weismann, in fact, was confronting one of the most difficult embryological problems: what determined the destiny of the cells and the germ-layers? In 1883 there could be only one non-teleological answer to this question. The answer was heredity or 'Vererbung.' We will return to this word and the concept that lay behind it toward the end of this paper. First, one last element of Weismann's study needs to be examined.

With most of the species he collected, Weismann found the germ site in the endoderm. This location resulted in the migration route, whether short or long, existing for the most part along the endodermal layer until the germ cells reached the ectodermal entocodon. Did the endodermal germ site indicate that the germ-cells were themselves endodermal in origin? The question forced Weismann to confront one of the most vexing embryological issues of his day: did the germ-layers possess integrity and specificity? Suffice it to say for the moment that Weismann had been very much impressed by the recent studies of Oscar and Richard Hertwig on the germ-layer theory. Although they had vacillated in their commitment toward the doctrine of specificity, the Hertwigs had spoken approvingly of the ectodermal origin of the germ cells.[29]

Weismann revealed no such hesitation; his biology demanded it. On the basis of its simple and non-compound nature Weismann reasonably maintained that hydra was the most primitive extant hydroid. On the basis of the principle of the division of labour and the contemporary understanding of the alternation of generation as a mode of asexual reproduction, Weismann held that the medusae of the *Hydrozoa* must be a phylogenetic stage beyond the colonial hydra-like polyps. Many of these also possessed germ sites associated with the gonads in the ectoderm of the manubrium. Only when evolution reached the point when the medusa retrogressed into

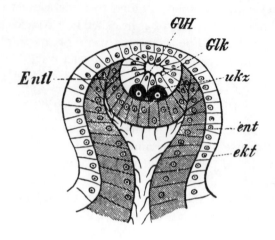

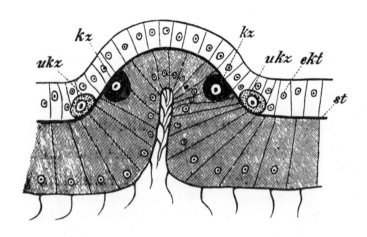

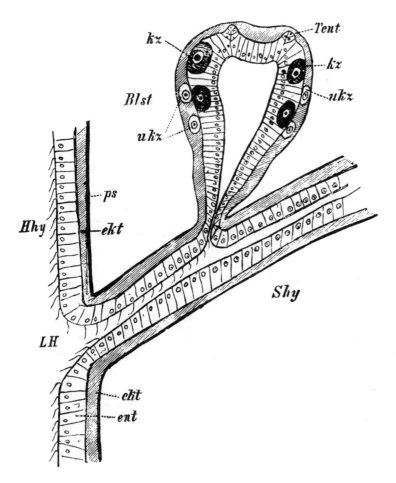

Fig. 3. Dislocation of the germ-cell site: early phyletic stages. These sketches represent three stages in the centripetal displacement of the germ-site. Sketch (a) depicts the second stage in Weismann's sequence. Two germ-cells appear in the entocodon. Sketch (b), representing stage three, is the point in hydroid phylogeny when the germ-cells first appeared in the endoderm. Weismann felt the specificity of the germ-layers dictated that the germ-cells originate in the ectoderm but that they could not be recognised as such. Thus he labels them as 'Urkeimzellen' or primordial germ cells. Sketch (c) indicates a more-advanced phylogenetic stage. Stages five and six may be seen in Fig. 4. German lettering indicated: Blst, blastostyle; ekt, ectoderm; ent, endodermal lamellae; Glk, Glockenkern or entocodon; Gll, Glockenkern Anlagen; Hhy, main hydranth; kz, Keimzelle or germ-cell; Shy, side hydranth; Tent, tentacle; ukz, Urmkeimzelle or primordial germ-cell. (From Weismann, *Hydromedusen*, pp. 231, 249.)

gonophores did this ectodermal association cease. Weismann argued that for the want of space and for the need of accelerating the reproductive process, the germ site was forced out of the ento-codon into the adjacent endoderm.[30] Similar pressures brought a further dislocation to the endoderm of the blastostyle, and then to the side and main hydranths.

Did this relocation to the endoderm of the germ sites imply that the germ cells now became endodermal in origin? Such a con-clusion would have violated too many of Weismann's working suppositions. But let us hear from Weismann himself:

How and by what means can it be possible that suddenly the endoderm cells differentiate into sex cells just as the ectoderm cells have done up to then? It is no exaggeration for us to regard this sort of thing as impossible. Thus if the cells in the pertinent [that is, endodermal portion of the] gonophore bud demonstrate the property of differentiating into germ cells, the conclusion is unavoidable that they must have [first] migrated from the ectoderm, WHETHER THIS CAN BE CONFIRMED BY OBSERVATION OR NOT.[31]

Weismann insisted that germ cells originated in the ectoderm and migrated into the endoderm whether this could 'be confirmed by observation or not.' To justify the lack of ocular proof, he spoke of 'Urkeimzellen' that were indistinguishable from the undifferentiated cells of the early ectoderm and that migrated into the coenosarcal endoderm and there became microscopically evident somewhere along their 'Marsch-route' to the ectodermal entocodon (Fig. 4). Only once did he secure tangible evidence of such a tortuous migra-tion.[32] In the female of *Eudendrium racemosum* Weismann observed germ cells in both the ectoderm and endoderm of the coenosarc. But which of these locations was primary? With one specimen Weismann caught, immobilised by his absolute alcohol treatment, two germ cells in the act of boring through the interstitial lamella on their way from the ectoderm to the endoderm where they then could proceed on their migration to the gonophores. He had confirmed by observation in one specimen the ectodermal origin of the germ cells and thereby rein-forced the prevailing doctrine of specificity of the germ-layers.

## General results

It is now time to stand back from these details of germ cell dislo-cations, migrations, and 'Urkeimzellen' and to ask about some

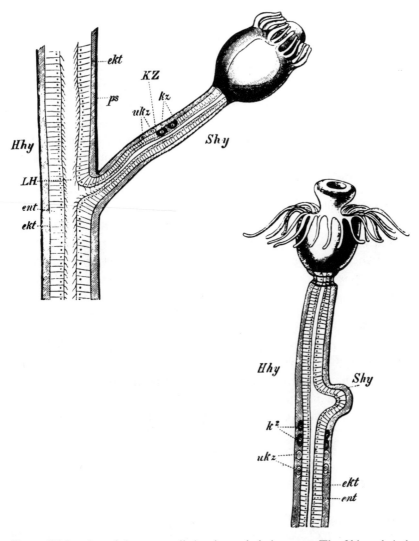

Fig. 4. Dislocation of the germ-cell site: later phyletic stages. The fifth and sixth stages (sketches (d) and (e)), of the phylogenetic dislocation of the germ-site provided the clinching moment for Weismann's argument that the ectodermal primordial germ-cells must be present in all hydroids although they could be recognised as such only in stage VI of *Eudendrium*. In one specimen, Weismann felt he had observed the germ-cells migrating from the ectoderm to the endoderm of the stem hydranth. Since the ontogenetic migration route of the germ-cells helped Weismann determine the phylogenetic sequence of the species of tubularians he examined, the biogenetic law in this case acted as the guiding principle. For explanation of the labelling, see caption to Fig. 3. (From Weismann, *Hydromedusen*, p. 250.)

general features of Weismann's biology – and about the embryology of the period.

If you have followed the argument about germ-cell migration, it will not surprise you that Weismann found his discoveries leading directly to his germ-plasm theory. Centripetal dislocation of the germ sites, a predetermined migration route from coenosarc to the gonophore, and 'Urkeimzellen' all added up to the conception of the continuity of germinal material. In 1881 Weismann delivered an address before the Versammlung of German Scientists and Physicians in which he introduced his idea of a separation among metazoans of the germ cells from the soma. During the same Spring of 1883 in which he published his full-length study of hydromedusae, he also delivered his famous Pro-Rector's lecture on heredity in which he fired the opening salvos in his war against the inheritance of acquired characters and where he first referred to 'the continuity of the protoplasm of the germ cells.'[33] The following year in an 'Autoreferat' of his *Hydromedusen*, he repeated the same expression, now placing it explicitly within the context of the diverse and alternating generations of his hydroids.[34] Modifying the expression in 1885 to the more familiar 'continuity of the germ-plasm', he incorporated it into the title of an extensive monograph.[35] But let us return to the hydromedusae and the nesting of suppositions which helped Weismann come to the conception of continuity.

Over thirty-five years ago Berrill and Liu stressed Weismann's dependence '... on the doctrine of recapitulation, in its original form now greatly discredited.'[36] Their observation has some merit. On two occasions in his *Hydromedusen* Weismann mentioned the biogenetic law, and he continued to subscribe to it throughout his life. But to pay nominal tribute is not the same as to incorporate the principle into the creative machinery of one's active science.

Weismann had come a long way from the days when he had used the biogenetic law to describe the phylogeny of caterpillar markings. We have seen how Weismann modified the biogenetic law while interpreting the development of the entocodon, and we have noted his functional concerns in conjuring up neo-Darwinian scenarios for the evolution and degeneration of the medusa. Both speak against a blind obsession with the biogenetic law. More telling, we have followed how Weismann's germ-layer analyses established the retrogression of the medusae and the phylogeny of the *Hydrozoa*. These determinations were essential for mapping out the migration route

of the germ cells and ultimately for arguing that there existed unobservable 'Urkeimzellen' that guaranteed a continuity of germ-plasm
into the endoderm and back into the ectoderm. As historians we
need to unravel the bond between the biogenetic law and the germ-
layer doctrine, two discredited nineteenth century conceptions.
Berrill, Liu, De Beer and Russell all recognised a connection, but
did not pursue it; Gould completely ignores it. We must turn to
Professor Oppenheimer's classic essay of 1940 on *The Non-Specificity of the Germ-Layers*[37] to find any historical analysis of the linkage
between the two.

You will recall that Oppenheimer graphically describes the 1870s
as a decade when the microscopic techniques and germ-layer
analyses of Kowalevsky and Metchnikoff became widespread. It was
only then that the germ-layer doctrine became bonded to the
biogenetic law. Haeckel, who in his *Generelle Morphologie* of 1866
could only deduce recapitulation from reductionist principles and,
as Oppenheimer points out, failed even to mention the germ-layer
doctrine, found by 1872 that the germ-layers gave him the analytic
and empirical proof that he desired.[38] Kleinenberg, Lankester, the
Hertwigs, Balfour, and a host of others found in this analytically
fashioned linkage a powerful guide to morphological and phylogenetic determinations. I believe Oppenheimer's double emphasis
is correct. To treat historically one concept without the other is to
miss the combined impact of both. Weismann, anyway, certainly
considered the two principles as inseparable. But why?

This brings me back to Weismann's use of the expression
'heredity' or 'Vererbung'. Today this term has been conditioned by
classical genetics. It conjures up the process of transmission genetics, meiosis, Mendelian patterns of trait distribution, and the
discovery of DNA. In the nineteenth century it encompassed more
than a transmission between generations. It denoted the genetic
connection between alternating generations in a complex life cycle.[39]
With the acceptance of the Virchowian cell theory, the term also
included the genetic connections between cell generations, and with
the success of the germ-layer doctrine in late sixties and seventies it
guaranteed a specificity between the primitive germ-layers and their
derivatives. Cell divisions, germ-layer derivatives, and alternating
generations expressed the laws of heredity just as regularly as did the
succession of sexual individuals, or for that matter, the succession of
species. As a consequence, 'heredity' was as much a part of the

developmental phenomena as it was a designation of transmission between traditional parents and offspring.

This more general use of the term 'heredity' is illustrated by a passage in Weismann's *Hydromedusen*. It consists of a bold statement about heredity and an image of development. We will turn to the image first. With this Weismann asserted the impossibility of endodermal cells of the gonophores suddenly assuming the 'inherited properties [Anlagen] of the ectodermal cells.' He asks you to picture that:

A long lineage of cell generations separate the two cells which lie close together, one of which is on the ectodermal side, the other, arising as a genuine endodermal cell, is on the endodermal side of the supporting lamella. They are connected only at the root of the entire polyp stalk, that is at the cleavage of the egg, out of which the first hydranth of the colony arises. How and by what means can it be possible that suddenly the endoderm cells differentiate into sex cells just as the ectoderm cells have done up to then? It is no exaggeration for us to regard this sort of thing as impossible.

It would be easy to argue that this passage reflected the pervasive and corrupting influence to the biogenetic law. In his bold statement, however, which precedes this passage by two lines, Weismann placed the emphasis elsewhere. 'What does this mean', he queried, 'other than that certain cell lineages [Zellengenerationen] alone possess the capacity to produce the sex cells, that consequently A STRICT LAW OF HEREDITY MUST RULE HERE AND NOTHING LESS, such as arbitrariness and accident.'[40] I believe that the prevailing notion of heredity was even more fundamental to nineteenth century biology than the biogenetic law and the germ-layer doctrine. The first drove the second and third.

In its expanded meaning the process of heredity played a different role in the understanding of life. Heredity was the conservative process that guaranteed that like produced like, that son copied father, that daughter cells were identical to the mother cell, that the germ-layers maintained their integrity, and that a phylogenetic 'Stammbaum' could exist in the first place. Countervailing this conservative force of heredity were the processes of change: the laws of variation and adaptation.

In the post-Darwinian period one commonly finds biologists referring to these two opposed principles as they tried to explain the irregular results of the biogenetic law. Haeckel wrote chapters on the processes of heredity and adaptation and then brought them together as the principles of stability and change in the *Generelle Morphologie*. In 1872 he got more quickly to the point. 'All the phenomena met with in the morphology of the Calcispongiae', he concluded, 'may be completely explained by the reciprocal action of two physiological functions, inheritance and adaptation; and we need no other causes to comprehend their production.'[41] On this side of the channel Balfour was equally forthright: 'There are, according to this theory, two guiding, and in a certain sense antagonistic, principles which have rendered possible the present order of the organic world. These are known as the laws of heredity and variation.... The remarkable law of development enunciated above...', Balfour continued, referring to the biogenetic law, 'IS A SPECIAL CASE OF THE LAW OF HEREDITY.'[42] We could easily find similar comments by Fritz Müller, Huxley, Gegenbaur, and Lankester, but this is enough to demonstrate that there is more behind Weismann's study of hydromedusae than simply the biogenetic imperative. Which brings me to my last point.

I have tried to present Weismann's study of hydromedusae as a complex of nineteenth century concepts: evolution theory, the concern for the alternation of generations, the process of the division of labour, the biogenetic law, the germ-layer doctrine, and now the dialectic between heredity and variation. These concepts interacted with and reinforced one another. They acted as a constellation of principles, and to a significant extent they arose out of the contemporary interest in embryology. There is, however, an irony to all this. The conception of the continuity of the germ-plasm is one of several developments of the last decades of the century that helped bring about the destruction of nineteenth century biology and rendered obsolete the magnificent edifice of descriptive embryology.

As Weismann developed his germ-plasm theory, he helped refashion our understanding of heredity. He was not the only biologist to take this route. In an enormously useful book Gloria Robinson briefly noted that Brooks and Carl Nägeli also reforged a new definition of heredity by means of continuity theories.[43] But Weismann's impact was more important. In the first place, Weismann became famous for using the continuity theory to pummel the hitherto unchallenged assumption of the inheritance of acquired characters. Since Lamarckian inheritance was a necessary corollary to the biogenetic law,[44] his crusade provided an ironic twist, indeed! In the second place, and I believe more important for the history of biology, Weismann began expanding upon his germ-plasm theory. As he did so, he quickly realised that heredity was not an antagonistic principle to variation. Heredity, instead, became the source of variation and soon the core to his neo-Darwinism.[45] This redefinition leads us directly into the Mendelian period.

If, then, we disregard the 'rhetorical' flourishes concerning the biogenetic law and look instead at how the law operated in conjunction with the rest of nineteenth century descriptive embryology, we find more than simply an obstructive principle. This study suggests that the notorious biogenetic imperative, fortified by the germ-layer doctrine and a nineteenth century conception of heredity, assured its own demise. Weismann, who never abstained from taking of its waters, became an unwitting agent in its destruction.

This paper was written with support from an NSF grant (FP SES 82-06331). The author would also like to express his thanks to Sears Crowell with whom he conferred about the biology of hydroids.

## Endnotes

1. E. S. Russell, *Form and Function. A Contribution to the History of Animal Morphology* (London: John Murray, 1916), pp. 312–13.
2. Gavin de Beer, *Embryos and Ancestors* (Oxford: Clarendon Press, 1958), p. 174. De Beer first published this work in 1930 under the title *Embryology and Evolution*.
3. *Ibid.*, pp. 76–7. For a more elaborate discussion of de Beer's conception of biology, see: Frederick B. Churchill, 'The Modern Evolutionary Synthesis and the Biogenetic Law' in *The Evolutionary Synthesis. Perspectives on the Unification of Biology*, ed. Ernst Mayr & William B. Provine (Cambridge, Mass.: Harvard University Press, 1980), pp. 112–22.
4. Stephen Jay Gould, *Ontogeny and Phylogeny* (Cambridge, Mass.: Harvard University Press, 1977), p. 202.
5. *Ibid.*, p. 102.
6. In a companion paper to this one, delivered first in Freiburg i/Br. on 30 May 1984 as part of the sesquicentennial celebrations of Weismann's birth (17 Jan. 1834), the author has examined the development and implications of Weismann's continuity concept following the events described here. Frederick B. Churchill, 'Weismann's Continuity of the germ-plasm' in Historical Perspective' in *August Weismann (1834–1914) und die Theoretische Biologie des 19. Jahrhunderts. Urkunde, Berichte und Analysen*, ed., Klaus Sander, pp. 107–24. The entire conference appears as a double issue of the *Freiburger Universitätsblätter, 1985, Jg 24, Heft 87–88*.
7. Gould, *Ontogeny* (n. 4), pp. 102–9. Weismann's study appeared as two monographs in Part II of his *Studien zur Deszendenztheorie*, that is, as 'Die

Entstehung der Zeichnung bei der Schmetterlingsraupen' and 'Über den phyletischen Parallelismus bei metamorphischen Arten.' The entire collection was published in English as *Studies in the Theory of Descent*, transl., Raphael Meldola (London: Sampson Low, Marston, Searle and Rivington; 1882).

8. N. John Berrill and C. K. Liu, 'Germplasm, Weismann, and Hydrozoa', *Quarterly Review of Biology* (1948) **23**, 124–32.

9. Frederick B. Churchill, 'Sex and the Single Organism: Biological Theories of Sexuality in Mid-Nineteenth Century', *Studies in History of Biology* (1979) **3**, 139–77.

10. Robert C. Olby, *Origins of Mendelism* (New York: Schocken Books, 1966) and John Farley, *Gametes & Spores. Ideas about Sexual Reproduction 1750–1914* (Baltimore and London: Johns Hopkins University Press, 1982). The latter serves as the most complete examination of the mid-century views of sexual and asexual reproduction.

11. August Weismann, *Die Entstehung der Sexualzellen bei den Hydromedusen. Zugleich ein Beitrag zur Kenntniss des Baues und der Lebenserscheinungen dieser Gruppe* (Jena: Gustav Fischer, 1883), pp. 12–14.

12. Nicholaus Kleinenberg, *Hydra. Eine Anatomisch–Entwicklungsgeschichtliche Untersuchung* (Leipzig: Wilhelm Engelmann, 1872).

13. William Keith Brooks, The Life-History of the Hydromedusae: a Discussion of the Origin of the Medusae and the Significance of Metagenesis, *Memoirs of the Boston Society of Natural History* (1886) **3**, 359–430.

14. August Weismann, Zur Frage nach dem Ursprung der Geschlechtszellen bei den Hydroiden, *Zoologischer Anzeiger* (1880) **3**: 226–33; Beobachtungen an Hydroid-Polypen, *Ibid.* (1881) **4**, 111–14.

15. At the end of his *Hydromedusen* Weismann explicitly addresses this issue (n. 11), pp. 293–5. It is worth noting that Brooks wrote his monograph also with an eye to commenting upon the issue of the alternation of generations.

16. Alexander Goette, Ein neuer Hydroid-Polyp mit einer neuen Art der Fortpflanzung, *Zoologischer Anzeiger* (1880) **3**: 352–9; Nicholaus Kleinenberg, Über die Entstehung der Eier bei Eudendrium, *Zeitschrift für wissenschaftliche Zoologie* (1881) **35**: 326–32.

17. Weismann, *Hydromedusen* (n. 11), pp. 228–9.

18. *Ibid.*, pp. 229–30, 244–54.

19. *Ibid.*, pp. 225–54. Weismann's argument in this section is at places difficult to follow because he jumps from one sub-order to another before coming to a conclusion. See also pp. 285–7.

20. Carl Claus, *Grundzüge der Zoologie zum Gebrauche an Universitäten und Höheren Lehranstalten sowie zum Selbststudium* (Marburg and Leipzig: N. G. Elwert, 1876), pp. 220–1.

21. Gottlieb von Koch, Vorläufige Mittheilungen über Cölenteraten, *Jenaische Zeitschrift für Medizin und Naturwissenschaften* (1873) **7**: 464–70, 512–15. Otto Hamann, Der Organismus der Hydroidpolypen, *Ibid.* (1882) **15**: 473–504; Oscar and Richard Hertwig, *Der Organismus der Medusen und seine Stellung zur Keimblättertheorie* (Jena: Gustav Fischer, 1878), pp. 64–6.

22. Eugen Korschelt and Karl Heider, *Text-book of the Embryology of Inverte-brates*, transl. Edward L. Mark & W. McM. Woodworth (London: Swan Sonnenschein & Co., 1895), Part I, pp. 47–8.

23. C. K. Liu and N. John Berrill, Gonophore Formation and Germ Cell Origin in Tubularia, *Journal of Morphology* (1948) **83**: 39–60. At the same time that Berrill and Liu were involved in their historical critique of Weismann's *Hydromedusen* (n. 8), they were also revising the traditional assumption about the ectodermal origin of the entocodon. After examining *Tubularia crocea*, they concluded, 'With regard to entocodon formation there is no doubt that in this species the epidermis of the gonophore bud plays no part. No sign whatever can be seen in any way suggestive of its proliferation, invagination, or isolation'. In an interesting historical aside they pointed out that their findings contradicted over sixty years of histological research. The previous misconception was due, in part, to the difficulty of catching the entocodon at its earliest stage of formation, but they also suggested that '... one should not underestimate the overwhelming influence and authority of Weismann, who stated categorically that the entocodon and germ cells were formed from the epidermis (ectoderm) throughout the Hydrozoa.' (p. 52). It is curious that they chose not to emphasise this aspect of Weismann's work in their historical paper.

24. Weismann, *Hydromedusen* (n. 11), pp. 254–6; Libbie Henrietta Hyman, *The Invertebrates: Protozoa through Ctenophora* (New York and London: McGraw-Hill, 1940), pp. 427–31.

25. Weismann, *Hydromedusen*, p. 256.

26. *Ibid*., pp. 258–62. Quotation appears on p. 264.

27. *Ibid*., pp. 249–53.

28. *Ibid*., p. 275.

29. Oscar and Richard Hertwig, *Organismus* (n. 21), pp. 31–7; *Studien zur Blättertheorie. Heft I. Die Actinien* (Jena: Gustav Fischer, 1879), pp. 166–8; Jane M. Oppenheimer, The Non-Specificity of the Germ-Layers, reprinted in Oppenheimer, *Essays in the History of Embryology and Biology* (Cambridge, Mass.: The M.I.T. Press, 1967), pp. 272–4.

30. Weismann, *Hydromedusen* (n. 11), pp. 232–3, 264.

31. *Ibid*., p. 288–9; translation modified from Berrill and Liu (n. 8), p. 127. Emphasis is Weismann's.

32. *Ibid*. This is best seen in Plate III, fig. 7.

33. August Weismann, Über die Vererbung, reprinted in *Aufsätze über Vererbung und verwandte biologische Fragen* (Jena: Gustav Fischer, 1892). Not having been translated until 1889, the English version uses the then-current expression 'continuity of the germ-plasm'.

34. August Weismann, Die Entstehung der Sexualzellen bei den Hydromedusen, *Biologisches Zentralblatt* (1884) **4**, 12–31.

35. August Weismann, *Die Continuität des Keimplasmas als Grundlage einer Theorie der Vererbung* (Jena: Gustav Fischer, 1885).

36. Berrill and Liu, Germplasm (n. 8), p. 129.

37. Oppenheimer, 'Non-Specificity' (n. 29).

38. Ernst Haeckel, (1872). On the *Calcispongiae*, their Position in the Animal Kingdom, and their Relation to the Theory of Descendence. *Annals and*

*Magazine of Natural History* (1873) S. 4, **11**: 241–2, 421–30. In an important passage Haeckel indicated with his characteristic flourish this scientific advance: 'In my *General Morphology* I sought to demonstrate synthetically that all the phenomena of the organic world of forms can be explained and understood only by the monistic philosophy; and now this demonstration is furnished *analytically* by the morphology of the Calcispongiae.' (p. 430.)

39. For example, see Richard Owen, *On Parthenogenesis or the Successive Procreating Individuals from a Single Ovum* (London: John van Voorst, 1849).

40. Weismann, *Hydromedusen* (n. 11), pp. 288–9. The translations of these two passages are modified after Berrill and Liu (n. 8). The first forms the pivotal point in their argument that the biogenetic law was the operative principle in Weismann's embryology. The emphasis in the second is mine.

41. Haeckel, Calcispongiae (n. 38), p. 430.

42. Francis Maitland Balfour, *A Treatise on Comparative Embryology*, 2 vols. (London: Macmillan and Co., 1885), vol. I, pp. 2–3. Emphasis is mine.

43. Gloria Robinson, *A Prelude to Genetics. Theories of a Material Substance of Heredity: Darwin to Weismann* (Lawrence, Kansas: Coronado Press, Inc., 1979), pp. 122–3.

44. De Beer, *Embryos* (n. 2), pp. 15–18; Gould, *Ontogeny* (n. 4), pp. 80–1. At this point Gould also notes that the notion of heredity was different in the pre-Mendelian era: 'Haeckel's definitions of heredity and adaptation do not follow modern usage. Heredity refers to characters that an animal receives from its parents, adaptation to those acquired during its lifetime. The adaptation of one generation may, of course, be the next generation's inheritance.' p. 81, ftn. Unfortunately Gould does not pursue the consequences of this older meaning.

45. August Weismann, The Significance of Sexual Reproduction in the Theory of Natural Selection (1886) in *Essays upon Heredity and Kindred Biological Problems*, ed. Edward B. Poulton, Selmar Schönland & Arthur E. Shipley (Oxford: Clarendon Press, 1889). The change in Weismann's understanding of heredity is the subject of my subsequent paper, see n. 6.

# Embryology and classical zoology in Great Britain

MARK RIDLEY

New College, Oxford, UK

## I

The aims of classical zoology are two: the description of the diversity of animal life, and the explanation of the relations between the main kinds of animals. Classical zoology pre-dates Darwin, but was stimulated more than any other field of biology by Darwin's theory; and the golden age of classical zoology begins, in Great Britain, soon after 1859. The beginning can be dated with the foundation of chairs in Zoology and Comparative Anatomy in Oxford (1860) and Cambridge (1866). Biology, from then on, has fragmented, but in gradual stages, which makes it less easy to choose a closing date. Classical zoology retains its influence, and interest, down to the present day. My narrative will finish in the early nineteen thirties.

I shall not, in this essay, give a balanced account of classical zoology between the 1860s and 1930s. I am going to concentrate on its embryology. Nor shall I be presenting a conventional history of ideas. Previous historians of embryology during this period have, very properly, concentrated on Germany. In Germany were almost all the great men: Von Baer, then Müller and Haeckel, then His and Roux, Gegenbaur and Weismann. Germany, too, was the breeding ground of nearly all the main embryological ideas. Great Britain was, in embryology, provincial; all the main ideas did reach Great Britain, in the end, but always after a delay. If the embryological criterion of homology received its first important statement in Britain in Huxley's Croonian lecture of 1858, it had been invented by Von Baer in 1828; if populist recapitulationism was taken up by Lankester in 1873, how much more influential and popular were Fritz Müller's *Für Darwin* of 1863 and Ernst Haeckel's *Generelle*

35

*Morphologie* of 1866; if Wilfred Jenkinson, just after the turn of the century, was the first major experimental embryologist in Britain, His had invented it in 1871; E. B. Ford and Julian Huxley's work on rate genes in the 1920s followed Goldschmidt's; and when Huxley, in 1920, treated an axolotl with extract of thyroids bought from the local butcher and successfully induced metamorphosis, the *Daily Mail* might expostulate that 'Young Huxley has discovered the Elixir of Life', but Huxley was only (known to himself) repeating the experiment of Laufberger in 1913 and Jensen in 1916. No wonder the historian of the ideas concentrates on Germany! But those long delays, of up to thirty years, can be made to demonstrate another point, which is more important here: the British zoological community was relatively isolated. Of course there were contacts. There was correspondence, there were visits. The visits, made by young zoologists, were always in the same direction: always from Britain to Germany. The young German zoologist had nothing to learn over here. And the British, too, knew where the action was. Lankester went, after graduating, to Leipzig. Balfour dropped in on Kleinenberg, in Messina, at Christmas in 1881. While Bourne was an undergraduate at Oxford, he spent a summer vacation working with Weismann in Freiberg; while Geoffrey Smith was an undergraduate, he visited Richard Hertwig in Munich; Julian Huxley went to Heidelberg in 1906 before coming up to Balliol, and went back there in 1913 (to learn comparative biology) before taking up a Professorship at Rice University in Texas. During 1913 Huxley, too, went to visit Richard Hertwig in Munich. He found the research in Heidelberg and Munich almost unintelligible, and not because of linguistic difficulties: an ability to read German was as essential a qualification for a zoological researcher as the use of the microscope. Many (no doubt) did not read German as readily as English; but then they were not all perfect microscopists either. The contacts between the zoologists of Great Britain and Germany, although they did exist, are little more than a trickle. They were not enough to break down the character and momentum of British zoology, and it is possible for the historian to treat British zoology by itself, taking relatively little notice of the events in Germany. This becomes even more possible if we shift our attention from ideas, to people. I do not doubt that the invention of ideas is historically interesting. But it will not provide the theme of this essay. We shall examine the ideas of British embryologists, but in their social and geographical setting.

My main concern will be with the most flourishing groups of embryologists of the period, with their rise and fall, and dispersal in space.

The geography of British classical zoology is simple. We need hardly go outside Oxford and Cambridge to write the entire history of embryology during this period. We shall occasionally go out, to London, and (for D'Arcy Thompson) to Dublin and Edinburgh. We should go to the continent as well; but not to Germany itself. We must go instead to that great Mediterranean clearing house of German zoology, the Stazione Zoologica, at Naples. The Stazione Zoologica was founded by the German embryologist Anton Dohrn in 1873. From then on it was through Naples that Germany exerted its strongest influence on British zoology. Nearly all British zoologists paid several, even regular, visits to Naples. Both Oxford and Cambridge provided biological scholarships which enabled the best recent graduates, in a late Victorian equivalent of the grand tour, to round off their zoological education in Naples. Lankester's visits, and his meetings there with Dohrn, are of the utmost importance for his changing views on recapitulation; Balfour, MacBride, Goodrich, Geoffrey Smith, Julian Huxley, Gavin de Beer all went to Naples. The earlier visitors met the phylogenetic embryologist Dohrn; the later ones such experimental embryologists as Meyer, who led the young Julian Huxley to experiment on regeneration and dedifferentiation, at first in sponges.

We have now defined a subject and its geography. We are thus in a position to trace the history of embryology during the golden age of classical zoology. Four main phases can be distinguished. We start as the introduction of the theory of recapitulation leads to an enormous flourishing of embryological research on phylogeny. Cambridge, with Balfour's school, is the main centre of this phase. Balfour's school dispersed after his death; some of its members left embryology, some stayed in it, but none of them kept the phylogenetic tradition alive. With the break up of Balfour's school, the centre of gravity shifts to Oxford. A new, causal embryology, which both grew out of and reacted against the older phylogenetic embryology, developed after the turn of the century. J. W. Jenkinson was the main figure. He, with Geoffrey Smith, trained up Julian Huxley. Both Jenkinson and Smith were killed in the Great War; but Huxley survived, and continued the tradition of experimental embryology afterwards. Under his influence embryology moves into a third, and then a fourth phase. The third phase sees a diversification of the causal approaches to embryology to include allometry and Mendelian rate genes, as well as traditional experiments on external and internal factors. Finally, towards the end of the twenties, Huxley and his student Gavin de Beer reacted against causal embryology. They turned their attention back to the original, traditional questions of classical zoology. They approached these questions through the theory of natural selection, and thus brought the causal approach (in a different form) to bear on classical, phylogenetic analysis. The result was a revolution in our theoretical understanding of the relation between embryology and evolution which has persisted to this day.

## II

The origins of embryology are in classical zoology. Classical zoology, as we have seen, seeks to understand the relationships between the different kinds of animals. It does so by distinguishing the homologous from the analogous similarities of species. Homologies are good indicators of real relationships, analogies are not. This was as true in the pre-evolutionary classical zoology of Richard Owen as it was afterwards. Instead of family trees, Owen was after Plans, Types, *Embranchements*. But, in his research, Owen (like his teacher Cuvier) had no use for embryology. He sought homologies solely in the older, comparative-anatomical criterion: resemblances of dead adult specimens. Embryology became a potential source of evidence when Von Baer stated his laws. Traits common to wider groups of species developed earlier; by tracing back the development of an organism in different species one could discover their successive, hierarchical relationships to each other. By the middle of the nineteenth century embryology was being introduced as a second criterion of homology, and throughout the second half of the century it was a matter of controversy which of the two, comparative anatomy or embryology, provided the more reliable criterion. Owen himself accepted that embryology provided evidence, but did not doubt that comparative anatomy was the more powerful method: homologies are 'mainly, if not wholly, determined by the relative position and connection of the parts, and may exist independently of ... similarity of development.' But his younger contemporary and great rival, Thomas Henry Huxley, thought exactly the opposite. In his Croonian lecture of 1858 'On the theory of the vertebrate skull'

Huxley said 'it were unwise to exalt either of these methods at the expense of its fellow' but 'the study of gradations of structure presented by a series of living beings [i.e. comparative anatomy of adults] may help the utmost in suggesting homologies, but the study of development alone can finally demonstrate them'.[1] Owen thought evidence from adults more reliable; Huxley evidence from embryos. Both gave reasons for their preferences, but neither is very clear.

The star of embryology had thus already begun to rise before the *Origin* was published. It would, we can assume, have continued to even had there been no Darwinian revolution: embryology was a little explored area, and had already demonstrated some relationships previously unproved. One example is the demonstration, by John Vaughan Thompson in 1830, that barnacles were crustaceans not molluscs. But there was a Darwinian revolution. And it undoubtedly gave a powerful stimulus to embryology. Now, it may appear strange that embryology, of all subjects, should benefit the most from Darwin. It appeared strange even to Arthur Shipley, who was sucked into embryology at that time; and frustrating to E. B. Poulton, who was not.[2] But reasons can be found. We need only look in the zoological textbooks of the period, and observe where the embryological evidence of homology is discussed. It has three typical settings, each of which reveals its importance for Darwinism: proofs of evolution, reconstruction of phylogeny, and the study of heredity. Darwin thought that embryological homologies provided good evidence of evolution; so too did Huxley; Romanes even thought that embryology provided the best kind of proof available. This does not, however, appear to be the main reason why people studied embryology. The main reason is to be found in the other two settings. Today genetics and the study of phylogeny are very different areas of biology. But this was not so in the second half of the nineteenth century. The theories of recapitulation and of heredity are inseparably linked: phylogeny and heredity were not distinct parts of biology. This need not surprise us: phylogeny and heredity are both ancestry. There is a strong connection between ancestry, inheritance, and development, in both of the two main kinds of theory of heredity: blending and the inheritance of acquired characters.[3] The blending theory, as formalised in Galton's Law of Ancestral Inheritance in 1889, supposed that an organism inherited a blend of $1/2$ its parental traits, $1/4$ its grand-parental, and so on. A common version of the theory of the inheritance of acquired characters supposed that the more, over the generations, a line of organisms acquired a trait, the more it became fixed in its inheritance, and the earlier it would be expressed in development. Earlier ancestral stages would then develop earlier. Only let the ancestral stages be adult ancestral stages, and we have reached the theory of recapitulation. And it was the theory of recapitulation which provided the greatest stimulation of all to embryology.

Huxley himself (and Darwin) had recapitulationist tendencies. But two other men were mainly responsible for introducing recapitulation into embryology, and so embryology into classical zoology. They were friends and contemporaries. Their names were Edwin Ray Lankester and Francis Maitland (called Frank) Balfour. We shall look at them in turn, taking Lankester first. He was trained up under George Rolleston in Oxford. Rolleston himself belonged to the older comparative-anatomical tradition: he compared adults, not embryos. But Lankester soon turned to embryology. In 1873 he published a frankly recapitulationist paper, which has moved a stage further than Huxley in its enthusiasm for embryological evidence: embryology is now the 'chief means' of detecting homology.

Lankester starts by remarking that 'A "natural" classification in modern zoology ... is a genealogical tree', and continues:

The chief means which the naturalist at present possesses of making out the genealogical tree of the animal kingdom lie in the fact that the individual animals living at the present day, in the process of reproduction, revert to the original simple condition (or nearly so) from which they have in the course of long ages been evolved as specific forms ... In passing from this simple condition to its adult form the individual goes through a series of changes, which are now explained by what may be termed 'the recapitulation hypothesis', which supposes that the individual organism in thus developing repeats more or less completely the successive series of forms which its ancestry has presented in the course of past ages; in fact the development of the individual is an epitome of the development of the species.

This is familiar stuff: recapitulation enabled genealogies to be determined embryologically. That was a good reason to use embryological evidence, but why should it be preferred to traditional comparative anatomy: why is it the 'chief means' of detecting homologies? Lankester did have a reason, although it is not well known; it also contains the key to so much of the future debate that it is worth examining. The classifications of that time, 'the four Cuvierian types or the six or seven types now generally adopted (Lankester wrote) are confessedly groupings based upon the anatomy of the adult organism; and therefore there has necessarily been a tendency in forming them to attach great importance to distinct plans of structure due to a secondary adaptation': Cuvier's *embranchements* are but four types of mechanical adaptation which 'might be assumed, as, indeed, in some cases they are, by organisms exhibiting divergent characters of an earlier and more fundamental character.' Similarity among adults, in short, is adaptive, and likely to be convergent. Embryology is more reliable because the developmental traits are (he thought) not adaptive, but ancestral. Lankester was aware that adaptations may occasionally be intruded into development, and conceal the recapitulation. It was (he said) 'the task of modern embryology' to distinguish adaptive from ancestral developmental stages.[4]

We have not heard the last of Lankester, but we should leave him there, in his biogenetic enthusiasm, for the moment. He did little embryological research, and built up no research school. To find these we must go to Cambridge, to that other great British comparative embryologist, Frank Balfour.

# III

Balfour entered Trinity College, Cambridge, in 1870, at the same time as Michael Foster arrived as praelector. Foster is a man of outstanding importance in the history of biology at Cambridge.[5] He was an experimental physiologist: not himself a great research worker; but a great builder of a research school. Foster and Balfour developed a close relationship, and Balfour stayed up, after graduating, to do research. From then on the pattern was set. One after another, brilliant undergraduates came up, were attracted to Balfour as they were repelled by the 'desperately dull and very formal' Professor Newton, and then sucked into comparative embryology. They were spellbound by his charming personality. Everyone who knew him sang the same tune, and whenever they described his personality that word 'charm' would come out. 'According to the degree to which their intimacy with him grew, those who got to know him were charmed with his kindly courteousness, fascinated with his brilliant, cheerful, often playful, companionship, held fast by his warm-hearted, steadfast friendship.' That was Michael Foster. Now listen to Stanley Gardner: he was 'one of the most brilliant and charming personalities that . . . zoology has known.'

Adam Sedgwick, Gardner tells us, 'conceived an affection [for Balfour] almost dog-like in its devotion'; for D'Arcy Thompson, Balfour was 'the man who meant most to him'; and Shipley later wrote that Balfour 'without doubt was the most attractive man I have ever met. He had to a peculiar degree that elusive and indefinable quality, charm, and he charmed us all . . . under Balfour and his lieutenant Adam Sedgwick, we all became comparative embryologists.'[6] And so they did. In the late seventies and early eighties he built up a superb school: there was William Bateson, working on *Balanoglossus*; W. F. R. Weldon and D'Arcy Thompson (who were friends at St Johns); A. E. Shipley, working on the lamprey; Walter Heape and (soon afterwards) Richard Assheton, working on mammals; Adam Sedgwick on *Peripatus*. We shall come back to them shortly.

The greatest work from the school was by Balfour himself, his immense *Comparative Embryology* (1880–1). Professor Gould has described Balfour as 'England's staunchest recapitulationist'[7], but this is misleading. It might lead one to think of the unbending, unthinking, fanatical recapitulationist of legend. The embryology of Balfour, however, was not the unthinking application of recapitulation to reconstruct phylogenetic trees: it was an exhilarating intellectual exercise combined with minute embryological description. Nor was he Britain's staunchest recapitulationist: his enthusiasm was much less than Lankester's. At the beginning of *Comparative Embryology* he distinguished between Von Baer's Law and recapitulation, and wrote that 'Von Baer was mistaken in thus absolutely limiting the generalisation, but his statement is much more nearly true than a definite statement of the exact similarity of the embryos of higher forms to the adults of lower ones.' He certainly thought that phylogeny informed ontogeny. That was the justification of the subject. But he did not underestimate the difficulties that result from adaptive intercalations, and the loss of ancestral stages. 'The satisfactory application of embryological data to morphology (he wrote) depends upon a knowledge of the extent to which the record of ancestral history has been preserved in development.'[8] This knowledge, or rather the obtaining of it, we have already seen Lankester describing as the 'task' of embryology. But Lankester never gave it much thought. Recapitulationists in general, as Gould demonstrates,[9] were willing to admit any number of exceptions to recapitulation; but they usually just proceeded as if the problems could be ignored. They had no techniques to distinguish

the ancestral from the non-ancestral developmental stages. But Balfour did not ignore the difficulties. He thought about them, and
applied his thoughts in the reconstruction of phylogeny. *Comparative Embryology* impresses most by the intelligent manner in which it
takes on these difficulties. Let us see, then, how Balfour tried to distinguish ancestral from non-ancestral developmental stages. The important chapter is number 13, in volume 2: general conclusions.

He proceeds in two stages. First he distinguishes between different
types of development; and then he asks how, in each type, natural
selection will intrude new adaptive traits, or remove old ancestral
ones. Then the truly ancestral traits can be recognised by elimination.
Such is the method in summary: now let us see how Balfour expresses
it. First, the developmental types: 'In a general way (he writes) two
types of development may be distinguished, viz. a foetal type and a
larval type.' In the foetal type the embryo develops within the parent
or on yolk supplied by the parent; the larval type develops independently, and has to provide food for itself. And how will natural selection act on them? Take foetal development first. Natural selection will
in this case tend to favour two processes: the abbreviation of development to reach maturity sooner, and an increase in the amount of yolk to
increase the embryo's chance of survival. Abbreviation will result in
the loss of characters 'which are of functional importance during a free
but not during a foetal existence'. The embryonic organs of locomotion, nutrition, and muscles, which are not needed by the adult, will
all disappear; those of circulation and excretion will not. Then there is
the effect of increasing the amount of yolk. Here we have to turn back
to Balfour's ingenious discussion of (among other things) the mechanical effect of yolk on the cleavage patterns of elasmobranchs. He
shows how the phylogenetic connection between the elasmobranchs
and frogs might have been obscured because of the large amount of
yolk in the elasmobranch egg.

So much for foetal development. What of larval? Larvae, which
have to fend for themselves, will tend to acquire their own 'secondary
adaptations' for feeding and locomotion. In summary, then, 'there is
a greater chance of the ancestral history being *lost* in forms which
develop in the egg; and of being *masked* in those which are hatched as
larvae.'[10]

Always, when trying to reconstruct phylogenies, Balfour was
thinking how natural selection would have modified that particular
kind of embryo. That was how he distinguished between ancestral

and non-ancestral traits. Balfour was not a staunch recapitulationist: he was an intelligent embryologist.

Balfour's distinction between larval and embryonic development is important for two reasons. One is its influence. It was used by Sedgwick, by Marshall, and (as we shall see later) by Goodrich. They did use it in different ways, but they were all following the same tradition. They were not simply concerned with the question of whether recapitulation was true, as if it were either true of all animals or of none; they were trying to work out how, in different types of ontogeny, ancestral stages would be retained. The other reason why Balfour's distinction matters is that it mirrors the future history of embryology. The mechanical effect of yolk is exactly the kind of argument to be found in 'causal', experimental embryology; the adaptations of larvae became the interest of that other tradition now associated with Garstang and De Beer. If we wonder why marine larvae came to form so large a part of the case against recapitulation, the answer is that it is they, more than any other kind of life cycle, which do not show it.

Now we can see how the embryological criterion led into causal embryology. Balfour used causal arguments in his reconstructions of homology. The homology of cleavage in elasmobranchs and frogs was obscured by a mechanical effect of the yolk. However, although Balfour's embryology may have contained the seeds of causal embryology, it was never itself causal. His purpose, and that of all his students, was to reconstruct phylogenies.

Balfour's school did not last long. In 1882 Balfour died in a mountaineering accident in the Alps. This provided the occasion for many of those outbursts of emotion which make so clear, despite the occasion of their delivery, the devotion of his students and colleagues. But much as they all loved him, after he had been taken from them by the Alpine rocks, they all combined their expressions of devotion to his person with rebellion from his science. They all, in their different ways, moved on. Many of them moved on to causal, experimental science. Such was the fate of Bateson, of Weldon, and (in a unique form) of D'Arcy Thompson. The collective biography of Balfour's school shows a recurring pattern. Just as they had all marched into comparative embryology under the charming instruction of Balfour, now, without his personal magnetism to hold them together, they all marched out again.

Let us look at Sedgwick first. He was Balfour's favourite pupil, and after Balfour's death took his place (as Reader in Animal Morphology). He kept up the tradition for a while, but increasingly came to distrust it. In 1894 the break came.[11] He then published a paper criticising Von Baer's Laws, mainly on empirical grounds: the laws simply were not true. He also had theoretical objections. Darwin had explained the recapitulatory pattern in development by heredity. A new variation, expressed at a certain age in development, would be inherited at that age. But in fact variations of an organ at, for example, the adult stage are not only expressed in the adult of the offspring; they affect all stages. If the organ is reduced in the adult, it is usually reduced at all stages. This line of argument we shall meet again when we come to Goodrich, and it was eventually to undermine the theory of recapitulation completely. Sedgwick was thus led by the Balfourian tradition into an increasing interest in heredity; but he never in fact studied it. Having destroyed the theory by which he had worked, he was left with no research to do. When he wrote again on embryology in 1909 he said all the same things as he had said in 1894. In the meantime he had written a standard textbook on zoology and held two Chairs. Research had left him behind.

Two other pupils of Balfour, William Bateson and W. F. R. Weldon, eagerly took up that field of research which Sedgwick had been led to but had refused to take up. Bateson remained in Cambridge and moved on to genetics: first to particulate inheritance, and then (after 1900) to Mendelism. Even while conducting a Balfourian investigation of the origin of chordates, he was having his doubts; already he could see that there was a next phase to the research, a phase that would move on from phylogenetic, to genetic, embryology, to gain a 'fuller understanding of the laws of growth and variation.' He wanted something more rigorous, something that would give him conclusions of greater certainty. By the time he wrote *Materials for the Study of Variation* (1894) he had finally parted with Balfour's methods: now he thought that recapitulation, or Von Baer's Laws (he confounds the two), were wrong, useless beyond remedy, and should be discarded along with the science that stood on them. Embryology had been successful when, in a general way, it was used to prove evolution, but when it was applied to particular questions of phylogeny, it 'has failed.' 'The principle of Von Baer, taken by itself, is clearly incapable of interpreting the phenomena of development.'[12] This is all he has to say, at the outset of his book, before he gets on to the real science, the study of variation.

While Bateson marched off into the study of particulate inheritance, his former colleague, W. F. R. Weldon, was off into biometry. Weldon remained in Cambridge until 1890. He then sat in the Chair of Zoology at University College London, having just been converted to biometry, by a reading of Galton's *Natural Inheritance* (which was published in 1889). Exactly when he parted with the embryological study of homology in not clear, although a lecture, given in 1885, on 'Adaptation to surroundings as a factor in development' may be suggestive; it was unpublished but survived in Pearson's memory when he wrote in 1906. Weldon moved from UCL to Oxford in 1899; but by then he had become a fanatical biometrician. Now he would write terrible things (in the privacy of his correspondence with Pearson) about old fashioned zoologists. The reformation of the universities, to channel huge and completely imaginary sums of money into biometrical research, had become his main interest. He soon burned himself out, and went mad before he died, at the age of 46, in 1906.[13]

For Sedgwick, Bateson, and Weldon, heredity was the natural development of Balfourian embryology. Walter Heape moved on to a different kind of causal science: the physiology of reproduction. He started out with Balfour, and retained an interest in embryology, turning out the odd paper and editing that notorious *Textbook of Embryology* to which MacBride contributed the volume on invertebrates; but Heape's main interest was reproductive physiology. From there came his main discoveries – the oestrus cycles, and the techniques of embryo transfer – which did later prove useful in embryology, but which are better seen historically in a physiological context.

Sedgwick and Bateson had stayed in Cambridge; Heape maintained close contact; Weldon moved to London: all four were in centres of biological research. Let us turn now to the history of Balfour's school in the provinces. Let us look, in turn, at A. Milnes Marshall, E. W. MacBride, and D'Arcy Thompson. Marshall moved to Owens College in Manchester were he retained, to the letter, the biology he had learned in Cambridge. He defended the theory of recapitulation in an address to the British Association in 1891, and in his textbook *Vertebrate Embryology* in 1893: and he was defending what he believed to be the general theory, justification, and purpose of embryology: 'It is on this fact of recapitulation that the great value of embryology depends.' 'Embryology reveals to us this ancestry, because every animal in its own development repeats its history, climbs up its own genealogical tree.' Of course, there were difficulties; but they would be overcome: all that was needed was more research.

Marshall's textbook is more innovative in its explanation of recapitulation. He, like his fellow Balfourians, no longer believed recapitulation to be the passive, inevitable expression of heredity. Nor does he think that recapitulatory stages are adaptations: he thinks they are positively harmful; and the fact that animals go through them suggests that they must, in some way, be constrained to. But what is the constraint? His answer comes from Kleinenberg, and it is embryological: 'each historic stage in the evolution [that is, ontogeny] of an organ is necessary as a stimulus for the development of the next succeeding stage.'[14] Here, I believe, is the first British glimpse of what was later to become a common theory. Joseph Needham took it up, and applied to it the idea of organisers. Indeed, in Needham's *Biochemistry and Morphogenesis* (1942) the whole subject of recapitulation has been reduced to a mere mention in the section on organisers. Recapitulated organs, such as gill slits, were now thought to be retained because of their functional importance in development. Waddington had even done an experiment on the pronephros of the chick whose results supported the theory that organs were recapitulated if they acted as organisers, growth centres, or endocrine organs.[15]

Marshall is one kind of provincial biologist. He stayed with, and hardened in, the concepts he had learned in his youth. E. W. MacBride is a similar case. He was trained up under Sedgwick in 1888, before Sedgwick's real break with the embryological criterion. He left for McGill University, in Canada, in 1896 where he nurtured his beliefs in the inheritance of acquired characteristics and in recapitulation. After his return to London, in 1909, he wrote a series of neo-Lamarkian and recapitulationist books and papers. In 1930, when De Beer published his *Embryology and Evolution*, MacBride was the last, and most ossified, survivor of the Balfour school. His explosion was just what De Beer wanted, and it was predictable enough: it is just the best known of MacBride's many noisy and inconsequential Irish displays.[16]

For a quite different provincial development we can turn to D'Arcy Thompson. He had kicked up his heels in Cambridge for a while, in the hope of employment. But none was forthcoming there: and he moved first to Dublin, and then back to his birthplace, Edinburgh. He was left to develop his ideas alone, and he developed a correspondingly unique set of answers to the questions which all of Balfour's school were grappling with. He did not move on to experimental, causal science; but he felt the same dissatisfaction with the speculations of phylogenetic embryology as all his former colleagues. He had come to doubt the entire Darwinian theory,[17] and became interested instead in purely physical causes of development. And his approach to phylogenetic transformation, his 'theory of transformations', is unlike anything else. It is abstract, rather than causal in the sense of biometricians or later synthesis. Its idiosyncracy alone made it difficult for his former colleagues to understand. D'Arcy Thompson's biographer quotes from a series of letters from Michael Foster which probably illustrate the reaction of the Cambridge school to his new departures. 'I confess I am not very attracted by this line of work, and doubt if its likely to be very fruitful.' And, more interesting: 'I shall be willing to present your paper – . . . But does your result wholly destroy the diagnostic value of the spicules? If the form is constant in a group – it does not matter how the form is brought about.' For Foster, the interest in embryology was whether it revealed homologies. Embryology which was not directed to this problem was 'not likely to be very fruitful.'[18] That is exactly what you might expect Foster to think, given that embryology's special value as a criterion of homology had so often been claimed as its justification. We may well question whether D'Arcy would have moved along so peculiar a line if his ambition to stay in Cambridge had been fulfilled. With people like Foster around he would have been kept in line. For Foster, the first among builders of research schools, could sense unerringly when an unfruitful line of work was in the offing.

It is important to place D'Arcy Thompson in the context of the whole Balfour school. He was just one among many to grow disatisfied with phylogenetic speculation. He was not the first, nor the most important; though he was the most idiosyncratic. Sir Peter Medawar has written that 'Comparative anatomy is no longer thought of as the central discipline of zoology; D'Arcy Thompson was the first man in this country to challenge its pretensions and to repudiate the idea.'[19] We can now see, after looking at the whole school, that Medawar was wrong.

We have seen that, although Balfour's embryology was primarily phylogenetic, it made use of arguments like those of causal embryology. Before finally leaving Cambridge let us look at one last figure who illustrates this particularly well. He is Richard Assheton. He came up to Trinity College in October 1883. The Balfour school still existed, although it has just lost its leader nine months before, Assheton was attracted to embryology and trained up. His career shows less of a change from the descriptive, phylogenetic embryology of Balfour to causal, experimental science. He was interested in both from the start. He, like Balfour, studied causal embryology to answer phylogenetic questions; but Assheton's embryology is more analytic and less descriptive than Balfour's. Assheton studied, for example, the mechanism of growth in annelids and chordates. This, like most of his work, was inspired by the question of the ancestry of the chordates. Annelids were, at that time, a candidate ancestor. Was metamerism homologous in annelids and chordates? The question was answered by an analysis of growth mechanisms, and turned out negative.[20] The important point for us is that causal embryology was being used as a method of phylogenetic analysis.

We can now leave Cambridge. After Balfour's death it declined as a centre of embryology. His school had been dispersed, explosively, across the geography and concepts of British biology. With Balfour gone, Cambridge was left to the physiologists. They took possession of their inheritance and since then have never, for a moment, let go.[21] When the ecologist Evelyn Hutchinson came up to Cambridge in the 1920s he found that the golden age of zoology was over. In so far as embryology was taught at all, it was by Cresswell Shearer, a Canadian, who had been taught by MacBride at McGill, and then learned experimental embryology in Naples. Shearer, however, inspired little work at all after he came to Cambridge, and only a small proportion of it was embryological. 'As a formal lecturer, Shearer was not entirely successful; his sense of time was rather uncertain and occasionally he gave the impression of being bored': thus James Gray, in Shearer's obituary notice. By the twenties, Shearer had grown far more interest in the renaissance of southern Italy under the Emperor Frederick II than in embryology.[22]

If this essay went on to a later period we should have to return to Cambridge. In the 1930s it again becomes a centre of embryology, but not in the Department of Zoology. That building was closed to them by its Professor, whose ideal zoology building was comprised of 'three floors – a ground floor for molecular physics, a first floor for biophysics, and a top floor for cell mechanics.'[23] The embryology of Waddington and Needham was performed elsewhere, in the new Strangeway's Laboratory or in the Biochemistry department.

The splitting up of Balfour's school represents the end of the first important stage of embryology in Great Britain. The study of recapitulation in some ways (as we have seen) passed into experimental embryology, while it moved away from the investigation of phylogeny. As we move into the second stage, the centre of gravity shifts increasingly from Cambridge to Oxford.

# IV

In the nineteenth century Oxford had no flourishing school of comparative embryology to compare with Balfour's. The first Professor, George Rolleston, compared only dead adults. He was not much interested in research anyway. Lankester did hold a fellowship at Exeter College during his first enthusiasm for recapitulation, but he did not stay for long: he left to become Professor at UCL in 1874. On Rolleston's death Oxford did attempt to steal Balfour, but Balfour declined; Cambridge sensed the danger and promtly secured him to an *ad hominem* chair. The Oxford Chair was also offered to Huxley, but in vain. It lapsed to Henry Nottidge Mosely. Mosely was a naturalist and taxonomist; he did no embryology. The two decades when recapitulation flourished most passed with no contribution from Oxford.

On Moseley's death Ray Lankester sat in the Linacre chair. When we last met Lankester, in the early 1870s, he was a fervent recapitulationist; but that stage did not last long. In 1873, it may be recalled, he preferred embryology to comparative anatomy as a criterion of homology, because adult anatomy was more liable to convergence: in 1879, his preferences had reversed. Now he would write that 'we are justified in assuming as a general law that animals or plants of like structure have descended from common ancestors – that is to say, that the same kind of organisation (especially where a number of elaborate details of structure are involved) has not been produced by natural selection.' Comparative anatomy had become the first method; and embryology but 'a remarkable aid to the correct building up of a pedigree.' He was still a thorough-going recapitulationist, saying for instance that 'It appears probable that the Nauplius phase is the recapitulative representation of an ancestor common to all [the Crustacea].'[24] He gradually mellowed further. His writings from the nineties onwards betray no fervent recapitulationism. As the embryological criterion had been absorbed by classical zoology, it had also been absorbed by Lankester.

Lankester, then, would inspire no school of embryological research. His successor, W. F. R. Weldon, had also started out as a phylogenetic embryologist, but also had lost his interest in it before coming to Oxford. At most he would encourage the new experimental embryology, from a distance. The most important contributions of Lankester and Weldon to embryology in Oxford were to plant, respectively, Edwin Stephen Goodrich and Wilfred Jenkinson in the department. The fruits did not develop until after Weldon's death. They grew up under his successor, the tolerant and affable Gilbert C. Bourne – 'Beejar' Bourne of the towpath.

Bourne himself was a pure comparative anatomist, not an embryologist. His own work therefore need not concern us, but his department must. He was Professor from 1906 to 1921, while important embryological work was being performed by Jenkinson and Goodrich; and Bourne himself had in turn introduced a third man, Geoffrey Watkins Smith. Jenkinson was a purely experimental embryologist, Goodrich purely phylogenetic, and Smith was interested in both. Let us see what these three thought of embryology, and of the theory of recapitulation.

We shall take Jenkinson first. He actually read Greats as an undergraduate, but while doing so he attended Lankester's class. After graduating he went to Weldon's laboratory in UCL; he practised experimental embryology there and with Hubrecht, in Holland. From his return to Oxford until the outbreak of the War he was the outstanding experimental embryologist in Oxford, and probably in the whole country. He worked intensely, and experiments poured out. Bourne called him 'the most thorough man I ever met.'[25] His *Experimental Embryology* (1909) was the first comprehensive textbook of the subject in English. In the pages of that erudite book we can see the content of experimental embryology, with its rigid devotion to experimentally determined causes and its exclusion of all other kinds of embryology. It divides causes into external factors and internal factors, a distinction first made by Lankester and passed down through successive generations of embryologists at Oxford. The external factors in Jenkinson's experiments were such things as gravitation, electricity, light, heat, pressure, and chemical composition. Experiments determined the effects of each of these on such developmental processes as the rate of cell division, of abnormal differentiation, and of other kinds of more precisely defined abnormality.

But Jenkinson had not confined himself to experiments. In 1906 he published a descriptive paper on the homologies of the germinal layers. Its subject was classical, phylogenetic embryology; its method descriptive; and its conclusion completely destructive. It works through the development of groups of vertebrates to see whether the relations between one developmental stage and the next are the same in different groups. 'The Vertebrata [he concluded] afford no support to the Gastreae or to any other theory which attaches a morphological, phylogenetic, or recapitulatory significance, a value for the determination of adult homologies, to the germinal layers.'[26] When Jenkinson, in *Experimental Embryology*, dismissed the biogenetic law in a few sentences, the dismissal stood on a firm foundation of fact.

Jenkinson's biography suggests a history, in Oxford at least, very different from that presented by Professor Gould. Gould has argued that empirical work did not drive the historical change from phylogenetic to experimental embryology. There was a change of interest, one for which embryologists had their reasons; but facts (according to Gould) were not among them. Once a generation of experimentalists had lost interest in phylogeny, Gould believes, they simply ignored it, and he describes their attitude to phylogenetic embryology as one of 'benign neglect.'[27]

The neglect of phylogeny in Jenkinson's experimental embryology, as we have just seen, was not benign. It was determined by empirical destruction. He neglected the biogenetic law after a 'a more intimate acquaintance with the facts has made it abundantly clear that development is no mere repetition of the ancestral series';[28] because, in other words, it was wrong; and it was wrong in the sense of being empirically false.

Of course Gould was taking a broader view than am I. I am not going to try a world-wide generalisation of Jenkinson. I am, however, going to generalise it to the other embryologists at Oxford. Let us turn now to Geoffrey Smith. He was a zoologist of wider interests than Jenkinson, and worked on many aspects of the subject; but we shall confine ourselves to his embryology. He too rejected recapitulation, and, like Jenkinson, he had rejected it because it was empirically false.[29] He examined it empirically in his lectures on the Crustacea. There he would test the Biogenetic Law by taking a group whose phylogeny was well known from anatomy, and then looking at the development of its species to see whether it recapitulated phylogeny. He started with the krill (*Euphausiacea*). He could find no trace in its development of the flattened leaf-like phyllopodan limbs characteristic of primitive Crustacea. He went on to the other groups, piling up the exceptions. There can be no doubt that he actively rejected recapitulation, and that his action was precipitated by empirical inquiry.

What, then, would Geoffrey Smith do with embryology? He also considered development a causal problem in the interaction of internal and external factors.[30] But he considered external factors less important: they only released forms that were potential inside the embryo. The internal generation of pattern was a much greater embryonic achievement than its influence by external factors. Smith thus set out to investigate the internal generation of form. He realised that it, like an external factor, could be studied experimentally. He chose, as an exemplary problem, the correlation between the primary and secondary sexual organs. The correlation is caused by internal secretions, sex hormones. Smith therefore measured the growth of secondary sex organs in parasitically and in experimentally castrated crabs, and in experimentally castrated cock chickens. The details of the results need not concern us; but we should notice one further feature of his work that will later prove important. As John Baker has observed,[31] Smith's measurements of the sizes of secondary sex organs influenced Julian Huxley's later work on allometry. Huxley, for example, quotes the work of his old tutor in *Problems of Relative Growth*. Smith's work, in its turn, was probably influenced by the measurements of correlations in morphology made by Weldon during his biometrical phase.

# V

Before coming to Julian Huxley I should say a bit about Goodrich.[32] Goodrich had entered the Slade School of Art (in London) in 1888; but while there he also attended Lankester's lectures in UCL. He was captured, and turned round, and (as we have seen) when Lankester returned to Oxford as Linacre Professor in 1892, Goodrich came too. He entered Merton to read for an undergraduate degree; after graduating he ran through the usual round of travels, returned to Oxford, and was soon (in 1900) elected a Fellow of Merton. He stayed until his death. He appears a more and more lonely figure as the years pass, as he continues the tradition of classical zoology, raising it to its peak of perfection. Although a great zoologist, and the Professor for over twenty years, he was no builder of a research school. He was not (I suspect) an inspiring man; he could not express himself clearly (if at all) on anything other than his kind of zoology: he liked to talk as one expert to another. But he was the best kind of Professor. He did not really run the department at all: he allowed people to follow their own interests. Under his tolerant regime, an exceptional zoological diversity was able to flourish, much of it indeed in reaction to the phylogeny of dead animals that Goodrich himself taught. His lectures, I have been told, combined beauty of illustration with monotony of voice. Many undergraduates therefore found them very dull. But it was Goodrich that introduced the successive generations to classical zoology while they lapped up such fashionable subjects as heredity and experimental embryology from Jenkinson and Smith, or (afterwards) from Julian Huxley. It is important we should know what he thought of the Biogenetic Law.

We are immediately faced with a difficulty. Goodrich's published works hardly even mention the theory of recapitulation. In fact they do not discuss any questions of principle: his publications are rigidly descriptive. Nor did he discuss recapitulation with his colleagues; Sir Alister Hardy, for example, cannot recall Goodrich ever talking about it. But he must have thought about it. Phylogeny was his great interest, and comparative embryology his method. It is not possible that he could have used that method without having thought about recapitulation. In the absence of any direct discussion, we can still look at his detailed procedures, to see whether they were guided by the biogenetic law. We should find that they were not. But fortunately we have a better source as well: his notes for a lecture on the 'General interpretation of ontogeny [and its] relation to phylogeny' have survived.[33] I cannot date them with any certainty, but they are in the same bundle as some lecture notes on 'embryology' which are dated as 1917. We may take that provisionally as the date of the lecture on ontogeny and phylogeny as well. The exact date however is not important, because Goodrich's views probably did not change much: he might have given the same lecture to Huxley in 1910, or to de Beer in 1920. Let us see what Goodrich had to say about the biogenetic law.

'The theory of recapitulation' he begins 'saturates modern embryological work.' But it is completely wrong: it postulates 'phantastic impossible ancestors', so its influence is nothing but a nuisance: it 'brings discredit on Comp. Anat. & Embr . . .' Then he goes on to explain, in two main stages, why it is wrong. To start with, not all developmental stages are ancestral; many are adaptive innovations. 'To most obvious objections Haeckel distinguished Palingenetic [ancestral] fr. Caenogenetic [adaptive] characters, but choice arbitrary.' There is, in other words, no method of distinguishing adaptive intrusions from ancestral hangovers. And adaptation may take place at any stage and to any extent: '*repetition of ontogenetic stages* of immediate *ancestors* wh. may be distorted & altered by adaptation at any stage leading to divergence in development & ultimately violent metamorphosis – free moving to sedentary life', and he gives other examples. This is a familiar objection, although we may notice that for Goodrich it is a reason for throwing out the theory completely; it is not just a difficulty to be overcome: it could never be overcome by research because the distinction between ancestral and adaptive characters is arbitrary.

He then considers recapitulation as a causal theory. What of Haeckel's statement that 'phylogeny is the mechanical cause of ontogeny'? Goodrich takes this statement literally, and disagrees. Drawing the diagram reproduced in Figure 1 he remarked 'obviously ancestral characters not repeated[.] Nothing which appears in A series which is not predetermined in Germ-track[.] $A_4$ does not *start* fr. same point as $A_3$ $A_2$ etc & so cannot really pass through same stages [. . .] starting point continually shifting.'

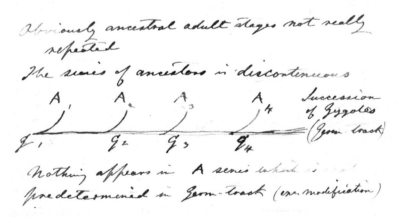

Fig. 1.

The same point was immediately made to me by Professor E. B. Ford when I asked him whether recapitulation was dead in Oxford in the twenties. 'Dying' he replied, and went on to say that its supporters did not really know what they meant by it. What they thought were re-capitulated stages were in fact produced by different genes: there were similar results but they were not the same genetically. I asked him if this was an orthodox criticism of recapitulation at that time (and mentioned Goodrich) and he confirmed that it was.[34]

At least two historians have remarked previously on the replace-ment of the theory of recapitulation by genetics in the twentieth century.[35] As with the replacement of phylogenetic embryology by experimental, Gould attributes this replacement to 'benign neglect', and to the theoretical undermining of recapitulation by Mendelism. Once again we can see that, in Oxford at least, the neglect is by no means benign; but Gould's second explanation does look more promising. When we look at it in more detail, however, we find that it does not fit either. He shows how Mendelism undermines the theor-etical explanation of recapitulation by 'terminal addition' and 'con-densation.' Philosophically it does; but, in Oxford, it was not the argument that people actually used. There the crucial point was that the recapitulated stages were not literally ancestral: they were gen-etically just as modern as any other stage. Ford, and Goodrich's lecture, do however provide detailed support for the general, if not the specific, explanation given by Gould. The biogenetic law was rejected more because it did not make genetical sense than because it under-mined two possible mechanisms of recapitulation. The story was not that simple, however, and the empirical exceptions to the law which Gould holds to be historically, or perhaps philosophically, unim-portant did matter in Oxford.

Having explained why recapitulation is wrong, Goodrich went on to explain why it sometimes appears to be right. 'Chiefly due (his notes say) to necessity of multicellular Metazoan always starting fr. *unicellular fertilized ovum.*' And then the 'correspondence between A[dult] series & ontogenetic stages only most general . . . Every stage in ontogeny really specific – gill arch of mammal not really like that of adult fish.' Recapitulation results mainly from the superficial simi-larities there are likely to be between different forms developing from a single cell to a multicell stage. Garstang made exactly the same point in his polemic of 1922, and Huxley a similar one in *Problems of Relative Growth*.[36]

Goodrich then gives a second reason. It continues a familar theme: are embryonic stages adaptations? 'Every stage really adaptive – but A[dult] must be self-sufficing & capable of reproduction – Ontogenetic stages need only be active self-sufficing in so far as they are free – not if much yolk or nourished by mother – hence larval development & embryonic. While A[dult] must be working machine O[ntogetic stages] need not.' He is putting Balfour's distinction to a slightly new use. He is using it to explain why recapitulation may take place. If embryos develop nourished by maternal yolk rather than their own exertions, or develop inside their mother, they do not need to be so closely adapted to the external environment. Ancestral characters may then be retained, and recapitulated. Free larvae are likely to evolve novel adaptations. Balfour (as we have seen) used the distinction to distinguish adapative from ancestral characters. Goodrich could have too, if he had wished, even though at the outset he had denied that it was possible. But let that pass.

Let us now compare Goodrich's explanation of recapitulation with the one that we met before when discussing A. Milnes Marshall. They provide two different kinds of explanation of recapitulation. I shall call Goodrich's theory 'ecological', and Marshall's 'embryological'; they could equally be called 'external' and 'internal'. The two kinds are interesting as they reflect an important question in modern biology, that of whether natural selection is all-sufficient, or whether the variation on which it acts is internally, developmentally constrained. As explanations of recapitulation, they both provide mechanisms of what Gould calls 'terminal addition' – the biogenetic law does require that new evolutionary stages are added on only at the adult stage. Goodrich's explanation is ecological. He suggests that, because the young stages are more protected from the external environment than are the adult stages, the adult is more likely to be altered by natural selection. It does not have to be the adult stage alone that undergoes evolutionary change; but, for ecological reasons, it is more likely to be.

The second kind of explanation is embryological. It has two forms, which are related, but distinguishable. The theory first suggested by Kleinenberg, taken up by Marshall and then Needham, and experimented on by Waddington, explains recapitulation by the importance of the recapitulated stage in development. There may be no external reason why (say) gill slits have to be recapitulated; they are not needed for breathing: but they may be retained because they are crucial for the development of a later important stage. The second form of the embryological explanation is more abstract. It makes use of the argument (made most influentially by Fisher in 1930) that evolution by macro-mutations is less likely than evolution in small steps. Changes in earlier stages may produce a big and disruptive change in the adult; changes at the adult stage may be smaller. Because evolution by smaller steps is more likely than evolution by large ones, evolutionary change in the adult is more likely than evolutionary change in earlier stages of development. The early stages will instead be repeated in descendants. This second embryological argument could be thought of as a general form of Kleinenberg's theory, or it could (in principle) be a separate argument. Waddington and Needham both mention it in the same breath as Kleinenberg's theory: the two arguments were not separate for them. Since then embryological arguments have become more and more popular. De Beer used them in 1938, and in our own time Gould's favoured explanation of recapitulation is the macromutation one – he does not even mention the ecological argument of Goodrich and Balfour.[37]

# VI

We have now examined the embryology of Goodrich, Jenkinson, and Smith. That embryology would have been what Julian Huxley heard when he came up to Balliol in 1906. He learned his embryology from Wilfred Jenkinson and Geoffrey Smith, supplemented by Naples and Germany. It was, in summary, a causal, experimental embryology, utterly divorced from the reconstruction of phylogeny. It was left to Huxley to continue the tradition after the war – Jenkinson fell at Gallipoli; Smith in the trenches at Pozières – for only he survived. With his return to Oxford, embryology moves into its penultimate stage. Huxley carried the experimental embryological tradition of Jenkinson and Smith into the 1920s.[38] With Goodrich playing Newton to Huxley's Balfour, Huxley now attracted the clever undergraduates: de Beer, Hardy, Ford, and many other important biologists, were all his pupils, and he turned many of them into experimental embryology of one form or another. Julian Huxley was of course a versatile biologist, but his embryology was if anything more rigidly experimental than Jenkinson's. He never worked on phylogeny. The first mention of recapitulation in his writings comes in popular books of the late 1920s, and it has there the appearance of an unfamiliar subject dimly recalled. Recapitulation is mentioned, uncritically but unenthusiastically, as evidence for evolution in both *Animal Biology* (a short textbook, of 1927, written with Haldane) and *The Science of Life* (which was written in 1927–8).[39] It was not until he wrote *Problems of Relative Growth* (1932) that he had much to say about recapitulation; and that was only to repeat de Beer. Experimental embryology, not phylogeny, was Huxley's interest while he was a tutor in Oxford (1919–25). Sir Alister Hardy recalls Huxley's tremendous enthusiasm for Child's experiments on planarians. And it was the work of Child, and Harrison and then Spemann, which was firing experimental embryology into its most confident phase. 'Students felt it to be the most rapidly advancing front of biological research' Medawar retrospected, and (according to E. B. Ford) 'It is probably not realized to-day how obsessed were those zoologists who had experimental leanings

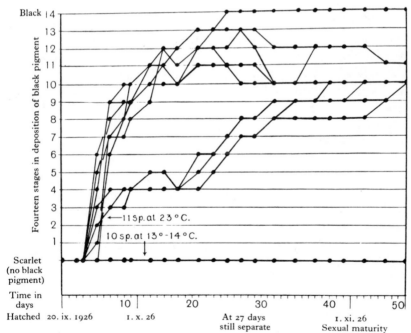

One dot = one or more individuals. (The mean values of this family are plotted in Fig. 1 *b*.)
*Gammarus chevreuxi*, a family of 21. Segregation for rapid and slow pigment deposition is visible at the higher but not at the lower temperature. The family is from a back-cross, and there are 6 "rapid" and 5 "slow" specimens, but several of them follow an almost identical course.

Fig. 2.

toward experimental embryology ... I remember more than one person speaking in the same terms, saying "Surely you are not going on with genetics when we are seeing what can be done in experimental embryology."[40] It came increasingly under the influence of the concepts of gradients and inducers (both almost absent from Jenkinson's book), to culminate in Huxley and de Beer's big book *Principles of Experimental Embryology* (1932). But we need not trouble ourselves with all that. Let us look instead in more detail at another kind of research that Huxley fostered: rate genes.

Mrs E. W. Sexton, who worked at Plymouth, first noticed the eye colour variants of *Gammarus*. The variants were taken in hand by E. B. Ford.[41] Goldschmidt had already worked on rate genes, but in moths which have a discontinuous development, through which it is difficult to follow the effects of a gene. *Gammarus* however has no metamorphosis. Ford & Huxley investigated two kinds of eye colour variant. The normal *Gammarus* had an eye with black facets and white interfacets: the two kinds of variant could have red facets instead of black, and non-white interfacets instead of white. Standard Mendelian crosses were performed, which (if I may simplify a little) demonstrated that each variant was controlled by a single locus with dominant wild-type. The phenotypes, however, were not static, they were actually the manifestation of a developmental process whose rate differed between the two genotypes. The development of blackness in the facets is shown in Fig. 2. As the Figure shows, even red-eyed *Gammarus* gradually develop black eyes; they just take much longer to do so than do normal *Gammarus*. The wild-type allele, it can be seen, has exactly the same effect as increasing the temperature: the gene, like heat, speeds up the rate of a chemical reaction. Ford and Huxley also reasoned that all, or most, genes could be considered as effecting developmental processes rather than as producing structures. This enabled Ford to rephrase Lankester's old distinction between external and internal factors.[42]

The *locus classicus* for the final stage of the embryological part of classical zoology in Great Britain is de Beer's *Embryology and Evolution*. I do not need to document its influence on the embryologists of the time and later; but we should consider the historical paradox it poses. The biogenetic law (as we have seen) was

completely dead in Oxford by 1905. The embryologists, the experimentalist Jenkinson, the phylogeneticist Goodrich, and Smith who
did both, unanimously pronounced recapitulation an error. If de Beer
(and Garstang) were only repeating the critique, whether or not they
thought they were saying anything original, they cannot have been in
fact.[43]

The key, I believe, is that Garstang and de Beer are not significant
so much in destruction as in theoretical innovation. Recapitulation
had indeed been slain long before 1922, or 1930. It had gone under,
found empirically wanting, by the first decade of the century. By then,
the theory of recapitulation was utterly extinguished among the leading embryologists in Oxford. I have not found anyone there who
believed it. It was still influential elsewhere, still that arrogant,
powerful figure, E. W. MacBride believed in it, still schoolchildren
learnt it, and (if de Beer and Hardy are to be believed)[44] then wrote in
their examinations that an animal, during its development, climbs up
its family tree. But in Oxford recapitulation was dead. Phylogenetic
embryology had been replaced by experimental embryology, with its
external factors and internal factors, internal secretions, and Mendelian genes. But the pure experimentalist had no interest in turning
to a causal analysis of phylogeny. As a causal theory of embryology,
recapitulation had been replaced by what we recognise as experimental embryology. As a method of reconstructing phylogeny embryological evidence had been reduced, as in the work of Goodrich, to the
status of just another character. But as a theory of the mode of evolution, recapitulation had been replaced by nothing at all.

It had been realised for some time that if the timing of sexual maturity could be brought forward in development, a descendant could be
derived from the larva of its ancestor. This was realised by Lankester
in an afterthought to his essay on degeneration of 1879; but Lankester
did not realise the importance of the idea. It was Garstang, impishly
amused by the way the process stood Haeckel on his head (descendant
adults from ancestral larvae rather than descendant larvae from ancestral adults), who made the most of it in Great Britain at least. He
christened the process paedomorphosis, and argued that it was the
mechanism by which most of the major groups of animals had
evolved. This theme was to be taken up by de Beer, and by Sir Alister
Hardy. Garstang did not visit Oxford often in the twenties (although
he retired there in 1934): his influence was exerted by his writings.

Rate genes provided a mechanism for paedomorphosis. Garstang himself was little interested by rate genes, or genetics, in general. I therefore do not agree with Gould's account, according to which Garstang's influence was due to his speaking the language of Mendelism. It was left to others to put paedomorphosis together with rate genetics to produce a general theory of evolution. Many biologists – and particularly de Beer – soon realised that a Mendelian account could be given of the possible changes in the timing of the development of organs; but Garstang was not among them.

Now we can see what de Beer was doing to the biogenetic law. If recapitulation was true, phylogeny must be generated only by changes in the adult: it must therefore be understood as changes in the adult. But if phylogeny could be generated by changes in the timing of developmental processes (heterochrony, as de Beer called it), then the old theory of recapitulation could be replaced by a newer theory of the mechanism of evolution. When de Beer blows his trumpet over recapitulation, he is not blasting at walls already fallen: he is replacing recapitulation by something new. At about the same time Huxley and Haldane both made the same synthesis, Huxley at least having been inspired by de Beer. Huxley, for example, after reviewing the work on rate genes in *Gammarus*, wrote that 'such phenomena are unintelligible on the Haeckelian doctrine. But they immediately fall into place when it is grasped that most examples of recapitulation constitute simply one side of a more general problem – the problem of altering the relative rates of growth and of other processes within the body.'[45]

# VII

There is one final element in the story. Let us turn to the course of de Beer's embryological career. He had been reared on Huxley's experimental embryology, which was not (like Jenkinson's) based on a refutation of the biogenetic law. Huxley simply ignored it, and got on with the experiments. De Beer's work through the 1920s all fits Huxley's pattern, and their joint book is the culmination of this phase.[46] But by the time he had written it he had already turned around and marched off, in the opposite direction, back into classical zoology. *Embryology and Evolution* marks the turning point. When de Beer came to evolution in the late twenties it was for the first time. He had the enthusiasm of a convert, and (having spent five years experimenting) he was behind the times in his new subject. He would never again do experiments, but moved on to detailed phylogenetic work, describing in intimate detail the development of the skulls of different species of vertebrates. He had returned to the inspiration of classical zoology.

Julian Huxley took a similar course. He, too, had been reared on experimental embryology; he too now returned to evolution. But he did not return to classical zoology: he moved on to the new, causal theory of evolution. He never spent his time working out genealogical trees, but became increasingly interested in the process of evolution. The change was, to some extent, influenced by his work on allometry. (Although the change from academic to populariser at that time must have been a bigger influence; as a writer without an institution he could not do experiments.) The work on allometry was not, of course, experimental embryology; but it started out under the influence of it. It has continued to interest embryologists – even Ross Harrison became interested in it – but has never yielded any embryological insights: as it developed it turned out to have more taxonomic than embryological use.[47] The allometric formula provided an easy way to describe form for classification. From here Huxley became increasingly interested in questions of species and speciation. He was soon at work on clines, and speciation is the most important topic discussed in his greatest book *Evolution: The Modern Synthesis* (1942).

It is too widely believed that experimental embryology was inevitably the next historic stage after the domination of the recapitulation theory. History has too often been told as the development of a descriptive into a causal, analytic science, with a replacement of one kind of science by another. I have shown how one tradition within recapitulation led to experimental embryology. I have shown too how experimental embryology, or at least two of its offshoots, Mendelism and allometry, led back into evolution. For de Beer at least experimental embryology was not the natural next stage after the study of phylogeny. Just as Balfour's school had broken up and its members moved on to causal and experimental biology, now, with Huxley's departure from Oxford, his school would break up and its members return to the study of evolution.

# Endnotes

1. R. Owen, *On the Archetype and Homologies of the Vertebrate Skeleton*, London, 1848, p. 6. T. H. Huxley, On the theory of the vertebrate skull, In M. Foster & E. R. Lankester (eds.), *The Scientific Memoirs of Thomas Henry Huxley* (1898) i, p. 541. On the criteria of homologies, see E. W. Russell, *Form and Function*,

London, 1916. Martin Barry was probably more important than Huxley in introducing the embryological criterion into Great Britain: see Jane Oppenheimer, An embryological enigma in the *Origin of Species*, In B. Glass, O. Temkin & W. L. Straus (eds.), *Forerunners of Darwin: 1745–1859*, Baltimore, 1959, p. 310–15.

2. A. E. Shipley, *Cambridge Cameos*, London, 1924, p. 143, 162. E. B. Poulton, *The Colours of Animals*, London, 1890, p. 286–7.

3. S. J. Gould, *Ontogeny and Phylogeny*, Cambridge, Mass., 1977, pp. 78–100. F. Galton, *Hereditary Genius*, London, 1889. For the close relation between heredity and recapitulation in contemporary textbooks, see, for example, T. H. Huxley, *Introduction to the Classification of Animals*, London, 1869, p. 42, and F. M. Balfour, *A Treatise on Comparative Embryology*, London, 1880–1, i, p. 3.

4. E. R. Lankester, On the primitive cell-layers of the embryo as the basis of a genealogical classification of animals, and on the origin of the vascular and lymph systems, *Annals and Magazine of Natural History* (1873) 4th series, xi, 321–38. I have quoted from pp. 321, 322, 323–4, 324, and 322, in that order. Lankester's Notes on the embryology and classification of the animal kingdom: comprising a revision of speculations relative to the origin and significance of the germ-layers, *Quarterly Journal of Microscopical Science* (1877) n.s., **17**, 399–454, is a follow-up.

5. G. L. Geison, *Michael Foster and the Cambridge School of Physiology*, Princeton, 1978.

6. M. Foster, Francis Maitland Balfour, *Proceedings of the Royal Society*, xxxv (1883) p. xxvii; J. S. Gardiner, Adam Sedgwick, *The Zoologist*, 4th series, xvii, 1913, p. 112, 111; R. D'Arcy Thompson, *D'Arcy Wentworth Thompson*, London, 1958, p. 52; A. E. Shipley op. cit. (2), p. 162–3, and p. 164 for the comment on Newton's lectures. Newton did temporarily inspire Walter Rothschild; but that impressionable youth was an exception to every rule: see M. Rothschild, *Dear Lord Rothschild*, London, 1983, pp. 73–6.

7. S. J. Gould op. cit. (3), p. 176.

8. F. M. Balfour op. cit. (3), i, p. 2n, ii, p. 298.

9. S. J. Gould op. cit. (3), p. 167–84.

10. F. M. Balfour op. cit. (3), ii, p. 299 italics original.

11. A. Sedgwick, On the law of development commonly known as von Baer's Law; and on the significance of ancestral rudiments in embryonic development, *Quarterly Journal of Microscopical Science* (1894) **36**, 35–52; The influence of Darwin on the study of animal embryology, In A. C. Seward (ed.), *Darwin and Modern Science*, Cambridge, 1909, pp. 171–84.

12. W. Bateson, On the ancestry of the Chordata, *Quarterly Journal of Microscopical Science* (1886) **26**, p. 548; *Materials for the Study of Variation*, London, 1894, p. 8, 9–10. (And see Hutchinson's remarks in his introduction to the reprint of *Materials*, New Haven, 1979.)

13. K. Pearson, Walter Frank Raphael Weldon 1860–1906, *Biometrika*, v (1906) p. 11 for the paper on adaptation; for his demented schemes of reform, *ibid* p. 22 and 37n, and D. J. Kevles, Genetics in the United States and Great Britain 1890–1930: a review with speculations, In C. Webster, ed., *Biology, Medicine, and Society 1840–1940*, Cambridge, 1981, p. 205.

14. A. M. Marshall, The development of animals, *Report of the British Association for the Advancement of Science* (1891) 826–85, *Vertebrate Embryology*, London, 1893, pp. 27, 26, 33.

15. J. Needham, *Chemical Embryology*, Cambridge, 1931, pp. 1635–8; *Biochemistry and Morphogenesis*, Cambridge, 1942, pp. 319–20. C. H. Waddington, The morphogenetic function of a vestigial organ in the chick, *Journal of Experimental Biology*, 15, 371–6, N. Kleinenberg, Die Entstehung des Annelids aus der Larve von Lopadorhynchus, *Zeitschrift fur wissenschaftliche Zoologie* (1886) **44**, pp. 212–24.

16. E. W. MacBride, Sedgwick's theory of the embryonic phase of ontogeny as an aid to phylogenetic theory, *Quarterly Journal of Microscopical Science* (1895) **27**, 325–42; *Textbook of Embryology. Invertebrata*, London, 1914; *Heredity*, London, 1921: these are the main works except for the de Beer controversy, on which see below.

17. D'Arcy Thompson, Introduction, In L. S. Berg, *Nomogenesis*, London, 1929, pp. xiii-xvi; and *On Growth and Form*, Cambridge, 1916, ch. 1 and passim.

18. R. D'Arcy Thompson op. cit. (6), pp. 89–90.

19. P. B. Medawar, Postscript: D'Arcy Thompson and *Growth and Form*, In R. D'Arcy Thompson op. cit. (6), p. 221.

20. R. Assheton, *Growth in Length*, Cambridge, 1916, esp. pp. 1 and 62.

21. A comparison of the Zoology Professors illustrates the general characteristic. At Cambridge they were A. E. Newton (1866–1907), A. Sedgwick (1907–9), J. Stanley Gardiner (1909–37), J. Gray (1937–59), C. F. A. Pantin (1959–66), T. Weis-Fogh (1966–75), G. Horn (1978–); Weis-Fogh, almost a century after the chair was founded, was the first Professor who had not spent almost his entire career in Cambridge: at Oxford the Professors were T. Rolleston (1860–81), H. N. Moseley (1881–92), E. R. Lankester (1892–9), W. F. R. Weldon (1899–1906), G. C. Bourne (1906–21), E. S. Goodrich (1921–45), A. C. Hardy (1945–61), J. W. S. Pringle (1961–78), T. R. E. Southwood (1978–); Rolleston, Moseley, and Goodrich were the only three to have spent most of their time in Oxford, and Rolleston hardly counts. The first three Cambridge Professors were comparative anatomists, the second four comparative physiologists; the first three at Oxford, and Bourne and Goodrich, count as comparative anatomists, Weldon a biometrician, Hardy a classical zoologist and ecologist, Pringle a comparative physiologist (from Cambridge), Southwood an ecologist. On the general difference, see H. R. Trevor-Roper, The acts of the apostles, *New York Review of Books*, 31 March 1983, p. 3, and N. Annan, *Leslie Stephen*, London, 1951, ch. 1, especially pp. 1–4, and 'The intellectual aristocracy', In J. Plumb (ed.), *Studies in Social History*, London, 1955, pp. 241–87.

22. G. E. Hutchinson, *The Kindly Fruits of the Earth*, New Haven, 1979, pp. 87–101; 138 on Shearer, on whom also J. Gray, Cresswell Shearer, *Obituary Notices of Fellows of the Royal Society* (1942) **4**, p. 16.

23. H. W. Lissmann, James Gray, *Biographical Memoirs of Fellows of the Royal Society* (1978) **24**, p. 62. The idea was Sir William Hardy's, but it gave Gray great pleasure to contemplate it.

24. E. R. Lankester, Degeneration, first published 1879, reprinted in Lankester's *The Advancement of Science*, London, 1890, from which I have quoted pp. 16, 17, 21.

25. G. C. Bourne to Mrs J. W. Jenkinson, 17 June 1915. The letter is in the possession of Mr R. Jenkinson, to whom I am most grateful for allowing me to see this and other material.

26. J. W. Jenkinson, Remarks on the germinal layers of vertebrates and on the significance of germinal layers in general, *Memoirs and Proceedings of the Manchester Literary and Philosophical Society*, 1906, p. 69.

27. S. J. Gould op. cit. (3), p. 186.

28. J. W. Jenkinson, *Experimental Embryology*, Oxford, 1909, p. 12.

29. G. W. Smith, *Primitive Animals*, Cambridge, 1911, pp. 58, 64, merely asserts the error of recapitulation. The full empirical critique on which I have based my remarks is in the G. W. Smith papers, MSS no. 235, lectures on Crustacea, preserved in the Archives of the Zoology Department, Oxford University (Henceforth: Oxford Zoology Archives). There is a rare memoir of Smith: *Geoffrey Watkins Smith*, Oxford (Printed for private circulation), 1917.

30. G. W. Smith papers, MSS no. 238, Experimental morphology and the theory of internal secretions (an evening address) and no. 240, Introduction to papers on the study of sex, Oxford Zoology Archives.

31. J. R. Baker, Julian Sorell Huxley, *Biographical Memoirs of Fellows of the Royal Society* (1976) **22**, p. 218.

32. I am most grateful to Professor E. B. Ford, Professor Sir Alister Hardy, and Sir Peter Medawar for talking to me about Goodrich, and zoology in Oxford in the twenties and thirties. I have also used A. C. Hardy, Edwin Stephen Goodrich 1868–1946, *Quarterly Journal of Microscopical Science* (1947) **87**, 317–55.

33. Goodrich papers, MSS no. 105, Embryology (lecture notes), Oxford Zoology Archives. E. S. Goodrich, *Living Organisms*, Oxford, 1924, pp. 145–7.

34. Professor Ford said this to me in conversation in All Souls, 7 March 1983. W. Garstang, The theory of recapitulation: a critical re-statement of the biogenetic law, *Journal of the Linnean Society, Zoology* (1922) **35**, p. 82, makes a related point, but see below for Garstang's genetics.

35. S. J. Gould op. cit. (3), pp. 202–6, and F. Churchill, The modern evolutionary synthesis and the biogenetic law, in E. Mayr & W. B. Provine (eds.), *The Evolutionary Synthesis*, Cambridge, Mass., 1980, pp. 112–22.

36. W. Garstang op. cit. (34), J. S. Huxley, *Problems of Relative Growth*, London, 1932, p. 239.

37. S. J. Gould op. cit. (3), pp. 231–4, who adopts a different classification. G. R. de Beer, Embryology and evolution, In G. R. de Beer, ed., *Evolution*, Oxford, 1938, pp. 57–78.

38. Huxley actually borrowed J. Gray's lecture notes to put together his first course on experimental embryology: J. S. Huxley, *Memories*, London, 1970, p. 125.

39. J. B. S. Haldane & J. S. Huxley, *Animal Biology*, Oxford, 1927, p. 225; H. G. Wells, J. S. Huxley & G. P. Wells, *The Science of Life*, London, originally published in parts, the whole first published late in 1930. Huxley op. cit. (38), ch. xii, tells when it was being written. I have worked from the 1938 edition, in which the relevant pages are 369–70. A. C. Hardy, Introduction, to

W. Garstang, *Larval Forms*, Oxford, 1951, p. 6, states that this passage was published in 1928.

40. E. B. Ford, Some recollections pertaining to the evolutionary synthesis, In E. Mayr & W. B. Provine op. cit. (35), pp. 340–1. P. B. Medawar, A biological retrospect, reprinted in Medawar, *Pluto's Republic*, Oxford, 1982, where the passage I have quoted appears on p. 294.

41. E. B. Ford & J. S. Huxley, Mendelian genetics and rates of development in *Gammarus chevreuxi, British Journal of Experimental Biology* (1927) **5,** 112–34. See also E. B. Ford, The inheritance of dwarfing in *Gammarus chevreuxi, Journal of Genetics* (1928) **20,** 93–102. There are some relevant recollections in Ford op. cit. (40).

42. E. B. Ford, *Mendelism and Evolution*, London, 1931, p. 30.

43. F. B. Churchill op. cit. (40), p. 120, 115. See also Gould on Garstang in op. cit. (3), 177, 183, 206, 428n7.

44. G. de Beer, *Embryology and Evolution*, Oxford, 1930, p. 9; A. C. Hardy op. cit. (39), p. 5.

45. J. S. Huxley op. cit. (36), pp. 239–40. The whole tone of Huxley's treatment suggests that he had learned it from de Beer: he even credits the term paedomorphosis to de Beer (p. 239). For J. B. S. Haldane: *The Causes of Evolution*, London, 1932, final chapter, which is very similar to: The time of action of genes, and its bearing on some evolutionary problem, *American Naturalist* (1932) **66,** 5–24. That Garstang was little interested in genetics or rate genetics was told me by both Hardy and Ford; the relevant page of Gould op. cit. (3) is 183. The Lankester essay is op. cit. (24), see p. 57.

46. D. Newth's statement (in his essay Michael Abercrombie: a Memoir, In R. Bellairs, A. Curtis & G. Dunn (eds.), *Cell Behaviour*, Cambridge, 1982, pp. 1–11) that de Beer was no experimental embryologist is inaccurate. For references, see J. Huxley & G. de Beer, *The Elements of Experimental Embryology*, Cambridge, 1934, and E. J. W. Barrington, Gavin de Beer, *Biographical Memoirs of Fellows of the Royal society*, **19,** 1973, 65–93. Newth would be an interesting source to follow the history of British embryology forwards from the period of this essay: his biases may amusingly be balanced against those of Sir Peter Medawar in his biography of Abercrombie: Michael Abercrombie, *Biographical Memoirs of Fellows of the Royal Society*, **26,** 1980, 1–15.

47. J. R. Baker op. cit. (32), p. 220; J. S. Huxley op. cit. (36), pp. 204–16.

# Was there a characteristic tradition in Britain determining the response to embryological issues?

Two things strike one about British embryology; the tenacity of involvement with recapitulation and the relative failure to establish any enduring schools of experimental embryology. What, if anything, might connect these features? Churchill and Ridley have shown how we must be careful to see beyond the notorious aspects of Haeckelian recapitulation, which are more remembered today, and to look for the positive contributions that the ideas might have played in earlier contexts. Even a superficial glance at the character of biology in Britain during the second half of the nineteenth century reveals an institutional structure which correlates naturally with distinctive scientific interests. British models of scientific success were private individuals like Darwin, Spencer or Galton, whose activities would necessarily gravitate towards the methods of the naturalist rather than the organised, collaborative and well financed structure of a laboratory. This was in marked contrast to the pattern in Germany where 'the microscopical art was rapidly becoming a national monopoly' (Churchill, 1979, *Stud. Hist. Biol.* **3**: 153). Certainly Britain was not noted for its contributions to cell theory, and when the professionalisation of science did come the lead was set by a mathematical, and model-building physics in Cambridge. If, then, the interests of British biologists inclined towards whole organisms, morphology, comparative anatomy and the study of populations, rather than towards chromosomes, recapitulation would have lain near their focus of attention.

Not inconsistent with this picture we see a strong undercurrent in Britain of what, at its peak in the 1920s and 1930s, would be known as 'organicism'. Russell, a characteristic representative of this approach, distinguishes between three schools of thought in his book

69

on the history of attitudes to morphology (*Form and Function*, 1916); 'functional' or 'synthetic' (which is the organicist version he himself favoured), 'formal' or 'transcendental' (which typified the earlier approach of the *Naturphilosophen* in Germany and seeks for unifying laws of body plan), and 'materialistic' or 'disintegrative' (which is the reductionist method of Weismann, which few British biologists followed). Despite the variety of their specific laboratory interests, scientists such as Hopkins, James Gray and Needham were united by their emphasis on 'organicism'. It was a concern which both linked and overreached particular methods. Behind the rich compilations of data and theory in Needham's writings, with their sometimes bewildering juxtapositions of liquid crystals, allometry, epigenetic landscapes and the biochemistry of the organiser, lay a search for an approach to that which would bring together all aspects of knowledge relevant to the centre target of explaining biological form.

The Theoretical Biology Club (aspects of whose aims are considered in this volume by Tennant and Olby) was part of this search; as Needham described it 'We were all devoted to a non-reductionist biological organicism which aimed at being at the same time non-obscurantist'. He quotes as the main influences on him, Teilhard de Chardin, Henderson, D'Arcy Thompson and Whitehead (M.-W. Ho & P. T. Saunders, 1984, *Beyond Neo-Darwinism*, p. vii). These then were the features that characterised one British contributor to experimental embryology. Julian Huxley is another, whose somewhat different, but equally diverse approaches would be hard to explain other than in terms of a tradition that occupied the middle ground of genetics, embryology and evolution theory. As an embryologist Huxley viewed the organism as a whole, from the perspective of growth rather than cells or their differentiation. Eventually Huxley abandoned embryology totally, and, as a leading figure in the integration of genetics with evolution theory in the 1940s, effectively disposed of embryology altogether in the new biological perspective of the 'Evolutionary Synthesis'. His book *Problems of Relative Growth* (1932) is dedicated to D'Arcy Thompson, another strongly enduring influence upon British biology. Thompson's impact cannot stem from any specific results or methods that he presented in his major work *On Growth and Form* (1917), since his approach was essentially based on analogy and has yet to give rise to a laboratory experiment. His appeal stems from the fact that he typifies the

British tradition in biology. As Olby discusses later, it was his primary concern with form, in a way that disregarded other considerations that might distinguish biological structures such as molecules, cells, organs or organisms, that typified an interest in macromolecules that was to lead to post-war molecular biology. How the British tradition responded to embryological issues after W.W. II is considered in the later paper by Yoxen.

The following three papers consider various aspects of the origin and characteristics of the newly emerging interest in biology in the USA over the comparable period.

# Preformation or new formation – or neither or both?

JANE MAIENSCHEIN

Department of Philosophy, Arizona State University, Tempe, Arizona 85287, USA

Preformation and epigenesis have seemed at many times, to many people, to represent clearly distinguishable positions on the issue of how organisms gain differentiated form. Preformationists traditionally have held that the form is already predetermined at the earliest stages of an individual's existence. Epigenesists traditionally have maintained that form emerges only gradually, during the developmental processes. Thus textbooks often refer to *the* preformation–epigenesis debate as if there were only one clear set of differences at issue. Recent historical literature has begun to establish, however, that 'the debate' actually has taken very different forms at different times. Shirley Roe, for example, has convincingly established the way in which the late eighteenth century debate between Albrecht Haller (a preformationist) and Caspar Friedrich Wolff (an epigenesist) took on its particular form because of the philosophical convictions of the participants.[1] 'The' debate then becomes a series of debates with changing emphases and tones as the philosophical and scientific climates changed.

American biologist William Morton Wheeler (1865–1937) recognised the persistence of concern about when form emerges but also some of the changing character of the specific points at issue. He postulated that there exist two different kinds of thinkers at any time. Some are most affected by the 'succession of phenomena, the ceaseless current of events, the changes that alter the complexion of the world, the great qualitative and quantitative differences produced by those changes in that which we call matter.' These observers note the rhythms and focus on the processes of change. Heraclitus, then Aristotle, Wheeler saw as first representatives of this first class of observers. The second class, represented by Parmenides and Plato, notes form

73

instead of process. These latter observers are impressed by 'similarity of the forms and conditions that recur from time to time and from place to place.'[2] Stability, not change, they see when they observe nature. When applied to development, the former represents a physiological and epigenetic view, the latter a morphological or preformationist view, Wheeler held.[3]

Wheeler suggested that these two types of observers constitute stable classes. Yet he also felt that the late nineteenth century had brought new data which required a reworking of the old preformationist–epigenetic distinctions. Thus, though Wheeler did not acknowledge the complexity of the shifts in earlier disputes and regarded the ongoing preformation–epigenesis debate as relatively fixed prior to his own time, he did see major changes by the century's end. Further new perspectives would probably emerge with more data, he felt. Neither a strict preformationist nor a strict epigenetic position would be likely to succeed. Rather:

The pronounced 'epigenecist' of to-day who postulates little or no predetermination in the germ must gird himself to perform Herculean labors in explaining how the complex heterogeneity of the adult organism can arise from chemical enzymes, while the pronounced 'preformationist' of to-day is bound to elucidate the elaborate morphological structure which he insists must be present in the germ. Both tendencies will find their correctives in investigation.

In fact, as Thomas Hunt Morgan (1866–1945) suggested, by 1901 it made sense to say that 'a process of pure epigenetic development, as generally understood nowadays, may also be predetermined in the egg.'[4]

It is on the changing perspectives of the late nineteenth century that this paper will primarily focus. In particular, I wish to demonstrate the American effort to undercut the German emphasis of the 1890s on distinguishing preformationist and epigenetic positions. Since I will focus on the American work ultimately, I have chosen to outline the German positions which were generally known and of concern to the Americans. Others also played important roles in stimulating American and European discussion, but the ideas outlined here represent the key theoretical positions. August Weismann and Oscar Hertwig represent respectively the revised predeterminist and epigenetic positions which focused on the extent to which form emerges due to inherited or physiological and developmental factors. Alternatively, Wilhelm Roux and Hans Driesch emphasised the extent to which the embryo acts as a self-differentiating mosaic of unstructured material which responds regulatively to external factors. Wilhelm His urged an intermediate position paralleled in significant respects to the American efforts to sort out the extent to which internal and external, inherited and developmental factors direct the emergence of form. The study of sex determination provides an example of one area where the discussion was played out.

## *The German debates:* predeterminism and epigenesis

Before evolution theory, both preformationists and epigenesists addressed the question: when does the embryo come to have the differentiated form that it is supposed to have? Or: at what point does the individual assume the characteristics of its type? Other concerns such as how that differentiation occurs also found expression, but all discussants began with the assumption of conformity to type. After the acceptance of evolution theory, there remained no type to which an individual could conform. Rather, the individual might feel pressures from the environment, hence external influences, as well as the effects of heredity, or strictly internal influences. An individual's form might resemble that of its parents but differ in some ways also. The biologist had to account for development of form with variations from as well as similarities to the parental forms. Since conformity no longer explained form, the embryologist had to provide an account of how that differentiated form appears. Preformation and epigenesis persisted, but with different emphases.

August Weismann (1834–1914) and Oscar Hertwig (1849–1922) served as primary spokesmen for these revised preformationist and epigenetic positions. Though others also contributed to the debates, Weismann and Hertwig provided a comparatively clearcut focus on what they consciously and explicitly regarded as the issues of preformation and epigenesis. They therefore serve to illustrate the state of the discussion in the 1890s. Weismann wrote of development in a number of essays, but his classic *Das Keimplasm* of 1892, translated into English in 1893, provided the focal point for supporters and critics alike.[5] In that work, Weismann explained that he had begun with sympathy toward epigenetic views and had rejected the 'evolutionary' idea that individual form merely unfolds from a pre-existent miniature state in the germ. In particular, Darwin's ideas of pangenesis had caused problems for him initially, Weismann proclaimed, but then he was forced into accepting such an evolutionary or preformationist position himself:

My doubts as to the validity of Darwin's theory were for a long time not confined to this point alone: the assumption of the existence of *preformed* constituents of all parts of the body seemed to me far too easy a solution of the difficulty, besides entailing an impossibility in the shape of an absolutely inconceivable aggregation of primary constituents. I therefore endeavoured to see if it were not possible to imagine that the germ-plasm, though of complex structure, was not composed of such an immense number of particles, and that its further complication arose subsequently in the course of development. In other words, what I sought was a substance from which the whole organism might arise by *epigenesis*, and not by *evolution*. After repeated attempts in which I more than once imagined myself successful, but all of which broke down when further tested by facts, I finally became convinced that an epigenetic development is an *impossibility*. Moreover, I found an actual *proof of the reality of evolution*, which ... is so simple that I can scarcely understand how it was possible that it should have escaped my notice so long.[6]

This 'proof' lay largely in explaining, within the context of cell theory, how body parts become so diversely differentiated. Cells become differentiated from each other but also differentiated themselves. This differentiation must result from different kinds of material particles (his biophores), Weismann asserted. This material must allow for transmission of independent structures which are not fixed but subject to variations. Thus *'The independently and hereditarily variable parts of the body therefore serve as an exact measure for determining the number of ultimate particles of which the germ-plasm is composed: the latter must contain at least as great a number as would be arrived at by such a computation'* of the number of variable parts.[7] The variable multicellular organism becomes so complexly differentiated, and thus requires so many different biophores, that even if those biophores were quite small there would not be sufficient space in the germ-plasm for them all. He could not envision a germ which contained a sufficient number of sufficiently variable biophores to explain an organism's situation, so any epigenetic account of development must fail. Instead of to biophores (which would yield varied characteristics by epigenetic development), then, Weismann turned to determinants (or material units which indirectly determine which characteristics will become manifest) as the basic unit of heredity. Some form of predeterminism must prevail.[8]

After making it clear that he held epigenetic views untenable and predeterminism necessary, Weismann then proceeded to detail his theory of the autonomous germ-plasm in heredity, development, and evolution.[9] Briefly, he maintained that 'Ontogeny, or the development of the individual, depends therefore on a series of gradual qualitative changes in the nuclear substance of the egg-cell.'[10] Nuclear changes, presumably with divisions of the chromosomes and accompanying reduction of the chromosome material, act to distribute the determinants for each body feature to different cells. Each cell becomes a particular cell type – such as nerve, muscle, epithelial, or whatever – simply because of the action of the determinants distributed to it.[11] Thus the cells are self-differentiating, 'that is to say, the fate of the cells is determined by forces situated within them, and not by external influences.'[12] It is *not* the case, as some epigenesists would say, that the cell's particular physical location in the embryo determines its fate. It is not the case, for Weismann, that conditions external to the cell itself can direct development.[13] Rather, under normal conditions, each cell has its predestined role because of the inherited determinants and their distribution to different cells during development. Internal factors alone determine development. Ontogeny is not a '"new formation" of multiplicity' or epigenesis, then, but the unfolding of multiplicity, or the evolution of previously invisible multiplicity.[14] There is no simple growth of pre-existing form as earlier preformationists had held. Instead Weismann's view was pre-determinist in a way that emphasised process as well as inherited structure, in that he made some concessions to the gradual manifestation, at least, of form. Determinants hold the key to that 'invisible multiplicity', and the nuclear divisions which distribute determinants to different cells explain development, according to Weismann.

Oscar Hertwig disagreed. Weismann's theory actually abandoned explanation of development, Hertwig felt, and it offered an unpromising closed system. In his classic work of 1894, *Präformation oder Epigenese*, Hertwig focused most directly on Weismann's preformation. In Hertwig's view, influenced by his education under Ernst Haeckel (1834–1919) and his subsequent move to a more direct concern with cytology and embryonic development, Weismann's theory merely transfers to an invisible region the solution of a problem that we are trying to solve, at least partially, by investigation of visible characters; and in the invisible region it is impossible to apply the methods of science. So, by its very nature, it is barren to investigation, as there is no means by which investigation may be put to the proof. In this respect it is like its predecessor, the theory of preformation of the eighteenth century.[15]

Instead of looking to fixed determinants as the source of cellular differentiation, the biologist must look to the complex of external and internal factors which affect cellular development. Instead of holding the developing embryo as a self-differentiating, pre-programmed system, the biologist should regard the embryo as also initially dependent on external conditions, Hertwig urged. Yet Hertwig's views did not represent a simple return to earlier epigenetic positions. He held, in agreement with Weismann, that the germ-plasm is already highly organised at the initial stages of development. From that organised germ, development then proceeds epigenetically to produce differentiated form. Thus, Hertwig believed that his theory should be considered evolutionary but also epigenetic.[16]

For Hertwig, the actions of individual cells and the interactions of cells to form a whole organism remained central. With his eye also on process rather than exclusively on pattern, Hertwig demanded an explanation of such phenomena as cleavage or gastrulation. How could such complex processes, which depend on close cooperation of cells, be explained by Weismann's theory which stressed differentiation and individuation of cells? If the embryo acted as a mosaic with the special determinants distributed to each cell, how could it respond to changing environmental conditions or to experimental conditions, he asked? Gravity, light, temperature, and such factors also manifestly affect development and differentiation, Hertwig felt certain, and Weismann could not explain that fact. Weismann's evolutionary or preformationist theory ignored nature, according to Hertwig, who saw different natural phenomena as demanding explanation.

For Hertwig, the shape of the whole organism and the distribution of materials in the egg played as important a role as the nuclear organisation. While Weismann focused on the nucleus and heredity, Hertwig concentrated on the cytoplasm as well and on development. The internal factors worked with external conditions to effect development. Thus:

I shall explain the gradual, progressive organisation of the whole organism as due to the influences upon each other of these numerous elementary organisms in each stage of the development. I cannot regard the development of any creature as a mosaic work. I hold that all the parts develop in connection with each other, the development of each part always being dependent upon the development of the whole.

Further: '. . . during the course of development, there are forces external to the cells that bid them assume the individual characters appropriate to their individual relations to the whole; the determining forces are not within the cells, as the doctrine of determinants supposes.'[17] Thus Hertwig's views of development clearly differed from Weismann's on at least these points.

It should also be clear that Weismann and Hertwig held different versions of preformationist and epigenetic views than their predecessors had. Though the terms remained, the character of the debate had changed. No longer did the preformationists appeal to the form of the type as already structurally embodied in the germ material and simply growing larger in any literal sense. Weismann's was instead a position of pre-determinism, where material determinants direct the development of differentiated parts. Still an evolution, development involved the self-guided unfolding of pre-determined multiplicity of parts instead of mere growth of preformed parts. The various parts become manifest only gradually as the determinants act in succession; thus Weismann's position moved somewhat toward including epigenesis even while separating heredity and development.[18] Hertwig's epigenesis did still hold form as emerging anew during development. But instead of conforming to type, the emerging organism begins as an organised germ cell, then responds to both internal and external conditions of the whole to grow and develop as a whole organism. The individual cells then work together cooperatively to produce the resulting differentiation. This position thus moved toward embracing some pre-organisation of the egg material. Weismann stressed heredity, Hertwig development, but the distinction was less extreme than in earlier debates.

Two distinct positions thus persisted with Weismann and Hertwig in the late nineteenth century, but they were neither as extreme nor were they as distinctly separated as previous preformationist or epigenetic views. Since each emphasised different problems and made different assumptions, it became unclear what besides rational argument might be considered evidence for one or the other point of view. The introduction of experimental results by Wilhelm Roux (1850–1924), Hans Driesch (1867–1941), and others helped point the way to a redefinition of issues and reorientation of discussion. Still within the general tradition of preformation–epigenesis debates, the exchange between Roux and Driesch exemplified a different emphasis, on the relative roles of internal and external factors in directing development.

## *An alternative emphasis:* self-differentiating mosaic or regulative totipotency?

The 1880s and 1890s brought heated debate about development. Not couched explicitly in earlier terms of preformation and epigenesis, or Weismann's and Hertwig's discussion of determinants and complex whole organisms, instead other contemporaneous debates centred on the extent to which the embryo constitutes a self-differentiating mosaic rather than a regulative body responsive also to changing internal and external conditions. Programmed mosaic or responsive system: which characterises the early embryo? The focus on this revised question represented an emphasis alternative to the more traditional questions about preformation and epigenesis addressed by Weismann and Hertwig.

Wilhelm Roux grappled with the complex problem of development. Eventually he and Weismann came essentially to agree on a theory of distributed determinants as causing development. Yet Roux seemed to the Americans to bring a different emphasis to his work. He elaborated Weismann's basic ideas into his own system of 'Entwickelungsmechanik' (developmental mechanics) and provided experimentally derived support for many of Weismann's claims. In early experiments on orientation of eggs, Roux decided that the orientation did not affect development.[19] As he rotated eggs about an axis, he found no change in development, which proceeded normally in both speed and detail. From this evidence, he concluded that eggs are self-differentiating and do not require external formative influences to guide their way to normal differentiated development.

In his classic, and much-discussed paper of 1888 on half-embryos, Roux set out to test what he saw as the two leading alternative theories of development. As he reported:

The following investigation represents an effort to solve the problem of self-differentiation – to determine whether, and if so how far, the fertilised egg is able to develop independently as a whole and in its individual parts. Or whether, on the contrary, normal development can take place only through direct formative influences of the environment on the fertilised egg or through the differentiating interactions of the parts of the egg separated from one another by cleavage.[20]

Working with frog's eggs, Roux cited assorted evidence for the self-differentiation of cells. Yet true to his rhetorical pronouncements for 'causal analytical experimental embryology', he maintained that only direct experimentation could establish that development acts as a sum of separate mosaic developments.[21] His experiments in this case consisted of puncturing with a needle one of the two blastomeres after the first cleavage, and a few examples of puncturing one or two of the four blastomeres after the second cleavage. He did not remove the punctured cells, and the early results remained unsatisfying. After modifying his technique, Roux succeeded in obtaining, in about 20% of the operated eggs, the ability of the unpunctured cell to survive and continue developing. These surviving and functioning cases formed his experimental sample, which he examined carefully through traditional histological staining and sectioning techniques.

Roux found a regularity in these abnormal cases. The experimental embryos produced only partial blastula or gastrula stages. The remaining blastomeres did not compensate for the injured blastomeres; rather it seemed that they developed as they normally would. From his results, Roux concluded that 'In general we can infer from these results that each of the two first blastomeres is able to develop independently of the other and therefore does develop independently upon normal circumstances.' Thus, 'All this provides a new confirmation of the insight we had already achieved earlier that developmental processes may not be considered a result of the interaction of all parts, or indeed even of all the nuclear parts of the egg. We have, instead of such differentiating interactions, the self-differentiation of the first blastomeres and of the complex of their derivatives into a definite part of the embryo.'[22] In short, the early developing embryo acts as a mosaic of independent parts. And probably the mosaic is effected by 'qualitative separation of materials', though he acknowledged that proof of that would require further research. From this conclusion of stable developmental pattern followed the suggestion that further study could reveal the exact role played by the particular first blastomeres, a suggestion that the Americans pursued in quite different ways with their cell-lineage work, as discussed later.

Roux did *not* go on in his paper of 1888 to conclude that every step of development produces independently developing cells and hence a perfect mosaic. Though he certainly did not reject such a possibility, he cautioned that 'How far this mosaic formation of at least four pieces is now reworked in the course of further development by unilaterally directed rearrangements of material and by differentiating correlations, and how far the independence of its parts is restricted, must still be determined.'[23] In this work, therefore, he clearly supported the cause of mosaic development, as a Weismannian sort of predeterminism. Yet he did not argue adamantly for an extreme form of predeterminism in which no regulative action could occur. And he left open the possibility that later stages, as the embryo becomes more complex, might exhibit increased dependent differentiation, where cells respond to other cells and to external conditions.

His continued research on half-embryos produced some cases where a whole embryo did result. By that time, though, Roux was sufficiently committed to his mosaic interpretation that he did not question his theory. Instead he generated an auxiliary hypothesis that there exists a reserve idioplasm which is called into action when regeneration or post-generation (following injury) occurs.[24] Despite such modifications, Roux maintained a predeterministic mosaic position. Other studies called Roux's interpretation into question. Of those pointing to contradictory conclusions, the experiments of Hans Driesch received most attention in Germany and the United States. Driesch began, like Roux, with the assumption that embryonic development depended on mechanistic processes and could be explained in mechanistic terms. His experimental results pushed him to different conclusions, however.

In his first of a series of 'Experimental Studies', Driesch followed Roux's lead and tested the potency (or ability to adapt) of cells after the first cleavage stage. Using echinoderm eggs instead of frog's eggs, Driesch expressly sought to repeat Roux's experiments with what he regarded as more durable, available, and easily observed material. Instead of puncturing one of two blastomeres with a needle, Driesch shook the two cells apart, following a technique of Oscar and Richard Hertwig. Apparently Driesch chose a different organism and method because he felt they would offer more easily obtained experimental results, for as Oscar Hertwig had shown, sea urchins presented a particularly favourable embryological subject because of their transparency and resiliency. Experiencing numerous problems with obtaining samples which survived and continued to develop, as Roux also had, Driesch nonetheless generated sufficient cases to discover that they produced half-sized normally structured blastulae. Thus, any theory of predetermination or of the pre-existence of 'organ-forming germ regions' must be wrong. Instead, the cells each retain totipotency, or the ability to adapt to the needs of the organism and become any part, at least until the four-celled stage, Driesch concluded.[25]

In his paper, Driesch did not go further to state that no pre-determination occurs or that Roux's theories of cellular self-differentiation or qualitative mosaic division were impossible. He did not endorse such an extreme, definite position at first, but only gradually became convinced of the continued totipotency of cells and hence of the absolutely critical importance of regulation in response to the changing environment during even later stages of development. Continued experiments made him increasingly dog-matic, increasingly anti-mosaic and anti-predeterminism.[26] His work of 1892, with its restrained interpretive comments, served best to underline the central issues in these empirically accessible debates, namely the extent to which development represents an unfolding or self-differentiation according to prescribed pattern or a continued regulated response to conditions.

What had seemed at first a compelling support for Weismann and Roux from Roux's experiments, Driesch undercut with his different results. The question of mosaic predeterminism or regulative epigenetic development remained open. Roux admitted that the mosaic undergoes development and hence is not strictly prestructured, and Driesch recognised that there might be some material arrangement in the germ. The positions were thus less extreme than those of earlier centuries but still clearly distinct, as were those of contemporaries Weismann and Hertwig.

Both Roux and Driesch became increasingly polemical and extreme with further experimental work appearing in print on each side. That tendency to emphasise one position or the other, to support either a preformationist or epigenetic view exclusively, may have stemmed from the researchers' different interests, as Wheeler suggests often happens. Roux, like Weismann, did focus on form, on structure and morphological changes during develop-ment. Driesch, like Hertwig, sought instead to understand the processes which take the embryo from one stage to the other. The former looked more to stable, internal factors, the latter more to changing external conditions. The apparent discrepancies of Roux's and Driesch's results received considerable attention from biologists. But the situation was a bit like a Gilbertian comic opera, as Herbert Spencer Jennings later suggested, with all singing 'For you are right and I am right and he is right and all is right.'[27] Historians have perhaps tended to overemphasise Roux's and Driesch's positions and their importance. Roux and Driesch did focus attention on heredity and development, and the apparently contradictory results were important, yet the American biologists at least looked to other alternative accounts of development as well as to elements of both Roux's and Driesch's. What Roux's and Driesch's debate did for the Americans was to underline the need for further evaluation of internal and external influences on development.

## A *different perspective:* Wilhelm His

An alternative to Roux or Driesch, to Weismann or Hertwig, came with the more centrist position put forth by the atypical German anatomist Wilhelm His (1831–1904). He served as an important alternative for the Americans, especially for Charles Otis Whitman (1842–1910), so plays an important part in our story. Though his most well known work appeared earlier (1874), His continued to publish and apparently particularly appealed to some of the Americans because of his cellular point of view and his emphasis on early cellular organisation.[28]

Beginning with the problem of how and when the organism gains its differentiated organisation, His rejected Ernst Haeckel's response, which was becoming a convenient starting point for discussion at the time. Haeckel said organisation came only with the germ layers, with his Gastraea Theory as the most accessible articulation of that view. Haeckel published various works emphasising that the essentially mere vegetative division of cells in early cleavage gives way to the organised gastrula stage.[29] Here, for the first time, the organism has identifiable separate layers which correlate with later parts of the body. For the first time with the gastrula, then, can the organism be considered organised and subject to the adaptive responses through natural selection. Haeckel believed that the earlier stages hold no interest for researchers, serving instead simply to multiply the primary nutritive material. Furthermore, after gastrulation, ontogeny follows phylogenetic patterns of development, or ontogeny recapitulates phylogeny as stated in Haeckel's biogenetic law.[30] Thus, evolutionary history will explain embryonic development, with a few exceptions which Haeckel fitted into a neatly modified version of the biogenetic law.

In contrast to Haeckel, His felt that evolutionary history explained nothing about development. An individual organism does not pass through the stages of ancestral organisms, as Haeckel said. No such primordial ancestor as the Gastraea had ever existed, in fact. Such suggestions add nothing and mislead understanding of developmental processes, which remain essentially physiological processes. Furthermore, His insisted, the earliest cell divisions do not simply separate material in preparation for the all-important gastrular separation into germ layers. The organism is organised from the beginning, he maintained. The fertilised egg has definite organisation, in the form of organ-forming germ regions.

In his most well known work of 1874, His asserted that study of the embryo must begin with the important principle of the organ-forming germ-regions. His convictions expressed in this early work found expression in his later work as well. 'The principle, according to which the germinal disk contains the preformed germs of organs spread out over a flat surface and conversely that every point of the germinal disk is found again in a later organ, I call the Principle of Organ-forming Germ-regions.'[31] In other words, organs exist pre-localised by the time the germinal disk is formed, and each pre-localised germ-region correlates with a later structure, and, indeed, gives rise directly to those later body parts. Developmental differentiation proceeds as the external manifestation of complex, interacting physiological processes internal to the organism. He called for a physiology of development. Look at structure, but also understand the physiological processes which bring about developmental change, His urged.

His's view might seem preformationist. But he denied such an interpretation. Instead, he believed that theories of development fall into four categories rather than the usual two of preformation and epigenesis. His own favourite was neither strictly preformationist nor epigenetic but the rather mysterious 'theory of transmitted movement.'[32] Development of form is the external manifestation of a complex, interacting, lawlike physiological process, His insisted. As with a telegraph message, we cannot see the process but do witness the result. And the message must have had a concrete organised beginning, as development does.[33] It is not the form itself nor the specific building material of the body which begins development, but 'the stimulation of form-producing growth.'[34] Production of form involves both the material medium, or the egg, and the processes of change which are stimulated by fertilisation, according to His.

Notwithstanding His's denials, some did see His as a preformationist. Oscar Hertwig did. In responding to Hertwig in 1901, His reiterated that the attempt to classify researchers into the old two categories of preformation and epigenesis did not make sense any longer. While many found difficulties in explaining the apparently contradictory results of Roux, Driesch, and others, that was partly because of their adherence to such old-fashioned positions, His felt. He found no problem, since he believed that his organ-forming germ regions could explain all. In Driesch's case of the half-sized regular embryos, for example, the separated blastomeres compensate for the missing parts because development involves interactive processes as well as initial organisation. The germ regions are not absolutely predetermined or in any sense preformed, that is, but retain some flexibility to respond to conditions. Form and process work together.[35]

Though Whitman and the other Americans at the Marine Biological Laboratory probably did not read all of His's theoretical work, and though they certainly quoted only very selectively from his writings and avoided some of his more difficult abstractions, they did assume a position similar in some ways to his on the preformation–epigenesis issues.[36] Thus His provided support if not initial inspiration for the Americans. They too held that neither preformation nor epigenesis was adequate in its old form. Nor did the new versions as put forth by Weismann and Hertwig or Roux and Driesch settle for them the issue of how form is achieved. Instead, the question of whether an embryo unfolds or emerges by new formation required a revised approach, they felt. Perhaps organised germ regions and regulative response could together explain development. Supporting this position was the American work on cell-lineage and cytoplasmic localisation that strongly sustained claims for a morphologically differentiated germ.

## *The Americans:* Whitman and the MBL workers

Wheeler's assessment of the situation typified the reaction of those American biologists working at the Marine Biological Laboratory in the 1890s.[37] The M.B.L. offered a unique setting for biologists from various American universities to spend their summers in research, and the work there concentrated in the early years on studies of development, including problems of development, inheritance, and physiology. Epigenetic and preformationist positions admittedly persisted, but neither was any longer regarded as fully in the right. Whitman served as initial spokesman for the preferred intermediate point of view. As first director of the M.B.L. in Woods Hole, Massachusetts, and as head of the important new biology program at the University of Chicago, Whitman exerted considerable influence over the younger group of professional biologists beginning their research careers. In particular, his contribution to the popular series of Friday night lectures at the M.B.L. would have been widely read and discussed.

Whitman had turned to Germany for his training in biology and had continued to read the German biological literature. Exposed while pursuing his doctoral degree in Germany to work in cytology as well as to concentration on phylogeny characteristic of Haeckel and others, Whitman had developed a deep interest in the early stages of development. He was particularly aware of the controversies surrounding the significance of the early stages for later development. The work on cell lineage which Whitman inspired at Woods Hole brought further attention to those early stages. This work also focused interest on questions about the extent to which those early stages exhibited regular, and hence probably inherited, or variable, and hence probably regulated, patterns of cell division. Thus Whitman brought questions about preformation and epigenesis to the attention of American biologists at the Marine Biological Laboratory, and they played a central role in transmitting a controversy active in Germany to an American context.

In his doctoral dissertation work, Whitman considered development of the leech *Clepsine*, exhibiting his conviction that eggs undergo early organisation, and hence some prelocalisation even if not preformation. So, as Whitman concluded, 'In the fecundated egg slumbers potentially the future embryo. While we cannot say that the embryo is predelineated, we can say that it is predetermined.

The "histological sundering" of embryonic elements begins with cleavage, and every step in the process bears a definite and invariable relation to antecedent and subsequent steps . . . .'[38] Bilateral symmetry finds precocious expression in the blastula or gastrula. Whitman considered the various relevant theories and evidence for each, and left many questions open for further discussion – many of which he pursued in a series of papers on *Clepsine*. He operated firmly within the tradition of studies of embryology *and* evolution and certainly did not reject evolutionary concerns.[39] One suggestion which Whitman did follow most closely in those later studies was that of an egg which experiences early organisation.

In another, short, paper of 1878, Whitman pointed out that the traditional view held that a pre-embryonic stage occurs and prepares the egg for its significant cleavage. Growth, impregnation, changes in the nucleus, and formation of the double-walled blastular sac occur during this stage. Then begins serious development with gastrulation, at which point the later stages are more or less sketched out. Only in the previous seven or eight years, Whitman pointed out, had that first pre-embryonic stage begun to receive serious attention. Careful study of the early stages reveals early changes. Yolk-spheres congregate and the nucleus changes, producing an amphiaster which resembles the pattern of iron dust near a magnet. Clearly, with these coordinated nuclear and cytoplasmic changes, 'we have here a life-phenomenon of very great importance.'[40] Other changes also occur in the cytoplasm. The male and female nuclei meet and blend. Complex changes these, presumably holding significance for later development. The early so-called pre-embryonic stage could no longer be ignored, Whitman definitely suggested, though he developed the implications no further in 1878.

By 1887, Whitman had spent two years at Harvard and resided at the Allis Lake Laboratory in Wisconsin. Two papers which appeared in that year demonstrate his growing conviction that the embryo already has some organisation at a very early stage and that researchers should focus attention on the early stages, well before the gastrula appears. In particular, in one paper he asked about the origin and fate of the germ-layers of *Clepsine*, and of the germ bands in particular. How the three germ-layers form and what significance the germ bands, or germ ring, holds remained open questions, which Whitman addressed. Comparing results from different organisms revealed parallels and homologies which went far toward detailing development, Whitman felt. In particular, he found Edmund Beecher Wilson's (1856–1939) studies illuminating. Wilson's *Lumbricus* and his own *Clepsine*, as well as other leeches exhibited a 'complete parallel' in origin and development of the germ bands in particular.[41] In the early cleavages, one finds a forecast of the future organism. Thus,

Although there is scarcely anything in the external appearance of the eight-cell stage to indicate the relation of its parts to the future embryo, yet we know by what follows that an immense work has already been accomplished. All those fundamental conditions and relations implied in the terms anterior and posterior, right and left, dorsal and ventral, are now definitely established. The ground-plan of the future structure is there, and the segregation and distribution of the building material have advanced far toward completion.[42]

For example, the annelid teloblasts, those particular cleavage products which undergo rapid and frequent division, those 'specialized centres of proliferation', give rise to the germ bands.[43] Clearly some important prelocalisation occurred and a later epigenetic or regulative development could not explain all the phenomena. Whitman sought to investigate these localisations further.

The comparative studies showed that when germ layers cannot be effectively identified and homologies remain unclear, one must look to the 'precise genealogy of the cells', or cell-lineages as such studies were later called. Germ-layers may not always give rise neatly to the same adult parts, and thus the phylogenetic relations cannot be discovered in the germ-layers. The germ-layer theory which finds organisation as beginning only with the germ-layers thus causes problems. 'When, as in the case under consideration, we find an organ arising sometimes from the ectoderm, and at other times from the mesoderm, we have to admit that there is no fixed and impassible boundary line between these two layers; and that its association with this or that germ-layer is not an infallible guide to its morphological identity.'[44] Look, then, to cell lineages, to the way in which particular cells correspond to particular parts of the germ-layers. Only then will you be able to sort out difficult questions of which parts are homologous to which, and which thus share ancestral, evolutionary relationships. Traditional concern with evolution and homologies of germ layers continued in Whitman's work, but the focus had begun to shift to earlier, cellular changes.

A second paper of 1887 underlines this new focus. Examining

maturation, or preparation of the egg to receive the 'spermatic element', and fecundation, or 'those attending changes in the protoplasm which form the concluding steps in the premorphological organization of the egg', Whitman admitted that the majority had come to regard the nucleus as the 'primum mobile' of development.[45] The movement of the two pronuclei was widely regarded as evidence that the nucleus controls developmental changes. But no, said Whitman. The pronuclei move through the cytoplasm. Indeed cytoplasmic forces direct the nuclear movements, as seen with the formation of astral lines through the cytoplasm. The egg cytoplasm exerts just as much influence as the nucleus, even at these earliest stages. Cytokinesis (movement of the cytoplasm) and kinokinesis (movement of the nucleus) work together, cooperatively.

With this study, Whitman had embarked on a series of efforts pursued over the next several years to establish definitely not only that the embryo undergoes early organisation, hence that it is preorganised, but also that cytoplasm exerts influence as well as the nucleus, hence that no complete nuclear determinism occurs. This work brought him directly into the revived debates over preformation and epigenesis as well as disputes over the significance of the cell theory.

Whitman rejected all the extreme positions. Those who argued for either preformation or epigenesis alone missed part of the story, he felt. Instead, 'The truth appears to me to lie on both sides, the error consisting only in unduly exaggerating the relative importance of one or the other factor'. Clearly nuclear changes occur, and they seem to play a role in directing cellular development at least. But the nucleus likely has arisen only by secondary adaptations and cannot be the 'seat of formative and regenerative energies.'[46] In some cases the cytoplasm directs the nucleus, as in the case of fertilisation which he had discussed the year before.[47] So both the nucleus and the cytoplasm exercise some directive control. But not all. The organism as a whole also exhibits a physiological unity, it responds with regulative reactions to changing conditions. A full explanation might require a move beyond such mechanistic accounts as biology traditionally uses. Perhaps new forces enter the process of development. Thus, none of either the preformationist or the epigenetic theories could offer a full account of development, Whitman insisted by 1888. Further study of development and the relative importance of predeterministic and epigenetic regulative factors remained an important problem for biologists.

In two lectures at the M.B.L. in 1894, Whitman returned to the debates. His comments especially in the 'Prefatory Note' indicate his disagreement with the current of opinion. In particular, the apparently contradictory experimental results of Roux and Hans Driesch echoed by experiments from some of the M.B.L. crowd, probably stimulated Whitman's concern. Roux's mechanism and Driesch's both seemed exaggerated and mistaken. Claiming that he had no particular view of his own to defend, Whitman concluded that he sought instead 'to have well-defined standpoints and clear ideas of guiding principles.'[48] Recognition of the initial orientation of cytoplasm and the whole organism, would bring the drift of opinion toward a new standpoint, more akin to His's emphasis on compromise and new definitions. Instead of the old extreme positions which sought to explain development as *all* due to preformation or *all* due to epigenetic responses, ask a new question: 'How far is post-formation to be explained as a result of pre-formation, and how far as the result of external influences?' In other words, what are the relative contributions of internal and external factors in directing development? As Whitman concluded, agreeing with a contemporary, 'This statement compasses the whole situation: "the successive formation of parts not previously existent", represents the accepted verdict on the old issues and the expressions, "imposition from without", and "generation from within" define the new issue, which lies wholly this side of the old . . . .'[49]

In a lecture of 1894, Whitman explicitly addressed the preformation–epigenesis debates. Effectively, he argued that anyone espousing solely one side or the other exhibited arrogance, for these two views remained complementary. Each had components of both truth and error.[50] What biology needed, he urged, was a clear formulation of the relevant standpoints and then investigation to discover the relative validity of each. Biology was actually ready to move to a new standpoint, once the issues became clarified. Instead of argument about whether development occurs by preformation or post-formation, then, the modern biologist should explore how far post-formation, or differentiation during later developmental stages, can be explained by the existence of preformation, or early organisation of the egg. How much depends on the developmental response to external conditions rather than on programmed internal unfolding?[51] This new question comprised such a very different problem than the old dispute, Whitman felt, that to an old preformationist,

someone like Weismann would appear as an epigenesist since his
system included some new generation during development of parts
which had not previously existed preformed. Instead of on preformation or epigenesis, Whitman clarified, the new issue centred primarily on whether generation occurs from within or is imposed
from outside.[52] And surely, the prudent biologist must acknowledge
that some of both internal and external influences act together and
should set out to study their relative importance and the way in
which the two sets of factors work together.

Whitman did not believe that pre-organisation explained everything; for example, cleavage planes were not pre-set. Whitman rejected as inadequate any sort of developmental mechanics which
sought to explain development while ignoring the initial historically
derived organisation of the egg. He took a moderate position
instead: some germinal organisation occurs, but later processes in
response to conditions also play a role. Evolution and epigenesis, but
neither in the old form.

In succeeding essays, Whitman pursued the same themes and
developed one of his favourite prescriptions for American biology:
cooperate! Biologists should cooperate among themselves, they
should use a variety of methods where more than one could prove
productive, and they should explore a variety of standpoints on a
problem. Do not choose one over-arching theoretical framework to
guide your work, he urged the young biologists. Do not become set
in your ways or in your ideas. Consider alternatives, avoid buying
into dogmatic statements of extreme positions.[53] For example, move
beyond preformation–epigenesis debates to new understanding of
heredity and development.

At least some of the young biologists at the M.B.L. agreed with
much of Whitman's view at first. Beginning with work on cell lineage, they settled to work to ascertain the extent to which the egg is
initially organised. Their study of the earliest developmental stages,
which carefully traced the origin and fate of each cell through the
early cleavage stages, received inspiration from His (and others) by
way of Whitman.[54] This concern with tracing cell fates might suggest a strong preformationist position, but that concern was
moderated by a conviction that regulation also occurs. Regeneration
and transplantation studies argued for regulation rather than mechanistic determinism. And work by Jacques Loeb (1859–1924), for
example, suggested that external conditions and accidents such as

where the cleavage planes occur exert more important influences than a preformationist position would allow. Thus, Wheeler for example, recognised the old preformation–epigenesis distinction as no longer useful and acknowledged that only further research would advance understanding. And researchers such as E. B. Wilson and Edwin Grant Conklin (1863–1952) took up cell-lineage studies and addressed the preformation–epigenesis issues. They also began with the conviction that the previous statements of the distinctions no longer proved useful. But each went on to draw his own distinction.

## *Variations on a theme:* Wilson and Conklin

Wilson agreed with Whitman that neither the old preformationist nor the old epigenetic position made sense any longer. After completing his own study of the marine worm *Nereis*, he concluded that the egg definitely exhibits some early organisation. In his major text of 1896, *The Cell in Development and Inheritance*, Wilson emphasised that biologists still had no account for the orderly progression of development. Both Weismann and Roux, on the one hand, and Hertwig, on the other hand, demanded too many assumptions and explained too little. 'The truth is', Wilson concluded, 'that an explanation of development is at present beyond our reach. The controversy between preformation and epigenesis has now arrived at a stage where it has little meaning apart from the general problem of physical causality.'[55] Clearly, part of the germ is inherited. Presumably it reflects the conditions acting on the germ material in earlier generations, which suggests a form of evolution or preformation. Yet, against that is the power of regeneration, which suggests at least some ability of new formation, or epigenesis. Much remains to be explored, Wilson concluded in 1896.[56]

In 1905, Wilson again addressed explicitly the issue of preformation and epigenesis. Over the intervening decade, his researches had led him to more certain convictions about the nature of the inherited germ at least.[57] Yet he continued to maintain that some elements of what had been considered preformation and epigenesis both played roles in explaining development.

Alice Levine Baxter has considered Wilson's work of that period and has concluded that Wilson 'was more of a preformationist than an epigenesist', yet 'that his definition of the term "epigenesis" was rather narrow', and that his particular views of preformation in fact helped Wilson to accept the chromosome theory which caused problems for some of his contemporaries such as Thomas Hunt Morgan (1866–1945).[58] Baxter's very useful studies of Wilson do show him as one of those Wheeler would have identified with morphology and with explaining changes in form rather than in processes of change. Baxter is convincing in establishing the importance which Wilson assigned to chromosomes. Wilson's concern with inheritance *and* development found focus in a nucleus of inherited chromosomes which he saw as programmed in some way to direct development. Yet it is perhaps misleading to label Wilson as a preformationist at all. Really he embraced some elements of both – while rejecting both in their previous or more extreme senses. The important point remains that Wilson saw the debates over development as having changed, away from the traditional arena in which preformation–epigenesis debates took place. Instead of concentrating on determinants or the complex functioning whole organism, self-differentiation or regulation, intrinsic or extrinsic factors, per se, Wilson returned to the traditional developmental question about how production of form occurs, by unfolding or new generation. He then set out his new response.

For Wilson, 'whether the embryo exists preformed or predelineated in the egg from the beginning or whether it is formed anew, step by step, in each generation': that was the traditional form of the basic question of preformation and epigenesis.[59] Recent research showed that in at least some organisms the early cells are directly and regularly correlated with parts of the future body. Other cases do not exhibit such regularity. No paradox here, Wilson said. The existence of 'organ-forming stuffs', some of which are actually visible as differentiated regions of the egg, could explain both phenomena. 'The evidence is steadily increasing', he wrote,

that such stuffs exist, that they have a definite arrangement in the egg, and that in cases where the form of cleavage is constant they are distributed in a definite way to the cells into which the egg splits up. The cleavage-mosaic is accordingly to be conceived as an actual mosaic of different materials that are somehow causally connected with the development of particular parts.[60]

Thus, for Wilson, to the extent that such material stuffs exist and are distributed in a regular, definite way correlated with late development, to that extent only can the origin of differentiated form be regarded as mechanical and as due to prelocalisation. But that extent may not, in fact, extend far. The prelocalisation may remain only very general, with the majority of details of differentiation coming from non-prelocalised factors.

Confirming such a conclusion, other studies showed that only a few stuffs exist in the early stages 'and that as development goes forward new stuffs are progressively formed and distributed.'[61] To this extent, the actual progress of development is epigenetic and not evolutionary. The embryo emerges only gradually. Yet differentiation is guided by causes lying deep in the hereditary material, presumably in the chromosomes, though the chromosomal role remained not completely demonstrated. This nuclear chromatin does, therefore, control development and offers a sort of preformation, but development remains truly epigenetic in essence.[62]

Whether Wilson was really more a preformationist thus loses importance here. Much more significant is the clear way in which he set up the discussion in 1905. Both preformation and epigenesis operate, yet it nonetheless made sense to separate the two, and to discuss each separately. Moving away from His's and Whitman's concern with uncovering the way in which internal and external factors inevitably work together, Wilson at least rhetorically separated what he saw as the components of preformationist and epigenetic development. Preformation he regarded as linked with inheritance and with chromosomes. Epigenesis he linked with developmental processes and with the cytoplasm and the organism as a whole. Though still regarding problems of heredity and development as essentially related, and though continuing to consider both together, as in his revised *The Cell*, Wilson's discussion of the two made clear that two separable factors operated to guide development. A logical conclusion followed that researchers could address each separately. Wilson himself pursued a path deeper into the inherited material of the nucleus, into the chromosomes and their complex dances during cell division. This particular separation became increasingly important throughout the twentieth century as study of heredity became genetics, of development became embryology, and the two increasingly diverged. A logical and perhaps expedient divergence, given assumptions at the time, but one which brought loss as well as gain.[63]

Wilson's fellow cell-lineagist, Conklin, also began with Whitman's concern for revising the preformation–epigenesis issues. Conklin, however, disagreed with Wilson's increased emphasis on the nucleus as primary bearer of inherited determinants for development. Nuclear divisions could not explain differentiation.[64] An ardent 'friend of

the egg', Conklin urged instead the central importance of cytoplasmic factors. The egg is organised early, he agreed with Whitman and
Wilson. Yet rather than turning to the 'deep hereditary stuffs' in the
nucleus and chromosomes to explain that phenomenon, he urged, let
us examine the cytoplasm itself and identify relations between early
differentiated organised areas and parts of the later embryo.
Agreeing with Wilson's general view that some sort of formative
material represents the beginning of differentiation, Conklin continued to call for a different emphasis, which kept heredity and
development together by examining the relations of inherited stuffs
and later development. In a later paper, he neatly summarised his
continued convictions:

What is the formative agent in embryonic differentiation? . . . It is impossible to understand, i.e., to make intelligible, development except as a
result of the formation and localization of different material substances.
Indeed development consists in morphological division of substances and
physiological division of labor. Cytologists and geneticists have made notable advances in the study of the distributions of differentiated
chromosomes and genes in the germ-cells, but the cytoplasm of the egg cell
has frequently been regarded as mere foodstuff for these nuclear elements,
in spite of the fact that practically all differentiation takes place in the
cytoplasm. We are beginning to realize that the central problem of development lies in this relation between the genes and the cytoplasm, and that the
cytoplasm is something more than mere nutritive 'stuff'.[65]

For Conklin, one particularly accessible problem for development
lay with ascertaining the extent to which and the way in which early
cleavages of the egg correspond to later embryonic development.
Does the division of cells play a causal role in producing differentiated form? Three possibilities exist, he reported. Either (1) cleavage
planes separate already differentiated germ regions, (2) cleavage
planes cut randomly across germ regions, or (3) cleavage planes cut
through undifferentiated, homogeneous material and do not yield
differentiation.[66] Roux's mosaic theory represents the first. Driesch's
views represent the third. And such views as Whitman's hold the
middle position, Conklin felt. Like Whitman, Conklin sought a
compromise, but in a different way. Each of the three alternatives for
cleavage holds true for some cases, he concluded. Some organisms
exhibit relatively more, others relatively less determinate cleavage.
Determinate cleavage holds when the cleavage pattern is constant
and differential (regularly correlated with later differentiated parts).

Indeterminate cleavage is neither. Determinate cleavage occurs because of protoplasmic direction, while indeterminate cleavage responds more to extrinsic environmental factors.[67] No nuclear determinants or even deep nuclear directives for Conklin. The active, responsive and initially inherited cytoplasm holds the key to explaining development.

Though Conklin did not go on explicitly in 1905 to tie his analysis with preformation–epigenesis discussions in any detail, he was certainly aware of such discussions. Given the context of his other work and of his contacts at Woods Hole, it is fair to assume that he regarded his determinate–indeterminate cleavage distinction as addressing the same problems. Determinate cleavage represents something like preformation, in an organism which acts like a developmental mosaic. Indeterminate cleavage represents epigenesis, or a responsive organism. Both occur, in different measures in different organisms. For Conklin, further consideration of the issue would have to address the extent to which and way in which the separation of blastomeres correlated with organised germ regions as well as with later-differentiated parts. Mapping such relations could go far to demonstrate how form becomes differentiated and when, he felt. Conklin denied the distinction of heredity and developmental processes, which formed part of Wilson's working assumptions after 1905 and 1906. For Conklin, heredity and development remain inextricably connected.

In his Harris lectures of 1914, Conklin addressed preformation and epigenesis directly. Yet neither in the 1915 nor even by the 1930 edition did he mention the nucleus or chromosomes in his discussion. Modern studies recognise that neither the old preformationist nor epigenetic positions hold true, he summarised. Rather, the germ has organisation, and development brings a formation of 'new materials and qualities, of new structure and functions.'[68] Still emphasising both internal and external conditions as shaping development, and still ignoring the separation of heredity and development, Conklin concluded that

Increasing complexity, which is the essence of development, is caused by the combination and interaction of germinal substances under the influence of the environment. The organization of the oosperm may be compared to the arrangement of tubes and flasks in a complicated chemical operation; they stand in a definite relation to one another and each contains specific substances. The final result of the operation depends not merely upon the substances used, not merely upon the way in which the apparatus is set up, but upon both of these things, as well as upon the environmental conditions represented by temperature, pressure, moisture or other extrinsic factors.[69]

While researchers such as Whitman, Wilson, Conklin, and Wheeler pursued their particular pieces of the middle ground between pure determinism and pure regulation, there were others who took positions a bit closer to Driesch's regulative perspective than any of these more morphological researchers. Those of Wheeler's investigators inclined toward process, those physiologists, stressed external factors and indeterminate early cell divisions more strongly. Men such as Morgan, Jacques Loeb (1859–1924), Charles Manning Child (1869–1954), or Frank Lillie (1870–1947) emphasised responses to changing conditions as causes of differentiation. These men remained adamant epigenesists in that they believed that form emerges only gradually, only as the result of regulative responses to external conditions. Not accidently did both Morgan and Loeb undertake extensive studies of regeneration, where regulation remains most obviously at work. Nor was it accidental that both Morgan and Lillie became major participants in the discussions of sex differentiation and sex production. The question of when and how an individual comes to have one sex or the other became something of a micro focus for the macro concern over whether development as a whole results from preformation or a new formation.

## *The focus on sex:* heredity and developmental, predeterminism and epigenetics

Sex offers a relatively clear case for study of the roles of heredity and development, an inherently interesting case, and after 1902 an apparently accessible case for study. During 1905–10, debates over sex determination became increasingly central to discussions over the relative role of heredity and development, or predeterminism and epigenesis. Partly this occurred because an individual becomes either one or the other sex, rather than some intermediate, and the researcher can tell relatively easily which sex has been developed – unlike the case with many other characteristics. Ultimately, it was sex determination which led Wilson further into the nucleus, pushed Morgan to genetics, and illustrated the separability of the related problems of sex inheritance and sex development or production.

Complex problems sorted out for sex determination pointed the way to understanding more general problems of heredity and development. In fact, sex differentiation promised a productive research program, with a viable framework to advance the process of clarifying previously confused issues. This study was thus central to changing attitudes about preformation and epigenesis, about heredity and development, or genetics and embryology.

In 1891 Henking discovered the regular existence of an unusual body in the nucleus of insects. Further work by 1902 identified that body as indeed a nuclear chromosome. The American, Clarence McClung (1870–1946) termed it an accessory chromosome and suggested that presence of the accessory determined maleness.[70] He was wrong, in fact, since the accessory actually determines femaleness in his particular insect species. Yet his suggestion that there exist two different types of spermatozoa, distinguished by their chromosomes, and two resulting different types of fertilised eggs, focused active attention on sex determination. If sex is determined by a chromosome, it seemed, then a sort of predetermination occurs, but perhaps the chromosome and sex were connected only secondarily, not causally connected. Did the apparently inherited chromosome determine, direct, or only follow sex determination, was the question.

For over a decade, numerous researchers had argued about whether sex differentiation in particular is predetermined in the fertilised egg, due to factors internal to the organism, or whether it emerges only gradually, guided instead by regulated response to external factors. Temperature, food, light – such factors might possibly determine each individual's sex, and to many physiologically oriented researchers prior to 1910, such factors did seem to be decisive.[71] By 1903, the internalists had won.[72] But the internalist position had fragmented, forming a range of opinions with two reasonably defined views on either end of a spectrum. Internal inherited factors determine development, said one group. The other group insisted that internal physiological responses and definitely not inheritance direct development and differentiation. The disputes centred on the significance for development of those accessory chromosomes in the nucleus which Henking and McClung had identified. Whether development occurs by preformation or epigenesis had become largely an issue over inheritance of chromosomes *vs.* regulative physiological development.

By 1905 Americans had assumed the primary roles in debating the case of sex determination, with considerable input also offered in particular by German cytologist Theodore Boveri (1862–1915). Though many people took part, the chief advocates of nuclear inheritance for purposes of this discussion included Nettie Stevens (1861–1912) and Wilson.[73] Against the ability of nuclear inheritance to explain the significant facts of development sat Morgan, Conklin, and Lillie, along with Loeb and Child. Sex chromosomes *vs.* interactive gradients and complex hormonal activity as *the* primary factor determining sex: we all know which position gained at least apparent dominance.[74]

This is not the place to expand on details of the debates, but the central point is clear and directly relevant. Some researchers willingly saw the regularity of chromosome divisions as evidence for chromosomal significance in development. Rather than concluding only what that evidence showed, however, namely that some correlation occurred, some went on to claim that chromosomes actually cause differentiated development. Though cautiously stated, this is the clear implication of Wilson's work after 1905, and of *Drosophila* studies by the Morgan group after 1910. Weismann and Roux had already presented logical arguments for a system in which determinants actually direct a mechanical, pre-programmed succession of developmental events. Though the evidence accumulating to show a correlation of chromosomes and characteristics such as sex was not conclusive, it did tend to push researchers in the direction of such neat mechanical systems as Weismann's and Roux's. Even previous sceptics of any such deterministic system, such as Morgan, became converts not only to the idea of chromosomal significance but even to a stronger hereditarian position, whereby heredity serves as the starting point and later determinant to development. The ability of a hereditarian chromosomal theory to address a variety of phenomena made such theories attractive to many including Morgan. Furthermore, such broad suggestive theories could serve as foundations for productive research programs.[75]

In contrast, other researchers reacted vehemently against such hereditarian views, paralleling with greater force Conklin's response to Wilson's stress on the nucleus. Some remained hereditarian but stressed the organised cytoplasm as inherited. Development can never be explained by the action of inherited materials, they argued. The organism represents a dynamic, ever-changing whole which

responds to constant alterations in conditions both inside and outside the organism. Although the egg is manifestly inherited and
although it may already possess some pre-organisation at the
earliest stages, it is physiological regulation, the actions of
hormones and enzymes and chemicals which ultimately direct
development. Inherited chromosomes really explain nothing of
development. Response of the organism is key.

By 1910 these commitments had become embodied in divergent
research programs. Genetics and embryology emerged as relatively
defined research directions. Though biologists still accepted that
both heredity and development occur, they moved away from the
earlier versions of an intermediate position after 1910. Whitman's
urge for cooperation and inclusion of both heredity and development did remain in the background. Biologists understood that
something was inherited and also that actual differentiation was
produced only gradually. But effecting a research program to deal
productively with both proved difficult. Conklin tried, as did a few
others such as Richard Goldschmidt (1878–1958). But attention
focused instead on the relatively quick results of Morgan's genetics
of sex heredity or on physiological hormone studies of sex production. Biologists had moved beyond a central position in their search
for productive results through divergent research programs.

# Epilogue

Even today researchers have followed divergent, specialised paths.
Biologists have largely ignored the various urgings by His,
Whitman, and Conklin that differentiation represents an interaction of both intrinsic (hereditary) and extrinsic (developmental)
factors. The study of heredity, largely through genetics but also to
some extent through study of cytoplasmic inheritance, has taken
research in a different direction from embryological research on
development. A fairly typical modern view, as stated by the
Medawars, though ignoring cytoplasmic inheritance, holds that
'Genetics deals with the characters of the information transmitted
from generation to generation, "epigenetics" (Waddington's term)
deals with the processes by which this information is translated into
real life – into flesh and blood and distinctive behavioural characteristics.' Thus, '. . . there *is* some real meaning in the antithesis: the
genetic instructions according to which development proceeds are

indeed preformed, but their realization is *epigenetic*, i.e. turns upon influences acting upon the embryonic cell from the outside. . . .'[76] Though biologists today generally acknowledge with the Medawars that both heredity and development occur, most proceed with their research as if only one or the other really mattered for explaining development, while the other provides only background.

Why? Why did the search for compromise by the Americans in the 1890s, which seemed so sensible and obvious, give way to this new division? Does the answer lie with the living processes themselves, with the way biologists think, or with external demands and constraints on science? Can biologists move forward successfully to a satisfactory understanding of both inheritance and development, or ought they even to try? These complex questions cannot be answered easily, and here I can offer only suggestions.

The answer does not, I feel certain, lie in the organisms themselves. Life does not involve a period of heredity neatly separated from a period of development. The egg manifestly develops even before fertilisation, while inherited material obviously continues to exert effects long after. Heredity and development are much more intricately interconnected than most research programs in either genetics or embryology after 1910 have admitted.

The willingness to divide complex biological processes into categories comes from the biologists themselves, then, and their responses to external demands placed on them. By 1910 a fairly well defined sense existed of what constituted a productive research program in biology. Answerable questions, definitive results. Not grand umbrella theories which proposed to explain everything at once. Not descriptive tidbits about pieces of development. Instead, biologists sought to formulate questions which were addressable through observation and experiment, thus yielding positive results. How best to conduct productive research? Narrow the problems and tackle accessible questions with available materials. Thus division of larger problems into questions of heredity and of development resulted – and has persisted.

Undoubtedly, Wheeler is also correct in claiming that different kinds of people exist and work with different assumptions. Perhaps some do find stable patterns more accessible and more comforting and hence find genes, prelocalised cytoplasm, and hereditarian assumptions most acceptable. Perhaps others find changing patterns and the processes of differentiation more intriguing problems and find developmental approaches more acceptable than hereditarian assumptions. More probably, there are more than Wheeler's two clearly distinct kinds of thinkers. Biologists probably fill a range of different convictions, with some falling between the extremes.

That middle ground receives less attention for a variety of reasons. It has traditionally promised less than extreme positions in the way of those definite and exciting results that granting agencies and administrators like. A struggle for authority for research funds and jobs in a world of increasing specialisation may also foster or at least reinforce a division of labour.[77] As a result researchers have felt pressures to concentrate in order to realise results, and they have accepted the expediency of rhetorical emphasis on one factor to the exclusion of others. Thus what biologists study in the laboratory may obscure their more complex, but less accessible convictions about the way form emerges.

I have traced the transformation and fall of a dichotomy in late nineteenth century biology. The debates between Weismann and Hertwig bore a family resemblance but were not precisely identical either to their eighteenth century predecessors or to the later exchanges. In America the dichotomy of preformation and epigenesis itself was rejected, only to be revived in a different form in the alternative focuses of research in genetics and embryology. Functionally, at least, biologists have since 1910 operated as if heredity and development could be neatly separated. Presumably, however, the process of transformation is never over, and the distinctions of current practice will themselves move into the historical background.

Support for this research came from National Science Foundation Grant #SES-8309388 and research assistance came from Nancy Tribbensee.

# Endnotes

1. Shirley Roe, *Matter, Life, and Generation* (Cambridge: Cambridge University Press, 1981);
   Peter Bowler, Preformation and Pre-existence in the Seventeenth Century: A Brief Analysis, *Journal of the History of Biology* (1971) **4**, 221–44;
   Bowler, The Changing Meaning of 'Evolution', *Journal of the History of Ideas* (1975) **36**, 95–114;
   Alice Levine Baxter, Edmund B. Wilson as a Preformationist: Some Reasons for His Acceptance of the Chromosome Theory, *Journal of the History of Biology* (1976) **9**, 29–57, especially pp. 30–8 on the debate generally.

2. William Morton Wheeler, Caspar Friedrich Wolff and the *Theoria Generationis, Biological Lectures* (1899) 1898: 265–84.
3. Wheeler, p. 282.
4. Wheeler, p. 284.
   Thomas Hunt Morgan, Regeneration in the Egg, Embryo, and Adult, *The American Naturalist* (1901) **35**, 949–73; citation p. 968.
5. August Weismann, *The Germ Plasm*, translated by W. Newton Parker and Harriet Rönnfeldt (New York: Charles Scribners, 1893).
6. Weismann, pp. xiii–xiv.
7. Weismann, p. 54.
8. Weismann, p. 53.
9. For more on Weismann's ideas see especially Chapter 1 of *The Germ Plasm*. Also: Frederick Churchill, August Weismann and a Break From Tradition, *Journal of the History of Biology* (1968) **1**: 91–112.
   Churchill, Hertwig, Weismann, and the Meaning of Reduction Division circa 1890, *Isis* (1970) **61**, 429–57.
10. Weismann, p. 32.
11. Actually, the process is not simple but quite complex. See above sources for more explanation of the process.
12. Weismann, p. 134.
13. As some researchers had argued, most notably Eduard Pflüger, Über den Einfluss der Schwerkraft auf die Theilung der Zellen, *Archiv für die gesammte Physiologie des Menschen u. der Thiere* (1883) **31**, 311–18.
14. Weismann, p. 138.
15. Oscar Hertwig, *The Biological Problem of Today*, translated by P. Chalmers Mitchell (Oceanside, New Jersey: Dabor Science Publications, 1977; original translation, 1900), p. 140; Original: *Zeit- und Streitfragen der Biologie*. vol. 1 *Präformation oder Epigenese? Grundzüge einer Entwicklungstheorie der Organismen* (Jena: Gustav Fischer, 1894). Volume 2 was a critique of Roux: *Mechanik und Biologie. Kritische Bemerkungen zu den entwickelungsmechanischen Naturgesetz von Roux*.
16. Hertwig, p. 136.
17. Hertwig, pp. 105–6 and 138.
18. Frederick Barton Churchill, *Wilhelm Roux and a Program for Embryology*, Ph.D. dissertation, Harvard University, 1966.
19. Wilhelm Roux, Beiträge zur Entwickelungsmechanik des Embryo, No. 1, *Zeitschrift für Biologie* (1885) **21**, 411–524.
20. Roux, Beiträge zur Entwickelungsmechanik des Embryo. Über die künstliche Hervorbringung halber Embryonen durch Zerstörung einer der beiden ersten Furchungskugeln, sowie über die Nachentwickelung (Postgeneration) der fehlenden Körperhälfte, *Virchows Archiv für pathologische Anatomie u. Physiologie u. klinische Medezin* (1888) **114**, 113–53. Partly translated in B. Willier and J. Oppenheimer, editors, *Foundations of Experimental Embryology* (Englewood Cliffs, New Jersey: Prentice-Hall, 1964), pp. 2–37. The page numbers cited here refer to the more accessible translation, p. 4.
21. Roux, 1888, p. 8.
22. Roux, 1888, pp. 25–6.

23. Roux, 1888, p. 28.

24. Further consideration of Roux's theories here would carry the discussion too far afield. See Churchill, *Wilhelm Roux*, for further discussion.

25. Hans Driesch, Entwicklungsmechanische Studien. I Der Werth der beiden ersten Erzeugung von Theil- und Doppelbildungen, *Zeitschrift für wissenschaftliche Zoologie* (1892) **53,** 160–78. Also in Willier and Oppenheimer, pp. 38–50. The page numbers cited here refer to the more accessible translation.

26. For example, he had begun to shift by 1894. See Driesch, *Analytische Theorie der organischen Entwicklung* (Leipzig: Wilhelm Engelmann, 1894).

27. Herbert Spencer Jennings, Biology and Experimentation, *Science* (1926) **64,** 97–105; citation p. 99.

28. Though the Americans' awareness of His has been acknowledged by some, the importance of that connection has not generally been appreciated. In particular, the tradition of concern for cell development, embracing the work of Schwann and Schleiden, Remak, and Virchow, had an important impact on His. As demonstrated by their citations and the problems they considered important, it is clear that Whitman and Wilson also were influenced by that tradition. Thus His's influence on American biology calls for further examination.

29. Ernst Haeckel, *Monographie der Kalkschwämme* (Georg Reimer: Berlin, 1872);
Gastraea-Theorie, *Jenaische Zeitschrift* (1874) **8,** 1–55;
Die Gastrula und die Eifurchung der Thiere, *Jenaische Zeitschrift* (1875) **9,** 402–508;
On Haeckel and germ layers, see: E. S. Russell, *Form and Function* (London: John Murray, 1916), chapter 16;
Alice Baxter, E. B. Wilson's 'Destruction' of the Germ-Layer Theory, *Isis* (1977) **68,** 363–74; especially pp. 362–6;
Jane Oppenheimer, The Non-Specificity of the Germ-Layers, in *Essays in the History of Embryology and Biology* (Cambridge, Massachusetts: MIT Press, 1967), pp. 256–94.

30. Jane Maienschein, Cell Lineage, Ancestral Reminiscence, and the Biogenetic Law, *Journal of the History of Biology* (1978) **11,** 129–58.

31. Wilhelm His, *Unsere Körperform und Das Physiologische Problem ihrer Entstehung* (Leipzig: F. C. W. Vogel, 1874), p. 19, E. G. Conklin translated this passage, 'Cleavage and Differentiation,' *Biological Lectures* (1898) 1896–7: 17–43; citation 18–19.

32. His, pp. 131–44 and 145–55.

33. His, p. 147, 148.

34. His, p. 152.

35. His, Das Princip der organbildenden Keimbezirke und die Verwandschaften der Gewebe, *Archiv für Anatomie und Physiologie. Anatomische Abtheilung* (1901); 307–37.

36. Frequent references to His occur in the American work, mostly to the principle and to *Unsere Körperform*.

37. On M.B.L. history see Jane Maienschein, 'A Marine Laboratory for America:

Early History of the M.B.L.', paper delivered at the M.B.L., 1984, and on deposit in the archives there.

For more of the flavour of popular debates at the time see the Marine Biological Laboratory's *Biological Lectures*.

For examples of the different emphasis in England, see G. C. Bourne, Epigenesis or Evolution, *Science Progress* (1894) **1**, 105–26;

St George Mivart, Critical Remarks on the Theories of Epigenesis and Evolution, *Science Progress* (1894) **1**, 501–8.

38. Whitman, The Embryology of *Clepsine*, *Quarterly Journal of Microscopical Science* (1878) **18**: 215–315; citation p. 263.

39. On Whitman and his concerns, see the useful study by Jeffrey Werdinger, *Embryology at Woods Hole: The Emergence of a New American Biology*, doctoral dissertation, Indiana University, 1980.

40. Whitman, Changes Preliminary to Cleavage in the Egg of Clepsine, *Proceedings of the American Association for the Advancement of Science* (1878) **27**, 263–70; citation pp. 263–6.

41. Whitman, A Contribution to the History of the Germ-layers in *Clepsine, Journal of Morphology* (1887) **1**, 105–82; p. 107; E. B. Wilson, The Germ Bands of *Lumbricus, Journal of Morphology* (1887) **1**, 183–97.

42. Whitman, 'A Contribution,' p. 111.

43. Whitman, 'A Contribution,' pp. 138–40.

44. Whitman, 'A Contribution,' p. 169.

45. Whitman, The Kinetic Phenomena of the Egg during Maturation and Fecundation (Oökinesis), *Journal of Morphology* (1898) **1**, 227–52; citation, p. 227.

46. Whitman, The Seat of Formative and Regenerative Energy, *Journal of Morphology* (1888) **2**, 27–49; pp. 30, 38.

47. Whitman, Oökinesis.

48. Whitman, Evolution and Epigenesis, *Biological Lectures* (1895) 1894: 205–24; p. 209.

49. Whitman, Evolution, pp. 221–4.

50. Charles Otis Whitman, Evolution and Epigenesis, p. 211.

51. Whitman, p. 221.

52. Whitman, p. 224.

53. Whitman's contributions to the Marine Biological Laboratory's *Annual Reports* and his letters in the M.B.L. archives reflect this attitude, for example.

54. For more on Whitman and influences on him, see Jeffrey Werdinger, *Embryology at Woods Hole: The Emergence of a New American Biology*, especially Chapters 1 and 6.

55. Edmund Beecher Wilson, The Cell-Lineage of *Nereis*: A Contribution to the Cytogeny of the Annelid Body, *Journal of Morphology* (1892) **6**, 361–480.

56. Wilson, *The Cell in Development and Inheritance* (New York: Johnson Reprint Corporation, 1966; original 1896), pp. 328–9.

57. Wilson, 1896, p. 330.

58. For discussion of Wilson, see Alice Levine Baxter, *Edmund Beecher Wilson and the Problem of Development: From the Germ Layer Theory to the Chromosome Theory of Inheritance*, Ph.D. dissertation, Yale University, 1974;

Jane Maienschein, Shifting Assumptions in American Biology: Embryology, 1890–1910, *Journal of the History of Biology* (1981) **14**, 89–113; Maienschein, What Determines Sex? A Study of Converging Research Approaches, *Isis* (1984) 75: 457–80; Werdinger, Chapters 2 and 7.

59. Baxter, Edmund B. Wilson as a Preformationist, p. 30.
60. Wilson, The Problem of Development, *Science* (1905) **21**, 281–94; citation p. 282.
61. Wilson, 1905, p. 288.
62. Wilson, 1905, p. 288.
63. Wilson, 1905, pp. 290, 292. See Baxter for further interpretation of Wilson's views.
64. See Frederick Churchill and Garland Allen, this volume, for discussion of the separation of heredity and development. Also, Churchill, William Johannsen and the genotype concept, *Journal of the History of Biology* (1974) **7**, 5–30.
65. Conklin, Cleavage and Differentiation, p. 30.
66. Conklin, Mosaic *vs.* Equipotential Development, *American Naturalist* (1933) **67**, 289–97; citation pp. 295–6.
67. Conklin, Cleavage and Differentiation, p. 19.
68. Conklin, pp. 28–9.
69. Conklin, *Heredity and Environment* (Princeton: Princeton University Press, 1915), pp. 83, 84–5; (Princeton, 1930), pp. 57–60.
70. H. Henking, Untersuchung über die ersten Entwicklungsvorgänge in den Eiern der Insekten, *Zeitschrift für wissenschaftliche Zoologie* (1891) **51**, 685–741;
    Clarence E. McClung, The Accessory Chromosome – Sex Determinant?, *Biological Bulletin* (1902) **3**, 43–84.
71. For discussion of the externalist position, see Maienschein, What Determines Sex? A Study of Converging Approaches.
72. As Morgan convincingly maintained in Recent Theories in Regard to the Determination of Sex, *Popular Science Monthly* (1903) **64**, 97–116. On sex determination generally, see John Farley, *Gametes and Spores* (Baltimore: Johns Hopkins University Press, 1982).
73. Nettie Stevens, Studies in Spermatogenesis. II. A Comparative Study of the Heterochromosomes in Certain Species of Coleoptera, Hemiptera, and Lepidoptera, with Especial Reference to Sex Determination, *Carnegie Institution of Washington Publication* (1906), pp. 1–32;
    Wilson, The Chromosomes in Relation to the Determination of Sex in Insects, *Proceedings, Society for Experimental Biology and Medicine* (1905) **3**, 19–23; Wilson, Studies on Chromosomes. III. The Sexual Differences of the Chromosome Group in Hemiptera, with some Considerations of the Determination and Inheritance of Sex, *Journal of Experimental Zoology* (1906) **3**, 1–40;
    Stephen Brush, Nettie Stevens and the Discovery of Sex Determination, *Isis* (1978) **69**, 132–72.
74. For example, Morgan, A Biological and Cytological Study of Sex Determination in Phylloxerans and Aphids, *Journal of Experimental Zoology* (1909) **7**, 239–352;

Garland Allen, *Thomas Hunt Morgan* (Princeton: Princeton University Press, 1978);
Charles Manning Child, The Process of Reproduction in Organisms, *Biological Bulletin* (1912) **23**, 30–9;
Child, *Individuality in Organisms* (Chicago: University of Chicago Press, 1915);
Frank Rattray Lillie, Sex Determination in Relation to Fertilization and Parthenogenesis, *Science* (1907) **25**, 372–6.
Jacques Loeb, *The Organism as a Whole* (New York: G. P. Putnam's Sons, 1916).

75. For discussion of the role of research programs and basic commitments of American biologists, see Maienschein, Shifting Assumptions;
Maienschein, Experimental Biology in Transition: Harrison's Embryology, 1895–1910, *Studies in History of Biology* (1983) **6**, 107–27.

76. P. S. and J. S. Medawar, *The Life Science* (New York: Harper and Row, 1977).

77. On such struggles for authority, see Jan Sapp, *Cytoplasmic Inheritance and the Struggle for Authority in the Field of Heredity*, Ph.D. dissertation, Université de Montréal, 1984.

# Origins of the embryological tradition in the United States

Relatively speaking, in the United States there was no well defined pre-existing tradition in biological research when, towards the close of the nineteenth century, there occurred an astonishing burgeoning of achievement in that field, although it is true that we can see represented in the earlier generation most of the approaches we have described in Europe – the Darwinian tradition was strongly represented by W. K. Brooks at Johns Hopkins and Louis Agassiz at Harvard who emphasised Haeckelian issues, while C. O. Whitman of Chicago and Woods Hole addressed the issues of cell theory. The common scientific backgrounds of the leading figures of the new generation (Conklin, Harrison, Morgan, and Wilson) in the first graduate school at Johns Hopkins under Brooks and H. Newell Martin and their later associations with Whitman, makes it of some interest to look for characteristic features, as Maienschein has suggested, and as will be considered further in the two papers that follow.

It was Whitman who deliberately set out to expand, coordinate and professionalise American research on the German model. His organisational success, in starting journals and as founder of a marine station at Woods Hole to rival Naples, was a major factor in determining the direction of research. *All* four Hopkins graduates did their first research in lineage tracing in marine embryos. Although Whitman's ideas had once 'met a bad reception' (Lillie, 1944, *The Woods Hole Marine Biological Laboratory*, p. 124) – and Brooks remained sceptical: 'Cleavage is nothing but the duplication of cells. Morphology does not begin until you get to the germ-layers' (Conklin, 1937, *The Collecting Net* xvi, p. 156) – they provided the perspective that dominated at the start of the 1890s, as can be judged from the remarkable record (published as the *Biological Lectures*) of the lectures which formed a centre point for thought and debate at Woods Hole. Whitman's views were derived in a period (1875–8), during his training in Leipzig prior to any serious concern with distinctions between nucleus and cytoplasm or to the understanding of fertilisation and nuclear behaviour that led up to Weismann's theories. His paper of 1878 traced successive cell derivations throughout the early embryogenesis of *Clepsine*: 'It was the first time that the primordia of any ectodermal organs had been followed to individual cells, and that the cleavage process itself had been adequately interpreted as a process of "histogenetic sundering".' (Lillie, 1911, *Journal of Morphology* 22, xix). The influences on Whitman at the time are suggested by the references he quotes in the paper: Lankester, Haeckel, His, who also worked in Leipzig at the time, and Leuckart, his own teacher. His views in 1890 are already detectable: 'the order of axial differentiation is in harmony with the supposed phylogenetic order of development', 'preliminary segregation' occurs 'long before cleavage.' (Whitman, 1878, *Quarterly Journal of Microscopical Science* 18, 215–315).

As Maienschein has described, the impact in America of Driesch's 1892 overturning of the convenient theories and results of Roux was immediate, and served to expose the scientific foundations of Whitman's theories. In a lecture in 1893 entitled 'The inadequacy of the cell-theory of development' (*Biological Lectures at Woods Hole*, vol. 2, pp. 105–24), he expressed his strongest, but last, defence. He is reacting against the whole approach that cell theory represented, including the physiological considerations of cleavage and the reductionism of concern with nuclei, chromosomes, or Weismannian particles, and he was insisting on the direct and simple translations of heredity into a new adult by way of parental organisation built directly into the egg. Here, as evidence that biological organisation can be established quite independently of any basis in cellular units, he points out how organisation applies equivalently to protozoa as to multicellular organisations, and to the results of regeneration in identical fashion to cytologically different processes of embryonic development. Given his conception of the organisation of egg cytoplasm as the direct mediator of inheritance, it is not surprising to find him denying the possibility of epigenetic mechanisms in development on the grounds that they would interfere with inheritance.

Driesch's findings were soon duplicated at Woods Hole; in *Amphioxus* by Wilson (1893), and in sea-urchins by Loeb (1894). However, soon quite different results were being obtained in other species, using the same methods, but consistent with Roux's or Whitman's views on in-built cytoplasmic organisation; in *Ctenophores* by Morgan, working with Driesch (1895), in *Illyanassa* by Crampton (1896), in *Dentalium* and *Patella* by Wilson (1904) and *Cynthia* by Conklin (1905). Wilson, who struggled hard to find compromise solutions admitted 'It is not easy to gather (Whitman's) precise position regarding the theory of cytoplasmic localisation' (*The Cell in Development and Inheritance*, 1896, p. 300). Morgan simply described it as 'fallacious' (*Wilhelm Roux Archiv fur Entwicklungsmechanik* 2, p. 83, 1895). The introduction of simple methods that could so successfully challenge a set of beliefs based purely on observation unleashed a flood of new experimental approaches, spurred on by the fanatical Jacques Loeb, whose aim it was 'to reduce biological appearances, the so-called manifestations of life, to the status of physicochemical reactions' (Flexner, 1927, *Science* 66, p. 335). 'The radical epigenetic view of development was held strongly for a time by Driesch and Jacques Loeb, among others, but anathema to most embryologists at Woods Hole' (Lillie, 1944, op. cit., p. 125).

The situation *was* confusing – the apparent irreconcilability of the mosaic and regulating behaviour of the embryos of different species still confounds us now after all – and was part of a general atmosphere of rapid and fundamental change in biology. On these narrow issues hinged solutions to the doubts and controversies throughout evolution theory and heredity. There arises an irresistable temptation to seek explanations here for the emergence of that remarkable new generation of scientists which included Morgan, Wilson, Conklin and Harrison. One is led to wonder especially how the great diversity of their future biological work could have come out of such similar and closely-knit starting points: all four biologists had, for example, been initiated into research by way of lineage studies in embryos and the small scale and cooperative spirit of the Woods Hole group made for an unusual degree of cohesion. All the more intriguing then is the question of why it was in the United States that the 'split' between embryology and genetics occurred so prominently.

Morgan's transition over twenty years from microscopic observation of sea-spider embryos to breeding *Drosophila* flies can only be understood by following (as is now possible through the studies by Allen) all the minute steps taken along the way by this immensely hard-working and productive man. The unique and complex combination of preconditions and circumstances can usually be traced back in one way or another to the concerns of Woods Hole in the early 1890s: the primary focus on broad evolutionary questions; the encouragement to move freely between species and techniques; the close association with Wilson which prevented Morgan disregarding, as he was inclined to do, the implications and value of conclusions arising out of cell theory; the revelations concerning the power and necessity of the experimental method that he had learned from Driesch and Loeb. But there was undoubtedly a second, quite distinct element to the rapid success of genetics as a discipline in the United States, as Allen shows in the following paper. This was the national pragmatism that made the positive outcome of experimental methods so attractive and provided motivation and support for the new science as its practical implications became clear.

The very success of genetics cannot have failed to invite comparisons with embryology, which would have appeared increasingly to be in some disarray. Lillie, Child and Harrison adopted the most rigorous experimental methods but this could not compensate for the underlying gaps on the more theoretical side. Some, like Conklin, continued to promote Whitman's interest in cytoplasmic organisation but, with the growing emphasis on chromosomes, seemed, along with embryology itself, more and more isolated. After 1895 the problem raised by Driesch's experiments was approached in a somewhat different way from that about to be taken in Germany. Morgan had initially come nearest to closing the divide between Roux and Driesch's findings in his studies of postgeneration and gastrulation using the early amphibian embryo. Child confined himself essentially to the evidence from regeneration, to invertebrates and to the role of external influences on tissue organisation. Harrison, while focusing explicitly on the Drieschian concept of a 'harmonious equipotential system', concentrated his efforts on later stages of development and on discrete organs, such as the limb-bud. His modification of Driesch's phrase to 'self-differentiating equipotential system' shows how, without explaining the discrete localisations of the organ rudiments themselves, he could effectively detach himself from matters of overall organisation and earlier stages. There is one factor which may explain why the American embryologists did not feel the need to pursue the resolution of the Roux/Driesch issue back to its foundations in the earliest stages of regulating embryos. It was the 'automatic self-contained perfection of the developmental process that holds our interest', as Morgan put it (*Experimental Embryology*, 1927, p. vii). Familiar as they were with the visible orderliness of cytoplasmic segregation and cell division that is typical of marine invertebrate embryos showing mosaic behaviour, it was an easy step to assume that these early processes could be directly equated, without more ado, with the increasingly acceptable elementary units of biochemistry and genetics, and to assume that similar considerations applied to the invisible early events in regulating embryos.

# T. H. Morgan and the split between embryology and genetics, 1910–35

GARLAND E. ALLEN

Department of Biology, Washington University, St Louis, Missouri, USA

## Introduction

In the first decade of the twentieth century Thomas Hunt Morgan (1866–1945) attacked both the Mendelian and chromosome theories of heredity. He complained that the Mendelian theory was based on the existence of hypothetical particles (the *Faktoren* or *Anlagen* of Mendel's original formulation), while the idea that chromosomes contained the material substance of heredity was based purely on circumstantial evidence. Behind those specific objections lay an even stronger criticism. As an embryologist, Morgan argued that both theories smacked heavily of 'preformationism' – the notion that the embryo and all of its parts lie perfectly formed in the egg or sperm (Morgan, 1909: p. 509). Against preformation, embryologists had posed the theory of epigenesis, the idea that formed embryonic parts emerge during development from formless matter in the fertilised egg. As one of the world's leading embryologists and a committed epigeneticist, Morgan felt that by postulating hereditary units that controlled specific traits (height, eye colour, wings, or sex), the chromosome and Mendelian theories ignored the important and interesting question of how differentiation of parts actually occurs. Morgan's embryological training led him to define heredity in an integrated way, as both the transmission of elements from parent to offspring, *and* the 'translation' of inherited information into adult traits. As late as 1910, the same year in which he published his first major paper on sex-limited inheritance in *Drosophila*, Morgan could write:

We have come to look upon the problem of heredity as identical to the problem of development. The word heredity stands for these properties of the germ cells that find their expression in the developing and developed organism.

(Morgan, 1910*a*: p. 449).

To Morgan, concern for the mere transmission of particles, whether abstract factors or material chromosomes, was uninteresting because it focused on only one small part of the overall process of heredity – that is, how offspring come to be like their parents. Heredity thus meant both transmission of potentialities during reproduction *and* development of those potentialities into specific adult traits.

However, by the 1920s, after his major work on *Drosophila* was well established, Morgan reversed not only his position on particulate theories, but on the very definition of 'heredity' itself:

Between the characters, that furnish the data for the theory [that is, Mendelian theory] and the postulated genes, to which the characters are referred, lies the whole field of embryonic development. The theory of the gene, as here formulated, states nothing with respect to the way in which the genes are connected with the end-product or character. The absence of information relating to this interval does not mean that the process of embryonic development is not of interest for genetics ... but the fact remains that the sorting out of the characters in successive generations can be explained at present without reference to the way in which the gene affects the developmental process.

(Morgan, 1926: p. 26)

By 1926, Morgan had divorced the study of transmission (genetics) completely from the study of development (embryology), in the process restricting the meaning of the term 'heredity' to encompass only the study of the material entities that passed from one generation to the next.

How could Morgan effect such a dramatic shift in definition of a term so fundamental to his own background and work as 'heredity'? The question is all the more intriguing since most biologists, including Morgan, *knew* explicitly that heredity and development were intimately related processes. For most of the great nineteenth century biologists such as Ernst Haeckel (1834–1919) and August Weismann (1834–1914), heredity and development were inextricably connected. A theory to account for one had in some ways to account for the other. And yet Morgan was able to defy not only a strong tradition in the history of biology, but even his own training as an embryologist. What factors entered into Morgan's separation of genetics from embryology, and how did it affect both fields during

the present century? A number of historians have dealt directly or indirectly with this question in recent years (Hamburger, 1980; Churchill, 1974; Gilbert, 1978; Sapp, 1982; Allen, 1974, 1978*a*, *b*, 1983). In the present paper I want to draw upon these views, as well as the work of Morgan o explain an organism's situation, so any epigenetic account of development and his followers, to explore the ultimate
consequences on both fields of the redefinition of heredity and the split between genetics and embryology that took place in the period 1910–35. I want to bring all of these views together in a general and comprehensive discussion of how and why the synthetic theory of heredity was restricted in its scope – particularly by Morgan and his school – during these years.

I am focusing on the work of T. H. Morgan and his school partly because Morgan was one of the principal architects of the new definition of 'heredity', and partly because of my own familiarity with his work. My general thesis will be that Morgan's redefinition of heredity emerged from the influence of several factors between 1910 and 1915:

(1) Lack of significant progress in his own research on experimental embryology;

(2) Beginnings of successful research work with the fruitfly *Drosophila melanogaster* after 1910;

(3) The promulgation of the genotype–phenotype conception of Wilhelm Johanssen in 1911;

(4) The influence of the philosophy of mechanistic materialism of Jacques Loeb, with its emphasis on analytical techniques;

(5) And finally, socio-economic factors in the United States: including the rise of large-scale agriculture for profit, with consequent availability of funds for research on heredity; and value to Morgan as a researcher to stake out and delimit a field of research with distinct, and limited, boundaries – a focused research program, if you will – in order to obtain money and students.

For practical purposes in this paper, I will focus mostly on point (4) and suggest how the work of other investigators leads to (5). In this paper, then, I argue that T. H. Morgan redefined the concept of *heredity*, dividing the older, more inclusive meaning of transmission and development into two distinct components, in the process creating two fields, genetics and embryology, out of what had previously been a single, unified field.

# Morgan's background and the revolt from descriptive and speculative biology

T. H. Morgan was born in Lexington, Kentucky, on 25 September 1866 of a distinguished southern family. His maternal great-grandfather was Francis Scott Key (1779–1843) and his paternal great-grandfather was John Wesley Hunt (1773–1849) one of Kentucky's first millionaires. His uncle was Colonel John Hunt Morgan (1825–64), whose escapades as leader of the famous 'Morgan's Raiders' during the Civil War have become legendary in the south.

After finishing his undergraduate work at the State University of Kentucky (now the University of Kentucky) in Lexington, Morgan headed for graduate school at the then relatively young Johns Hopkins University in Baltimore. At Hopkins Morgan studied primarily under William Keith Brooks (1848–1908), a student of Louis Agassiz (1807–73) and later his son Alexander Agassiz (1835–1910) at Harvard. Like the elder Agassiz, Brooks was a morphologist, that group of biologists prominent in the post-Darwinian period who studied detail on both adult and embryonic structure as a key to understanding evolutionary (phylogenetic) relationships. In addition, however, Brooks was interested in a variety of broader issues within biology. He had a strong philosophical bent, and loved to discourse at length on topics such as the biogenetic law and evolutionary perfectionism (McCullough, 1969). Brooks also had the ability to relate the most minute observation of animal structure to broad biological principles, and to see major areas of biology as problems to be solved rather than fixed and final 'truths'. To Brooks the main problems of biology in the late nineteenth and early twentieth centuries were those of heredity, evolution, and especially embryonic differentiation. Above all, he thought that these problem areas could be probed 'without undertaking to resolve biology into physics or chemistry' (Brooks, 1900: p. 710). Morgan always found Brooks' tendency toward speculation and philosophical thinking exasperating and ultimately unproductive (Allen, 1978b: p. 45).

Morgan's doctoral work, under Brooks, was a morphological study of the embryology and phylogeny of the pycnogonids, a group of arthropods known as 'sea spiders' (Morgan, 1891). This work involved a detailed microscopical study of the very earliest stages of cleavage of sea spider embryos. Morgan challenged the then-dominant theory of Anton Dohrn (1840–1909), Director of the Naples Zoological Station, which claimed that the pycnogonids were derived from the annelids long before that group gave rise to the arthropods. In his thesis, Morgan argued from embryological considerations that the pycnogonids were descended from the arthropods, specifically the arachnids, *after*, rather than before that group split off from the annelids. Even as a doctoral student Morgan showed that propensity to challenge established ideas which remained with him throughout his life.

In 1891 Morgan left Hopkins for his first teaching job, filling the position at Bryn Mawr College recently vacated by another of W. K. Brooks' students, E. B. Wilson (1856–1938), who had moved to New York to head the Zoology Department of Columbia University. At Bryn Mawr the other member of the biology department was Jacques Loeb (1859–1924), newly arrived from Germany. Loeb was a thorough-going materialist, grounded in nineteenth century German materialistic philosophy, and a strong advocate of the mechanistic conception of life (Loeb, 1912; Fleming, 1964). Morgan and Loeb became close friends, and it was through this association that Morgan first began to question seriously the descriptive and morphological tradition in which he had been trained.

The second influence came from Morgan's work at the Stazione Zoologica in Naples in the summer of 1892 and for most of a full academic year, 1894–5. Here he met the young and enthusiastic experimental embryologist Hans Driesch (1867–1941) with whom he also became lifelong friends. Driesch was at that time involved in the heated controversy with Wilhelm Roux (1850–1924) over the mosaic theory of development. Despite the differences in interpretation of their experiments, Driesch and Roux agreed on one thing: the importance of experimental and mechanistic analysis of embryological problems. Both also agreed on the importance of studying embryological problems for their now intrinsic value, and not as mere avenues for reconstructing phylogenies. Morgan was greatly excited by the atmosphere of experimental and mechanistic work being carried out by Driesch and others at the Naples station. His experiences there became a profound turning point in his career, catalysing him to become an experimentalist. From that point on, Morgan waged all-out war against speculative and idealistic biology, such as that found in the writings of Ernst Haeckel (1834–1919), August Weismann (1834–1914), and even his old teacher Brooks. As he came to oppose speculative biology, Morgan increasingly championed not only experimentalism but at a deeper level the mechanistic philosophy of life.

Between 1891 and 1910 Morgan established himself as one of the foremost experimental embryologists in the United States, and the world. He carried out research on regeneration, early cleavage and differentiation, and wrote articles and books on evolution, heredity, sex determination, and the chromosome theory. Throughout these studies, Morgan demonstrated his conviction that the problems of heredity, development and evolution were intimately connected. For example, his first major work on evolution, *Evolution and Adaptation* (1903), grew directly out of his studies on regeneration. In the Preface to his book Morgan pointed out that his interest in evolution grew out of the question of how regenerative powers could have evolved by Darwinian selection. To Morgan, the adaptive value of regeneration could be realised only when fully developed; thus he did not see how such a process could evolve by the accumulation of many small intermediate stages, none of which would be adaptive in and of itself. *Evolution and Adaptation* is a lengthy critique of the Darwinian theory, in which Morgan concludes that the mechanism of natural selection is insufficient to explain the evolution of regeneration, and a whole host of other adaptive traits.

In those same years Morgan was also concerned with the relationship between heredity, evolution and development. Between 1900 and 1910 he was a strong critic of the Mendelian and chromosome theories of heredity, on the general ground that they could not encompass either developmental or evolutionary phenomena. Both theories to Morgan smacked of preformationism – that is, they referred to ultimate particles, or preformed elements and ignored the really interesting *process* in biology, namely, how adult structures develop out of undifferentiated germinal elements. Both theories were really structural, not functional. As Morgan wrote of the Mendelian theory in 1909:

The nature of Mendelian interpretation and description inextricably commits to the 'doctrine of particles' in the germ and elsewhere. It demands a 'morphological' basis in the germ for the minutest phase (factor) of a definitive character. It is essentially a morphological conception with but a trace of functional feature. With an eye seeing only particles and a speech only symbolizing them, there is no such thing as a study of a process possible ... It has been possible, I think, to show by means of what we know of the genesis of these color characters that the Mendelian description – of color inheritance at least – has strayed very wide of the facts; it has put factors in the germ cells that it is now quite certainly our privilege to remove; it has declared a discontinuity where there is now evident epigenesis.

(Morgan, 1909, p. 509)

In addition, Morgan felt that particulate concepts of heredity, whether applied to the more abstract Mendelian particles, or to the visible chromatin elements, ended, rather than initiated, further research. They seemed to solve a problem simply by identifying it with a structure. In the same year that he published his first account of the white-eyed male *Drosophila*, 1910, Morgan wrote critically of the finalism implicit in particulate theories:

It may be said in general that the particulate theory is the more picturesque or artistic conception of the developmental process. As a theory it has in the past dealt largely in symbolism and is inclined to make hard and fast distinctions. It seems to better satisfy a class or type of mind that asks for a finalistic solution, even though the solution be truly formal. But the very intellectual security that follows in the train of such theories seems to me less stimulating for further research than does the restlessness of spirit that is associated with the alternative [that is epigenetic or embryological] conception.

(Morgan, 1910, pp. 451–2)

From an embryologist's point of view, particulate theories shifted the focus from process to structure, and thus ignored the most basic and interesting problem: namely, how adult traits are developed epigenetically. In addition, of course, the existence of Mendelian particles was purely speculative, a feature which particularly distressed Morgan. He had come to his strong mechanistic and experimental bias in the 1890s by opposing just such particulate views in the work of Naegeli, Weismann and Haeckel, and was not prepared to backtrack with the appearance of a new particulate scheme (*i.e.* Mendelism). What was particularly telling to any embryologist was the propensity for those adhering to particulate theories to speak of adult traits as if they actually resided, in miniature at least, within the fertilised egg. This was preformationism at its worst. As Morgan emphasised, germ cells do *not* contain miniature wings, eyes or legs, but only the potential, under appropriate conditions, for producing these traits. Thus Morgan chastised his colleagues for using the term 'unit character' to refer to a Mendelian trait on the grounds that the particulate unit only contained the potentiality for generating the adult trait, and was not the trait itself. (Morgan did not use the term 'unit character', in fact, until as late as 1919, up to that time having used 'unit factor'.)

Thus to Morgan, as an embryologist in the early twentieth century, the problems of heredity and development were inseparable. Learning about transmission of information between parents and offspring was of no value without also learning about the development of the trait into its ultimate adult form. Quite literally, as Scott Gilbert has emphasised, the Mendelian theory developed by the Morgan school began with solid roots in embryology (Gilbert, 1978).

Morgan's general interest in the interrelationships between heredity, evolution and embryology is also evident in the very way in which, between 1908 and 1910 he hit upon use of the fruit fly, *Drosophila melanogaster*. As an opponent of the Darwinian theory of natural selection, Morgan became highly enamoured of Hugo deVries' mutation theory, first published in 1901–3. DeVries postulated that large variations, called 'mutations', occur in such a way as to produce new species in one generation. DeVries' theory was based on experimental results obtained with the evening primrose (*Oenothera*), and seemed to Morgan an excellent alternative to Darwin's theory of origin of species by selection of slight individual variations. In an attempt to demonstrate the occurrence of species-level mutations in animals, Morgan began breeding *Drosophila* in his Columbia laboratory around 1908 (Allen, 1975). When he discovered the white-eyed male fly in 1910, he realised it was not, after all, a new species but that indeed a single trait, eye colour, was transmitted in a discrete pattern as if it were actually a part of the X chromosome. Thus, Morgan's interest in *Drosophila* could easily change focus from evolutionary to hereditary questions because, in his mind, they were all so intimately connected. As he changed his focus, Morgan also ultimately changed his mind, abandoning his earlier scepticism of the Mendelian and chromosome theories and constructing a synthesis between them. It was this work for which Morgan became most well known, and for which he won the Nobel Prize in Medicine or Physiology for 1933.

From 1910 through the late 1920s, Morgan and a small group of enthusiastic co-workers, A. H. Sturtevant (1891–1971), Calvin B. Bridges (1889–1938), and H. J. Muller (1890–1968), among others, actively elucidated the relationship between genes and chromosomes of the fruit fly. As is familiar to most biologists, this was the period of the 'classical school' of genetics. Morgan and his group demonstrated that genes could be regarded as discrete units, arranged in linear fashion along chromosomes. They were able to map the genes, and eventually demonstrate a close correlation between genetic and cytological maps. They uncovered many facts about gene interaction, and demonstrated the complexity of organisation of the genome of diploid, sexually reproducing organisms. Along with work on plant genetics at such centres as Cornell (R. A. Emerson) and the University of Missouri (L. J. Stadler and Barbara McClintock) the Morgan school established clearly and unequivocally the material nature of hereditary units, and their structural organisation in the chromosome.

Ironically, the Mendelian-chromosome theory, for which Morgan became the acknowledged leader, is pre-eminently a *transmission* as opposed to translation, or epigenetic conception of heredity. Shortly after beginning his work with *Drosophila* Morgan shifted his concern from the embryological, developmental to the genetic, transmission view of heredity. Behind Morgan's changing conception of heredity lay an important philosophical issue: the growth and spread of mechanistic materialism in biology during the early twentieth century. In his career, Morgan played an important role in the development of mechanistic thought and its application to a variety of biological problems, particularly in embryology and genetics.

## Morgan's mechanistic materialism

Mechanistic materialism is a special form of general philosophical materialism that can be summarised as five propositions. (1) The parts of a complex whole are distinct and separate from one another: for example, the atoms in a molecule or the gears and levers in a clock. Each part is thus a distinct and separable entity. (2) The proper method for studying the whole is to break it down into its component parts, each of which can be investigated independently of its more complex involvement with other parts. This method of investigation is often referred to as *analysis*. (3) Behind the method of analysis lies the general assumption that the whole is equal to the sum of its parts and no more. There are no mystical or 'emergent' properties arising out of the association of the parts in the whole. Thus, if we know all about each part it should be possible to reconstruct the whole in its totality; nothing more is needed. (4) Systems change over time due largely to constant forces impressed on them from the outside. For example, the planets move in definable orbits because of the gravitational attraction of other bodies; populations evolve because organisms are constantly presented with challenges from an ever-changing environment. It is important to point out that the mechanistic view does not deny change in systems; the world does change constantly. However, that change arises out of influences of one system on another, rather than arising out of conditions existing within the system (for example, organism, population) itself, that is, part of its own internal nature. (5) The mechanistic world view is basically *atomistic*. Mechanists tend to see all phenomena in terms of a mosaic of separate, interacting, but ultimately independent parts.

Morgan's bias for mechanistic biology was directly related to his revolt against the descriptive and speculative side of the morphological tradition. Although his initial encounter with Jacques Loeb at Bryn Mawr in 1891–2 may not have presaged much, the friendship between the two men developed slowly over the next 20 years, so that by the time Loeb came to the Rockefeller Institute in 1910 each thought highly of one another's work. Loeb's paper, 'The Mechanistic Conception of Life' in 1911, and his collected essays, under the same title in 1912, present clear and impassioned arguments for a reductionistic and physicochemical approach to the study of life. Loeb championed the analytical approach of the physical sciences for biology – to make biology an exact science. This meant pre-eminently breaking complex problems down into their component parts and studying each separately, in isolation from the others. Morgan was primed for Loeb's views by his stays at the Stazione Zoologica in Naples and particularly his contact with Hans Driesch and the *Entwicklungsmechanik* tradition. Although he was a strong advocate of mechanistic biology, Morgan was never as rigid and hard-lined a mechanist as Loeb. He was too good a naturalist and too well rounded a biologist to accede to statements such as Loeb's claim that phototrophic insects were 'photochemical machines enslaved to the light' (Fleming, 1964: p. xxii). However, Morgan did join with Loeb, and physiologist W. J. V.

Osterhout, to edit a series of books for Lippincott and Co. under the title, 'Monographs on Experimental Biology', which focused on topics dealing with physicochemical biology. The significance of this series has been discussed elsewhere (Allen, 1966*a, b*) but I should emphasise here that it not only provided a forum for the publication of mechanistically conceived work, but also demonstrated by example the clear *advantage* that such work could have in exploring new directions in biology. Appropriately, Morgan titled his own volume for the series *The Physical Basis of Heredity* (1919), to indicate clearly the mechanistic, physicochemical basis for his own theories of heredity. Finally, the culminating work of Morgan's life, establishment of the Division of Biology at California Institute of Technology (CalTech), was a clear testament to his mechanistic philosophy. The buildings at CalTech were constructed under Morgan's guidance so that biology should be closest to organic chemistry, while the staff consisted almost exclusively of younger workers pursuing studies in physicochemical aspects of plant and animal physiology, biochemistry, genetics and embryology (Allen, 1978*b*).

It was Morgan's commitment to mechanistic biology and the analytical methodology it advocated, that provided the general backdrop against which his change of position on the problems of heredity can best be understood. With his mechanistic bias more clearly in hand, Morgan could take advantage of three factors that, stepwise, led to the separation in his mind between the problems of heredity transmission and embryonic development: (1) Wilhelm Johannsen's promulgation of the genotype–phenotype distinction, (2) the rapid progress of the *Drosophila* work, and (3) the agricultural climate around the turn of the century which emphasised genetic transmission rather than embryonic differentiation, as the crucial problem to be understood. Let us see how these various factors entered into Morgan's thinking in the period 1910–25.

# Morgan, heredity, and the genotype–phenotype distinction

Around the turn of the century many biologists were aware that the term heredity was used in several different and ambiguous ways. Danish botanist Wilhelm Johannsen (1857–1927) pointed out the ambiguities in his now-famous paper of 1911, 'The Genotype Conception of Heredity' (Johannsen, 1911), while William Bateson had voiced many of the same complaints some 15 years earlier (Bateson, 1894; also Coleman, 1970: p. 294). More recently, historians of science Jan Sapp at the University of Montreal and Scott Gilbert of Swarthmore College, have dissected out the various ways in which the term heredity was understood by workers in different fields of biology in the early decades of the present century. Sapp has identified at least five different contexts in which heredity was used, each with its own special slant and meaning: embryological, Mendelian, biometrical, evolutionary, and cytological (Sapp, 1982: pp. 27–33). Gilbert has also pointed out that with the rise of Mendelian genetics, and especially of the Morgan school, the term heredity evolved from its earlier embryological, to its more modern, genetic meaning (Gilbert, 1978).

From the mid-nineteenth century onward embryologists almost unanimously adopted the epigenetic view of development – that is, seeing form in an embryo develop out of formless ground substance rather than merely the growth in size of an already preformed individual. The epigenesis paradigm had shifted the attention of embryologists away from the problem of *transmission* (that is, the passage of preformed elements from parent(s) to offspring) to the development of form. In the older, embryological definition of heredity, transmission becomes but one part of the larger problem, namely, how offspring come to resemble (or, to a lesser extent, vary from) their parents. Much of the descriptive period of embryology, from the 1850s through the 1870s was at least in part aimed at showing in detail how and at what stage in development different organs and organ systems began to form, and how eventually specific adult traits emerged. To be sure, heredity meant transmission, since something material had to be passed from parent to offspring. But in the descriptive period embryologists recognised that what were transmitted were potentialities, not actual formed structures. It was the actualisation of these potentials during the growth of the embryo that formed the basic definition of heredity, and the focus of interest for most embryologists.

As cytological work in the 1870s and 1880s began to appear, the actualisation of potentials in the fertilised egg could be interpreted as largely a function of either the nucleus with its interesting array of chromatin threads, or of the cytoplasm, which microscopy was revealing as more highly organised than the mid-century term 'protoplasm' had originally suggested. Cytologically oriented workers such as Weismann, Boveri and Strasburger argued that hereditary potentialities resided in the nucleus, while embryologically oriented workers such as Morgan and E. G. Conklin argued that the structure of the cytoplasm determined the actualisation of potentials. While some lip service was given from both sides to the interaction between nucleus and cytoplasm, the emphasis of each group was clear and relatively exclusive. The debate between nuclear and cytoplasmic proponents of embryonic control has had a long history that persists in varying degrees to the present day.

In addition to the notion of nuclear or cytoplasmic control of development was the idea that embryonic differentiation was also influenced by environmental factors (for example, temperature, humidity, amount of food available, etc.). It was envisioned that the environment could act at various stages of embryonic development, from fertilised egg to older embryo, or on the formation of germ cells themselves, triggering one or another potentiality into actuality. Particularly prominent were environmental theories of sex determination that abounded around the turn of the century (Allen, 1966a). Since environmental factors always acted through the agency of the cell cytoplasm, environmental theories of embryonic development were thought of as allied to cytoplasmic, rather than nuclear control mechanisms. Morgan's lifelong friend Edwin Grant Conklin stated clearly the embryological view of heredity when, in 1908, he wrote:

Indeed, heredity is not a peculiar or unique principle for it is only similarity of growth and differentiation in successive generations ... In fact, the whole process of development is one of growth and differentiation, and similarity of these in parents and offspring constitutes hereditary likeness. The causes of heredity are thus reduced to the causes of successive differentiation and development, and the mechanism of heredity is merely the mechanism of differentiation.

(Conklin, 1908: pp. 89–90)

Conklin went on to say:

Heredity is today the central problem of biology. This problem may be approached from many sides – that of the breeder, the experimenter, the statistician, the physiologist, the embryologist, the cytologist – but the mechanism of heredity can be studied best by the investigation of the germ cells and their development.

(Conklin, 1908: pp. 89–90)

As we have seen, Morgan was initially imbued with the same view. How, and when did he begin to change his mind about the scope of heredity? Shortly after he discovered the white-eyed *Drosophila* in 1910, Morgan encountered the conception of distinguishing between genotype and phenotype promulgated by Wilhelm Johannsen from 1909 onward. Primed as Morgan was from his recent discoveries of sex-linked inheritance, Johannsen's ideas appear to have been a major catalyst in Morgan's change of mind about the inclusive meaning of the embryological definition of heredity. As Jan Sapp has persuasively argued, Johannsen's paper of 1911 was a virtual polemic *against* the older view of heredity, seeking to delimit and in a sense redefine the concept to mean only what we today would call genetic *transmission*.

In his 1911 paper, Johannsen separated out the several meanings of the term heredity which he found in current use among biologists (Johannsen, 1911: p. 130). As Johannsen emphasised, there were two components to the development of adult traits in organisms: the genotype, that is, the hereditary elements passed vertically from parent to offspring, and the phenotype, the development of those genotypic capabilities horizontally into adult traits. Heredity, as a field of study, Johannsen argued, should be solely concerned with the vertical process – that is, the passing on of the genotypic elements. The study of the development of phenotype belonged more properly to the realm of embryology.

The genotype, Johannsen emphasised, was purposely an *ahistorical concept*, whereas the phenotype concept, especially under the influence of Haeckel's 'biogenetic law', was a truly historical one (Johannsen, 1911: p. 139). To Johannsen the genotype conception was as ahistoric as a chemists's atoms combining to make molecules. $H_2O$ is always water, regardless of where the hydrogen and oxygen atoms have been prior to their combination. The hydrogen and oxygen atoms recovered from water can be recombined with other elements and behave exactly as every other hydrogen and oxygen atom. The same was true, in Johannsen's view, of Mendelian factors, or what he called 'gens.' The fundamental basis of heredity, he stated, was hidden deep within the gametes. Only through the analytical methods of Mendelian theory, coupled with experimentation and mathematical formulation, could the process of heredity be understood. The term 'phenotype' was coined by Johannsen as a derogatory word referring to the morphological, descriptive view of heredity characteristic of the old natural history tradition. Thus, Johannsen's view rigorously separated the field of Mendelian genetics from all earlier studies of heredity, that is, those originating in evolutionary, biometrical, or embryological theory. In so doing, Johannsen purposefully gave a new meaning to the term heredity.

Although I believe Johannsen's concept was a watershed for Morgan in demonstrating the value of a more limited definition of heredity, it is not so easy to prove that Morgan actually knew of Johannsen's work as early as 1911 or 1912. The evidence is circumstantial, but not definitive. First, Johannsen was to have spent part of the winter quarter as a guest lecturer at Columbia, as a major stop-off in his American lecture tour of the late fall and winter of 1910–11 (See letter from G. H. Shull to E. G. Conklin, Summer 1910; Conklin Papers, Princeton University). While no records survive to indicate that Johannsen actually did go to Columbia, his student and erstwhile biographer Otto Winge clearly states that Johannsen was at least preparing to depart for the United States in the autumn of 1910 (Winge, 1958: p. 86); all evidence points to the fact that Johannsen did come to the United States at that time. Second after 1911, Morgan did begin to change his views on heredity, as I will demonstrate in more detail in the next section. Third, Morgan was present at the American Society of Naturalists meeting in Ithaca, N.Y., on 28 December 1910, where he not only delivered his own paper, 'The application of the conception of pure lines to sex-limited inheritance and to sexual dimorphism', in a symposium titled 'The Study of Pure Lines of Genotypes' (Morgan, 1911), but also must have heard Johannsen himself deliver what was to become his important 1911 paper, 'The genotype conception of heredity' (Johannsen, 1911).

It is thus highly unlikely that Morgan did not meet Johannsen at least at the Naturalists Symposium, if not later at Columbia, and discuss heredity in general, and the genotype–phenotype distinction in particular. What I am suggesting in this roundabout way is how likely it was that Morgan knew about the genotype–phenotype distinction early enough in his *Drosophila* work for it to serve as an important stimulus for separating the problem of hereditary transmission (the genotype) from that of embryonic development (the phenotype).

What could the genotype–phenotype conception actually have accomplished for Morgan? It could have helped him redefine his view of heredity from a primarily phenotypic (that is, embryological) to a primary genotypic (that is, Mendelian) one. It could have helped him move from a process-oriented, epigenetic view of heredity to a morphological conception based on the transmission of material particles from parent to offspring through the germ cells. It could have helped draw his attention away from the cell cytoplasm, the realm of concern of most embryologists at the time, to the cell nucleus, the concern of Mendelian genetics. And finally, primed as he was from the point of view of mechanistic philosophy, Morgan most likely found the genotype conception to be the perfect analytical tool for breaking a complex problem down into its simpler, and experimentally accessible, components. The genotype conception focused the many questions of heredity which were emerging from Morgan's *Drosophila* work.

## The theory of the gene and the final split between genetics and embryology

With his mechanistic bias already well grounded, and armed with the genotype–phenotype distinction, Morgan proceeded to develop a more definite, rigorous and restrictive definition of 'heredity'. This can be seen more clearly in his changing definition of the most fundamental unit of heredity, the *gene*. Empirical work quite close at hand had shown Morgan that he could give up opposition to notions of particulate inheritance. His own studies with *Drosophila*, proceeding so well from 1911 onward, had convinced him that there was considerable value heuristically at least, in the Mendelian theory. At the same time, the work of a number of cytologists, particularly that of his friend, Columbia colleague, and department chairman Edmund Beecher Wilson (1856–1938), had suggested that there was possibly some relation between Mendel's factors and the chromosome theory of heredity. Morgan came to realise that hereditary units could be viewed as real, material components of cells, and not just as abstract particles.

The influence of Wilson on Morgan is difficult to assess, though clearly it had profound effects on his thinking. Partly because Morgan and Wilson worked just down the hall from one another in Columbia's Schermerhorn Hall (Zoology Department) during the winters, and lived across the street from one another at Woods Hole in the summer (both working at the Marine Biological Laboratory), there was little need for written correspondence. The evidence is clear, however, that both men admired each other: Wilson brought Morgan to Columbia in the first place in 1904 (from Bryn Mawr), and once remarked that the most favourable organism he (Wilson) ever discovered for advancing biological work was Thomas Hunt Morgan. Morgan, in turn, wrote Wilson's obituary for the National Academy Memoirs, and praised his work in clear, though unimpassioned terms. Both men saw biology very much alike. As former students of Brooks, both conceived of biology as a whole, the integration of problems of heredity, development and evolution with the methods of cytology and cell physiology. It seems highly likely that Wilson provided a strong influence on Morgan in many ways, not the least of which was, specifically, overcoming the latter's opposition to the chromosome theory as 'preformationist'. By 1905 Wilson, and Morgan's own former student at Bryn Mawr, Nettie M. Stevens (1861–1912), had concluded independently that chromosomes – specifically the accessory chromosomes – determine sex. Although Morgan initially rejected this interpretation, Wilson championed the idea so strongly, and with such good evidence, that by the time the *Drosophila* work was under way and his own discovery of sex-linked inheritance had been made, Morgan was ready to admit that chromosomes could determine not only general traits such as sex, but specific ones such as eye colour or wing shape as well. Indeed, Morgan's own analysis of the *Drosophila* work ultimately depended on being able to associate given phenotypic traits with given chromosomes, and with specific points, or loci, on these chromosomes. Certainly an important factor in that shift in Morgan's thinking was a result of Wilson's influence.

But Morgan was also primed on his own to make the association between Mendel's work and the chromosome theory. Chromosomes provided a *material* basis for Mendel's otherwise abstract 'factors'. This was important to Morgan, as he and his co-authors made explicit in the 'Preface' to *The Mechanism of Mendelian Heredity* in 1915. In justifying the constant application of cytological work on chromosomes to the Mendelian breeding results, Morgan *et al.* write:

Why then, we are often asked, do you drag in the chromosomes? Our answer is that since the chromosomes furnish exactly the kind of mechanism that the Mendelian laws call for; and since there is an ever-increasing body of information that points clearly to the chromosomes as the bearers of the Mendelian factors, it would be folly to close one's eyes to so patent a relation.

(Morgan *et al.* 1915: pp. viii–ix)

Morgan's materialism had led him to accept Mendelism when (1) empirical evidence from breeding experiments with *Drosophila* demonstrated it, and (2) Mendelism could be given a physical basis in cell structures, such as chromosomes. It was clear to Morgan by 1915 that chromosomes could be viewed as vehicles for transmitting Mendelian factors. Yet, at the time of the epoch-making *Mechanism of Mendelian Heredity*, Morgan had not clearly stated that Mendel's factors were in fact what Johannsen had called 'gens', nor that they were highly precise, discrete (molecular or atomistic) segments of chromosomes. (Morgan may have been less inclined to use Johannsen's term at this point, because Johannsen himself was something of a non-materialist and resisted thinking of Mendelian genes as physical segments of chromosomes. In fact, Johannsen never fully accepted the Mendelian chromosome theory as having material reality (Churchill, 1974; Sapp, 1984).)

By 1919, however, Morgan was willing to be more precise on the physical nature of the gene as well as its possible biochemical nature. On the gene's structure he wrote:

[The analysis of crossing over] leads then to the view that the gene is a certain amount of material in the chromosome that may separate from the chromosome in which it lies, and be replaced by a corresponding part (and by none other) of the homologous chromosome. It is of fundamental significance in this connection to recognize that the genes of the pair that interchange do not jump out of one chromosome into the other . . . but are changed by the thread breaking as a piece in front of or else behind them, but not in both places at once . . .

(*Ibid.*)

On the functional, biochemical nature of the gene, Morgan indulged in a little speculation and suggested (following his friend Jacques Loeb) that genes might function as molecular entities producing enzymes:

The hereditary factor in this case must consist of material which determines the formation of a given mass of these enzymes, since the factors in the chromosomes are too small to carry the whole mass of the enzymes existing in the embryo or adult.

(*Ibid*, p. 245)

Throughout the evolution of his attempts to understand the nature of the Mendelian gene more clearly, Morgan exhibited his mechanistic bias by viewing genes essentially as separate atomistic units. Again in the *Physical Basis of Heredity*, he wrote:

The essential point here is that even although each of the organs of the body may be largely a product of the entire germ-plasm, yet this germ-plasm is made up of units that are independent of each other in at least two respects, *viz*, in that each one may change (mutate) without the other's changing, and in segregation and in crossing-over each pair is separable from the others.

(*Ibid*, pp. 240–1)

Thus by 1919, Morgan not only used the term 'gene' explicitly for the first time (it did not occur, for example, in *The Mechanism of Mendelian Heredity* four years earlier), but also appears to have thought of genes as specific, discrete, and independent units acting in an atomistic way. The gene as a mechanistic element had come into its own.

At the same time that he was refining his definition of the gene, Morgan also began to sharpen the distinction between nucleus and cytoplasm of the cell on the one hand, and between embryology and genetics as fields of research on the other. In *The Mechanism of Mendelian Heredity* in 1915, Morgan had allowed that certain components of the cytoplasm (e.g. plastids) could self-perpetuate and thus might be important units of heredity outside the nucleus. However, he gradually abandoned this position over the next few years. In a letter to his friend Jacques Loeb in 1919, Morgan wrote that he was particularly anxious to counteract the notion that the cytoplasm transmits any fundamental characteristics to the next generation:

It is this point [that the cytoplasm transmits fundamental properties] that I am anxious to go for, because of its widespread belief among biologists in general for which I can find absolutely no real basis except an emotional one. It is for this reason mainly that I have not hesitated to hold up as examples two of my best friends and a very famous German investigator.

Morgan made the same point in his Nobel Prize speech, delivered in Stockholm in June, 1934 and published the next year (Morgan, 1935). The man who had once stated that the cytoplasm was the seat of the really interesting activities relating to heredity (Morgan, 1910: p. 453) now argued that nothing of fundamental importance for heredity occurred within the cytoplasm, and that it was the nucleus on which the geneticists' attention should be focused. The nucleus and cytoplasm, once considered so vitally linked in Morgan's mind, by 1919 had become separate and distinct.

Morgan made the distinctions between the realms of embryology and of genetics as fields of investigation even more sharply by the mid-1920s. In *The Theory of the Gene* (1926), Morgan put it explicitly: embryology was the study of the development of genetic potentialities into adult realities; genetics was the study of the transmission of hereditary elements from parent to offspring (Morgan, 1926: p. 26).

The practicality of such a split in directing future research was obvious to those who were anxious to get on with the study of genetic transmission. In 1914, William Bateson wrote that as a result of splitting off genetics from embryology,

... we at least can watch the system by which the differences between various kinds of fowls or various kinds of sweet peas are distributed among their offspring. By thus breaking the main problem into its parts we give ourselves fresh chances.

(Bateson, 1914: p. 289)

Thus, by the mid 1920s, the science of genetics as a new field was established on several grounds:

(1) As a materialist view of heredity, in which hereditary units are identified as discreet, material units, the genes.

(2) As a mechanistically established concept, in which the genes are regarded as independent units shuffled and reshuffled in successive generations; the fundamental character of each unit being unaffected by its recombination with other units.

(3) As an experimentally established science based on the correspondence of two distinct and independent lines of investigation: experimental breeding and the cytological observation of chromosome structure.

(4) As the foundation for a new field of research, called genetics, with its own problems and methods – related to, but set apart from, both embryology and evolution.

It would be erroneous to suggest, however, that all biologists, particularly embryologists, agreed with Morgan's redefinition of 'heredity'. Morgan's close friend and colleague from the Marine Biological Laboratory in Woods Hole, F. R. Lillie, was one who staunchly resisted the mechanisation of the hereditary process. Lillie argued that the gene theory did not speak to the problems of embryogenesis:

I do not know of any sustained attempt to apply the modern theory of the gene to the problem of embryonic segregation. As the matter stands, this is one of the most serious limitations of the theory of the gene considered as a theory of the organism. We should, of course, be careful to avoid the implication that in its future development the theory of the gene may not be able to advance into this unconquered territory. But I do not see any expectation that this will be possible, even in principle, as long as the theory of the integrity of the entire gene system [*i.e.*, that all genes are present, or at least active] in all cells is maintained. If this is a necessary part of the gene theory, the phenomena of embryonic segretation must, I think, lie beyond the range of genetics.

(Lillie, 1927: p. 366)

Lillie even went on to claim that he did not foresee a future synthesis possible – at least in the late nineteenth century sense (by which he meant Weismannian) – given the new view of geneticists about what constituted heredity (Lillie, 1927: p. 367). Another critic of the Morgan school was his old friend and colleague, from both Woods Hole and Johns Hopkins days, Ross G. Harrison. Harrison wrote in 1937 that the new gene theory, as prestigious as it seemed to be, was much too one-sided. It focused only on the problem of transmission, and failed to deal in any significant way with the embryological issues (Harrison, 1937: p. 372). Beyond this, Harrison probed more deeply at the philosophical foundations of the split. He argued that geneticists were too atomistic, while embryologists sought a more holistic interpretation of the hereditary process:

The embryologist, however, is concerned more with the larger changes in the whole organism and its primitive systems of organs than with the lesser qualities known to be associated with gene actions. As Just remarked ... he is more interested in the back [of a fruitfly] than in the bristles on the back, and more in the eyes than the eye color.

(Harrison 1937: p. 372)

In addition to classical embryologists such as Lillie and Harrison, another group, namely those concerned with the problem of cytoplasmic inheritance, reacted against the Morgan definition of heredity. As Jan Sapp has pointed out, both Herbert Spencer Jennings and more particularly his student, Tracy Sonneborn, found the definition of heredity associated with the Morgan school far too limiting. In his autobiography Sonneborn described himself as 'a lifelong critic of what seemed to be a blind and erroneous faith in the gene as a source of all heredity' (Sonneborn, 1978: p. 1). In the same paragraph he referred to the transmission conception of heredity as 'a stifling dogma'. Focusing as much attention as it did on the nucleus, and particularly the chromosomes, the Mendelian-chromosome theory, as defined by Morgan, left little room for nucleo–cytoplasmic interaction. Viktor Hamburger has pointed out that as late as 1951 the effects of this rigid separation (between genotype and phenotype and between nucleus and cytoplasm) were felt by Belgian embryologist Albert Dalcq. Arguing that classical genetics ignored the important role that cytoplasm as 'an organized system' played in the process of differentiation, he wrote:

This notion [of pattern in the cytoplasm and of the importance of the whole], so intimately tied to a pattern, is lacking in the system of concepts used by geneticists ... [These] are based on a particularistic, atomistic viewpoint which neglects, despite everything, this other factor which resides in the totality of the organization.

[Dalcq (1951): p. 135; translated by Hamburger]

And, interestingly enough, even so arch a mechanist as Jacques Loeb recognised that the atomistic gene concept posed a problem for embryologists. In 1916 he wrote:

The difficulties besetting the biologist in this problem [harmonious interaction of parts of an organism] have been rather increased than diminished by the discovery of Mendelian heredity, according to which each character is transmitted independently of any other character. Since the number of Mendelian characters in each organism is large, the possibility must be faced that the organism is merely a mosaic of independent hereditary characters. If this be the case the question arises: What moulds these independent characters into a harmonious whole?

(Loeb, 1916: pp. v–vi)

These objections all focused on two major issues: (1) the atomistic nature of the gene, including the separation of the nucleus from the cytoplasm, and as a consequence (2) the failure of modern geneticists to take a holistic approach to the problems of heredity. Those who objected to the Morgan definition of heredity saw in the genotype–phenotype distinction, and in the nuclear–cytoplasmic split a major obstacle to the future solution of one of the main problems of modern biology: how the fertilised egg differentiates into a multicell adult.

Yet, despite objections by some well respected members of the biological community, it was the more restricted definition of heredity that triumphed in the years after 1915. What additional factors made this triumph possible?

# The economic, philosophical, and sociological basis of the 'new heredity'

Primary factors contributing to the redefinition of heredity – and consequently the split between traditional embryology and genetics – included a changing economic, philosophical, and social *milieu*, in the United States between 1890 and 1930. In the post-Civil War period, especially between 1890 and 1914, American agriculture began to undergo a major revolution, part of a more general economic shift associated with industrialisation and urbanisation. With regard to just one of these, urbanisation, a critical shift was well under way by the 1880s. Cities were swelling rapidly, draining population away from the countrysides. Families that had once fed themselves while producing a small surplus on their family farms, were now living in urban areas and needed to be fed. Family-based agriculture was no longer productive enough for the needs of an increasingly urban-industrial work force. Agriculture had to be managed and planned, on a larger scale than before; and it had to be made more profitable as an area for business investment. But all this required managing and planning of a sort that was foreign to the principles of *laissez-faire*. After much debate and struggle, first the private sector, then government, adopted the notion of economic and social planning. This meant, among other things, the training and use of experts to advise and manage. 'Experts' often meant 'scientific experts' (broadly interpreted, including 'social scientists' after 1914 or 1915) who could help in what came to be known as 'rational planning'. Applied to agriculture, rational or scientific planning meant research into the management of plant and animal nutrition, the use of fertilisers and animal feeds, crop rotation and diversified planting, and especially by the turn of the century, animal and plant breeding (Rosenberg, 1976*b*). In the latter half of the century much effort had gone into the development of fertilisers, animal feed, and the physiological/nutritional aspects of agriculture. While these efforts had yielded some major increases in productivity, they had reached a limit in both extent and profitability. This was not so with breeding, however. The results of good breeding had a very different economic potential. In 1910 US Secretary of Agriculture James Wilson could argue strongly for the study of heredity as a way of opening up a new horizon in economically profitable agriculture. Writing in the opening pages in the newly founded *American Breeders Magazine*, Wilson noted that both fertilisers and animal nutritive feeds must be reapplied year after year to have the desired effect, whereas, the hereditary effects obtained through breeding were different:

Heredity is a force more subtle and more marvelous than electricity. Once generated it needs no additional force to sustain it. Once new breeding values are created they continue as permanent economic forces.

(Wilson, 1910: p. 5)

Wilson then goes on to make his point more explicit:

But the cost of improvements through breeding usually represents only a small fraction of the added values. The increase of products secured pays the price in a short time, and, since there is no further expense, the annual increase afterward is clear profit. The farmer will be able to retain a part of the larger production in the form of added profit and part will help to reduce the cost of living to those in the cities. Larger production on the farm will also give increased business to the transportation company, the manufacturer, and the merchant, and will provide the nation the larger product with which to hold our balance of trade.

(*Ibid.*)

Wilson's enthusiasm was not mere political rhetoric. There was a widespread belief, partly catalysed by the rediscovery of Mendel's laws in 1900, that the science of breeding was off to a new and momentous start. This is evidenced also by the fact that not only the United States government, but also private foundations, were becoming increasingly interested in scientific planning in general, and agricultural planning in particular. The Carnegie Foundation contributed significantly to the research of Luther Burbank, and after 1904 was the sole support of C. B. Davenport's extensive (for those days) Station for the Experimental Study of Evolution, at Cold Springs Harbor, Long Island (the station's function was to study scientifically the processes of breeding and selection). In addition, after 1915 the Carnegie Institution of Washington funded the work of Morgan and others on *Drosophila*. While the enthusiastic hopes of Secretary Wilson and others that the rediscovery of Mendel would have a payoff in immediate agricultural results was not forthcoming in the short run, the new genetics did eventually contribute profoundly to the development of agriculture in the United States. The work of investigators such as Donald Jones (Connecticut Agricultural Station), E. M. East and later Paul Mangelsdorf (Harvard's Bussey Institution), Charles Zeleny (University of Illinois), and L. J. Stadler & Barbara McClintock (University of Missouri, Columbia) and R. A. Emerson (Cornell) was all carried out in a specific agricultural context, and in many cases led directly to some practical agricultural gains (hybrid corn and wheat being amongst the most notable).

I suggest that around the turn of the century there was more than passing academic or intellectual interest in the study of heredity, defined as patterns of transmission of genetic elements from parent to offspring. It was this atmosphere that gave considerable impetus, even indirectly, to the redefinition of heredity effected by Morgan and his followers after 1910. The agricultural breeding context distinctly favoured a transmission as opposed to a translational concept of heredity.

On the sociological side, the splitting of transmission genetics from embryonic development had an important practical consequence within the scientific community. It allowed Morgan, his immediate followers, and others who took up the Mendelian-chromosome theory, to define a new and a separate field of investigation. Between 1910 and 1925 the Mendelian-chromosome theory became what Imre Lakatos calls a full-fledged *research program* (Lakatos, 1970). Lakatos has emphasised that research programs consist not only of concepts but also of methods of research (in this case, for example, breeding coupled with cytological observations), standard protocols (for example the use of pure strains for breeding, or the correlation between breeding and cytological data), and philosophical methods, including the notion of what is a proper explanation in a field (for example, mechanistic *vs.* holistic interpretation; or the role of quantitative and mathematical thinking in scientific explanation). By establishing a research program, Lakatos points out, scientific workers define their fields and problems, thus influencing the direction of future research and the development of field-wide methods for dealing with challenges to accepted orthodoxy. For example, Morgan and his group not only defined rigorously the problems which were to become the future focus of genetic research, but also established an orthodoxy which focused almost exclusively on the cell nucleus as the centre of heredity. Any attempts to discuss cytoplasmic inheritance (or what was sometimes referred to as 'maternal effects') were strongly discouraged. Jan Sapp has studied the history of this subject exhaustively, and has shown how those who sought to publish on cytoplasmic inheritance, such as Tracy Sonneborn and later his student, Donald Nanney, at first found their papers rejected by orthodox genetic journals (Sonneborn, 1978).

Sapp applied the idea of struggle for authority among competing fields, as developed by French sociologist Pierre Bourdieu (1975), specifically to the case of nuclear vs. cytoplasmic genetics in the early twentieth century (Sapp, 1984). Bourdieu's idea is that scientific fields, or research programs in the Lakatosian sense, are in competition with other fields for money, students, and the opportunity to control academic or research positions. The competition is most keen among closely related fields, but exists to one degree or another between all fields. In Bourdieu's model, scientific authority, or competence, is understood as the socially recognised legitimacy of the individual to speak and act on scientific matters. The content of scientific ideas is thus seen as related to the social reality of establishing a professional niche, that is, a research program. Thus, Bourdieu sees the choice for pursuing certain theories over others, as well as the manner in which the theories are put forward, as an integral part of the social context among competing fields.

Applied to the development of genetics and its separation from embryology, Bourdieu's idea suggests the following scenario: As Mendelian genetics began to have some success dealing with the process of transmission, it became advantageous to begin determining the boundaries of the new field – that is, to establish its problems and its scope. This meant for Morgan and his group that it became increasingly advantageous to eliminate from the study of heredity itself the knotty problems of embryonic development with which Morgan himself (and others) had had little experimental success. By so doing, Morgan was able to outline what appeared to be a successful and easily approached field of scientific endeavour. Had he insisted on working simultaneously with the problems of the development of phenotype and transmission of the genotype, it is doubtful that the field could have developed in any clear-cut way. Although he himself never renounced either his interest in embryology, or his belief that the Mendelian gene ultimately had to be interpreted in embryological terms, Morgan was pragmatic enough to see the advantage of pushing embryology aside for the time being.

Developing the new field of Mendelian genetics with a strong central focus (transmission, assortment and recombination), a set of research techniques, and most importantly some clear and immediate results, Morgan was able to attract attention, students, and (ultimately) research money in a way that would have been impossible had he insisted on studying heredity in the older, more holistic way. Morgan thus drew a boundary between the new field of genetics and the old field of embryonic development. Everything within the boundary was included in the new research program; those who would try to force upon the Mendelian-chromosome theory the burden of explaining embryonic development were told to become (or remain) 'embryologists'.

In putting forth this analysis I do not wish to suggest that Morgan or his group made such choices consciously, or were acting in a particularly ruthless or opportunistic way to exclude certain topics from their new research program. I do want to suggest, however, that the conscious and subconscious aspects of what it means to establish a new field of research and to gain the recognition, money, and students which can result, is not a negligible factor in what constitutes the formation of a scientific research program. After all, research programs are more than merely good ideas. They involve techniques, equipment, laboratories, people, the desire for individual recognition, and money, all of which have some direct influence on the content and direction of the scientific ideas themselves.

# Conclusion

Once developed, the split between embryology and genetics has persisted in various forms down to the present day. Those who attempted to effect any rapprochment in the 1930s and 1940s – such as Richard Goldschmidt or C. H. Waddington – only succeeded in raising interesting theoretical questions for which they could find few answers. When both T. H. Morgan (1934), and Hans Spemann (1924), tried to write about the relation of genetics to classical embryology, they achieved little more than to remind their readers that somehow genes had to direct the differentiation of phenotypic traits in the developing embryo. Although the provocative work of the Spemann School had opened a whole new approach to the problem of embryonic differentiation, it did not depend on any knowledge of

Mendelian genes or chromosomes. It was focused at higher levels of organisation, namely tissues and organ systems. At the cellular level the ideas of induction and morphogenesis had very little to say.

As the above essay has attempted to show, no serious biologist in the 1920s or 1930s doubted that the 'physiology of development', to borrow Goldschmidt's phrase, was an important area of study. No one doubted that genes were material entities which had a real chemical function during the development of the embryo. No one saw this more clearly than Morgan's colleague at Columbia, E. B. Wilson, who emphasised the relationship between genetics and embryology in the title of all three editions of his classic text, *The Cell in Development and Heredity* (1st edition 1896, 2nd 1900, 3rd edition, 1925). Yet even Wilson, visionary as his view of biology was, could not effect a reunification of genetics and embryology in the older, nineteenth century sense.

A final issue that I would like to raise is to assess Morgan's actual role in effecting the split between genetics and embryology. Was Morgan's role crucial? Did he, and the position he took with respect to the burgeoning *Drosophila* work, create the dichotomy, or was it inevitable anyway? In other words, how much was the split a matter of Morgan's personal style and influence, and how much was it a natural consequence of economic, social and intellectual developments within biology itself at the time? Of course, the answer to this question in some ways reflects one's particular style of interpreting history; and a major discourse on the historical role of the individual *vs.* external conditions is quite beyond the scope of this essay. However, it would be foolish to pretend that the facts – especially in the case presented here – speak clearly and unequivocally for themselves. There is still a great deal of research to be done – especially on the relationship between genetics and agriculture (in the United States and elsewhere) in the period between 1900 and 1940. However, I do want to suggest how the specific and general facts outlined in this paper can be interpreted in a logical and consistent way in the context of the broad history of biology and society in the early twentieth century.

In many ways I think the dichotomy between embryology and genetics was inevitable, and was not an idiosyncratic product of T. H. Morgan's own particular research. Of course, Morgan as an individual did make a difference, probably influencing both timing and the manner in which the split took place. But he was not the causal agent; the split was very much a product of its time, with Morgan being the unplanned agent.

Given the economic and social milieu of the period 1890–1930 it seems inevitable that a disproportionate amount of interest in and attention to the problems of hereditary transmission, as opposed to embryonic development, would manifest itself. Scientific agriculture inevitably made genetic transmission a more critical problem than embryonic differentiation, because it emphasised producing, through planned breeding, self-perpetuating strains of commercially important products. Thus, a level of interest – ultimately translated into financial support – existed for studies of transmission, thereby fostering a separation of those concerns from the broader issues of development. At the same time, the prevailing analytical method fostered by the mechanistic-materialist school of thought made such a separation seem more logical – even necessary, perhaps – than it would have seemed a generation earlier. The philosophical milieu was also favourable to a split, though it did not, in itself, stimulate research into genetics more than into embryonic development.

Morgan's use of *Drosophila* as a favourable organism, while important in the subsequent history of genetics, does not in itself seem crucial enough to have been a major factor responsible for splitting genetics off from embryology. For one thing, Morgan was not the only investigator using *Drosophila* for breeding purposes in the laboratory. In fact, he got his ideas from W. E. Castle, who bred the fruitfly for other purposes as early as 1905 (Allen, 1975: p. 325). While Castle himself may not have been able to exploit *Drosophila* to the extent that Morgan and his group did, there is no question that interest in breeding the small insect was already established before Morgan ever took up the issue. For another, quite different organisms, such as maize (corn), or guinea pigs, were also available, and became additional foci for genetic and cytogenetic studies by the 1920s. All the basic genetic principles illustrated by *Drosophila* can be elucidated with maize, wheat, and other plants, organisms of more direct agricultural use than *Drosophila*. Thus, I do not think the argument can be made that the split between genetics and embryology resulted primarily from Morgan's (or anyone else's) discovery of *Drosophila* as a highly favourable organism for breeding purposes. Breeding was already of powerful interest (as discussed above) by 1910. As Bateson pointed out in his visit to the USA as early as 1903,

American breeders were eager for knowledge of Mendel and his methods, suggesting that interest in transmission was 'in the air', and that had Morgan's *Drosophila* not come along to spur on the study of genetics in 1910, eventually the work would have been carried on by others, perhaps with the same or different organisms.

With regard to Morgan himself, the split that he personally made in his thinking was not inevitable, but resulted from the confluence of several factors in his own career. By 1909 or 1910, Morgan's own experimental work in embryology had not yielded much in the way of startling or original work, though he was certainly recognised as one of the most knowledgeable leaders of the field. So, in a sense, Morgan was 'primed' for a new direction in his research efforts. The discovery of *Drosophila* came at just the right time to help Morgan crystallise his own thinking about the hereditary process, and specifically to help overcome his earlier scepticism of Mendelism. At the same time the work of E. B. Wilson and Nettie Stevens, as well as his own cytological studies in 1909 and 1910 on the chromosomal basis of sex determination in aphids and phylloxerans, helped Morgan to overcome his scepticism of the chromosome theory. Probably as important as all of this was the incorporation into his laboratory after 1910 of Sturtevant, Bridges and Muller. These young and enthusiastic students not only pushed the *Drosophila* work ahead with great imagination and rapidity, but also they never had the doubts about either Mendelism or the chromosome theory that Morgan did, and often argued with their teacher to overcome his various misunderstandings and doubts. In a sense, they were important figures in paving the way for the ultimate synthesis of Mendelism and cytology that was the crowning achievement of the *Drosophila* work.

Morgan did, of course, contribute something, too! His own basic pragmatism and intellectual honesty allowed him to change his mind on fundamental issues when presented with new evidence. For example, in the face of Wilson's and Steven's work on chromosomes and his own work on *Drosophila*, he could come to accept theories he had previously rejected. He could also see the value in divorcing the problem of transmission from development, though he himself remained vitally interested in both. Morgan also brought to the *Drosophila* work a philosophical materialism that made him appreciate more than many of his generation the values tying the chromosome theory to Mendel's, thus giving the abstract 'gene' a real, material basis. This same turn of mind also prepared Morgan to appreciate Johannsen's genotype–phenotype distinction, and to recognise the value of studying the transmission and developmental aspects of heredity separately. Morgan's mechanistic bias emphasised the value of breaking complex biological processes – such as the old, inclusive concept of heredity – down into their component parts. Thus, the confluence of *Drosophila* as a favourable organism for studying transmission, the presence of an enthusiastic group of co-workers, the concurrent cytological studies of chromosomes and heredity by Wilson, Stevens, and others, and the publication of Johannsen's genotype conception of heredity, all helped push Morgan away from his older, inclusive, to a newer, more circumscribed, definition of heredity.

And where is the split between embryology and genetics today? The historical facts suggest that the breach has not yet been healed completely. Genetics has persisted in being the science of 'the gene', even if that 'gene' is now defined as a few thousand base pairs of DNA. Biologists still talk about *the gene* for eye colour or *the gene* for height, as if genes were blueprints, each acting separately, or atomistically, from each other. Genetics is still very much the study of patterns of transmission between parent and offspring, and is thus still heir to its mechanistic-materialist background. Embryology, for its part, while avoiding the pitfall of reductionism and atomism over the years, has often lapsed into a kind of holistic vagueness that has not been productive for experimental research. One need only think, for example, of Paul Weiss' 'field theory' as a well intentioned, but non-materialistic theory to understand how holistic views alone do not yield verifiable theories (Weiss, 1968). Yet, despite these disadvantages, embryologists have kept alive a holistic view of the development process which focuses attention on the interaction of components in a developmental system, and somehow recognises that the whole is indeed greater than the sum of its parts.

I would not want to risk being much of a prophet, or at any rate, being held to my predictions in the future. But if I may venture a guess, I would say that the old dichotomy between embryology and genetics is just now in the process of breaking down. The two fields are self-consciously aware of a need for reconciliation – in fact, something even more, the synthesis into a new theory of heredity, one that incorporates the notion of transmission *and* translation into a comprehensive whole. Yet Lillie was right: any future synthesis will not return to the Weismannian version of holism. Today, any holistic view will be held up to the test of experimental analysis. Molecular genetics, with its equal emphasis on transmission (DNA replication) and translation (protein synthesis) is playing a major role in breaking down this old dichotomy. Yet so far, molecular genetics has provided only a few clues to the basic problem of development: how segments of DNA are turned on and off in a coordinated way so as to produce an ordered sequence of differentiating parts. There is a glimmer of hope even for this knotty problem in recent work in immunology where studies of both vertical and lateral transmission are being integrated with the molecular biology of gene expression. I am too little of an expert on immunology to point out many of the specific aspects of this integration. But the mammalian $\alpha$-globulin genes may be a good case in point, where the line between vertical transmission of the Johannsen sort and horizontal translation of the molecular genetics school becomes so blurred as to be meaningless. If a gene's expression is related to its position in a genome, or in a particular cell line, then translation *depends* on transmission, and *vice-versa*. I suggest that as biologists look more closely at the whole cell, and study its genome in relation to overall cell processes such as protein synthesis, the title of E. B. Wilson's classic *The Cell in Development and Heredity* will come to have a new meaning – one that would have been applauded enthusiastically by Wilson himself, and, with only a touch of scepticism, by his friend and irreverent critic, Thomas Hunt Morgan.

# References

Allen, Garland E. (1966a). Thomas Hunt Morgan and the problem of sex determination. *Proc. American Philosophical Society* **110**, 48–57.

Allen, Garland E. (1966b). T. H. Morgan and the emergence of a new American biology. *Quarterly Review of Biology* **44**, 168–88.

Allen, Garland E. (1974). Opposition to the Mendelian-chromosome theory: the physiological and developmental genetics of Richard Goldschmidt. *Journal of the History of Biology* **7**, 49–92.

Allen, Garland E. (1975). The introduction of *Drosophila* into the study of heredity and evolution, 1900–1910. *Isis* **66**, 322–33.

Allen, Garland E. (1978a). *Life Science in the Twentieth Century*. New York: Cambridge University Press.

Allen, Garland E. (1978b). *Thomas Hunt Morgan: the Man and His Science*. Princeton, N. J.: Princeton University Press.

Allen, Garland E. (1983). T. H. Morgan and the influence of mechanistic materialism on the development of the gene concept, 1910–1930. *American Zoologist* **23**, 829–43.

Bateson, William (1894). *Materials for the Study of Variation*. London: MacMillan and Co.

Bateson, William (1914). Address of the President of the British Association for the Advancement of Science. *Science* **40**, 287–302.

Bourdieu, Pierre (1975). The specificity of the scientific field and the social conditions of the progress of reason. *Social Science Information* **6**, 19–47.

Brooks, W. K. (1900). The lesson on the life of Huxley. In *Smithsonian Institution Annual Report, 1900*, pp. 700–11. Washington D.C.: Government Printing Office.

Churchill, Frederick (1974). William [sic] Johannsen and the genotype concept. *Journal of the History of Biology* **7**, 5–30.

Coleman, William (1970). Bateson and chromosomes: conservative thought in science. *Centaurus* **15**, 228–314.

Conklin, Edwin Grant (1908). The mechanism of heredity. *Science* **27**, 89–99.

Dalcq, A (1951). Le problem de l'Evolution, est-il pres d'être resolu? *Annales de la Société Royale zoologique de Belgique* **82**, 117–38.

Fleming, Donald (1964). Introduction to *The Mechanistic Conception of Life* by Jacques Loeb. Cambridge, Massachusetts. Harvard University Press, preprint of the 1911 volume.

Gilbert, Scott (1978). Embryological origins of the gene theory. *Journal of the History of Biology* **11**, 307–51.

Hamburger, Viktor (1980). Embryology and the modern synthesis in evolutionary theory. In Ernst Mayr & William Provine (ed.), *The Evolutionary Synthesis*. Cambridge, Mass.: Harvard University Press.

Harrison, Ross G. (1937). Embryology and its relations. *Science* **85**, 369–74.

Johannsen, Wilhelm (1909). *Elemente der exakten Erblichkeitslehre*. Jena: Gustav Fisher.

Johannsen, Wilhelm (1911). The genotype conception of heredity. *American Naturalist* **45**, 129–59.

Lakatos, Imre (1970). Falsification and the methodology of scientific research programmes. In Irmre Lakatos & Alan Musgrave (ed.), *Criticism and the Growth of Knowledge*, pp. 91–196. Cambridge, England: Cambridge University Press.

Lillie, Frank R. (1927). The gene and the ontogenetic process. *Science* **66**, 361–8.

Loeb, Jacques (1912). *The Mechanistic Conception of Life*. Cambridge, Mass.: Harvard University Press (Reprint, 1964).

Loeb, Jacques (1916). *The Organism as a Whole*. New York: G. P. Putnam.

McCullough, Dennis M. (1969). W. K. Brooks' role in the history of American biology. *Journal of the History of Biology* **2**, 411–38.

Morgan, T. H. (1891). A contribution to the embryology and phylogeny of the Pycnogonids. *Studies from the Biological Laboratory, Johns Hopkins University* **5**, No. 1: 1–76.

Morgan, T. H. (1903). *Evolution and Adaptation*. New York: Macmillan & Co.

Morgan, T. H. (1909). Recent experiments in the inheritance of coat colors in mice. *American Naturalist* **43**, 494–510.

Morgan, T. H. (1910). Chromosome and heredity. *American Naturalist* **44**, 449–96.

Morgan, T. H. (1911). The application of the conception of pure lines to sex-limited inheritance and to sexual dimorphism. *American Naturalist* **45**, 65–78.

Morgan, T. H. (1919). *The Physical Basis of Heredity*, New York; Lippincott.

Morgan, T. H. (1926). *The Theory of the Gene*. New Haven: Yale University Press.

Morgan, T. H. (1934). *Embryology and Genetics*. New York: Columbia University Press.

Morgan, T. H. (1935). The relation of genetics to physiology and medicine. *Scientific Monthly* **41**, 5–18.

Morgan, T. H., Sturtevant, A. H., Muller, H. J. & Bridges, C. B. (1915). *The Mechanism of Mendelian Heredity*. New York: Henry Holt.

Rosenberg, Charles. (1976a). Science, technology, and economic growth: the case of the agricultural experiment station scientist, 1875–1914. In *No Other Gods*, pp. 153–72. Baltimore: Johns Hopkins University Press.

Rosenberg, Charles. (1976b). The social environment of scientific innovation: factors in the development of genetics in the United States. In *No Other Gods*, pp. 196–209. Baltimore: Johns Hopkins University Press.

Sapp, Jan (1982). The field of heredity and the struggle for authority, 1900–1931: some new perspectives on the rise of genetics. Unpublished paper. Quoted with permission.

Sapp, Jan (1984). *Cytoplasmic Inheritance and the Struggle for Authority in the Field of Heredity, 1891–1981* (Montreal: Institut d'Histoire et de Sociopolitique des Sciences, unpublished Ph.D. dissertation).

Sonneborn, Tracy M. (1978). *My Intellectual History in Relation to My Contributions to Science*. Unpublished autobiography, Lilly Library Archives, Indiana University.

Spemann, Hans. (1924). *Vererburg und Entwicklungsmechanik*. Leipzig: Akademische Verlagsanstalt.

Weiss, P. (1968). *Dynamics of Development: Experiments and Inferences*. New York: Academic Press.

Wilson, Edmund B. (1925). *The Cell in Development & Heredity*, 3rd edn. New York: Macmillan.

Wilson, James. (1910). The new magazine has a place. *American Breeders Magazine* **1**, 3–5.

Winge, O. (1958). Wilhelm Johannsen: the creator of the terms gene, genotype, phenotype and pure line. *Journal of Heredity* **49**, 82–8.

# Experimental technique in the rise of American embryology

'Today' wrote Morgan in 1927 (*Experimental Embryology*, p. 1) 'the need (of the method of experiment in problems of development) seems so obvious that we are apt to forget the strong opposition that the movement at first met from embryologists of the old school'. The following paper addresses this aspect of the 'Woods Hole School'. Harrison's break from the constraints of the old school is of particular interest because, even more than Morgan, his approach to embryology was dominated by scientific method, leading in the case of his use of the technique of heteroplastic grafting or tissue culture to experiments of unparalleled elegance. Such is the importance of his methods, it is sometimes difficult to see how they related to the long-term, underlying objectives of Harrison's embryology.

Although his first experience in research involved lineage tracing in the oyster at Woods Hole in 1890 and he was awarded a Ph.D. at Johns Hopkins in 1894, Harrison only spent time in Woods Hole on one occasion (1896) in the subsequent thirteen years. His thesis work was carried out in Bonn (1892–4) where he subsequently undertook medical training. His choice of the development of paired fins for study under Moritz Nussbaum clearly anticipates a theme running throughout his later work; an interest in individual organs, in bilateral symmetry and the implications of these for underlying organising mechanisms. Dr Witkowski's paper shows how the work on neuron theory evolved gradually from that on the fin. It is worth recalling the context of his interest in this subject; the neuron theory in this period formed part of controversies over the 'inadequacy of the cell theory' which itself was an issue of fundamental concern for more obviously embryological matters because on it depended the choice between cell-based and organismic approaches to the origin of differentiated

147

structure. The processes of neurons obviously represented fertile material for those seeking evidence either on the universality of the cell theory or on syncytial and non-cellular components or on levels of anatomical organisation extending well beyond the dimensions of single cell bodies. Besides this the nervous system already attracted a special fascination of its own. Sherrington's *Integrative Action of the Nervous System* (1906) hinted at its role as overall controller not only of behaviour, but also of structure. With growing interest in the role of stimuli and response in tropic and reflex situations, similar concepts were being carried directly over into the realm of development, as we have already seen. Thus Davenport could write (1895, *Bulletin Harvard Museum* **27**, 173) in 1895, 'It is a highly probable belief that no movement takes place in protoplasm except as a response to stimuli. The very fact that ontogenesis is a complex of actions indicates that there must be a large number of stimuli raining in upon the different parts of the developing protoplasm to which they respond'. At the same time Herbst and others were demonstrating situations (e.g. regrowth of ablated insect antennae) in which nervous tissue appeared to be necessary for developmental events such as regeneration and heteromorphosis. Both Loeb and Morgan took up this work after 1900. The continuity underlying Harrison's work is emphasised by the way in which he returned to his work on paired organs in his study of limb polarisation from about 1915. A quotation from a paper in 1898 could well stand as a description of the work he was to do forty years later in attempts to find polar molecules in limb buds using X-ray crystallography (1940); referring to the concept of polarity he says 'Fundamentally, a purely geometrical conception, it signifies more when used by the morphologist, implying not only symmetry, but also an internal cause for that symmetry, by virtue of which every particle of the organism has the same polar relations as the whole'. (*Wilhelm Roux Archiv Entwicklungsmechanik*, 1898, **7**, 469)

Harrison's attitude to, and employment of, the experimental method is significant, not only for its own interest, but also for the insights it gives us about beliefs and priorities shaping early American embryology. Is it not important that perhaps the most 'classic' of American embryologists should be best known for work which, as seen today, is only obliquely embryological, namely the neuron studies and tissue culture? Can we make out here another reflection – paralleling Morgan – of an American preference for tangible results and the prospects of practical applicability?

# Ross Harrison and the experimental analysis of nerve growth: the origins of tissue culture

J. A. WITKOWSKI

Muscle Research Centre, Department of Paediatrics, Royal Postgraduate Medical School, London W12 0HS, UK

## Introduction

In her opening address, Dr Oppenheimer referred briefly to what must be one of the most fundamental assumptions in embryology, that is the assumption that experimentation is a valid procedure for studying the developing embryo. While the emphasis placed on the differing roles of observation and experimentation in pre- and post-1880 embryological research may obscure other equally important differences (Oppenheimer, 1957), it is clear that experimentation was an issue of great theoretical and practical importance in embryology at the turn of the century. I would like to illustrate what it was like to work in embryology at this time by describing Ross Harrison's development of tissue culture. (For biographies of Harrison see Abercrombie, 1961; Nicholas, 1961; Wilens, 1969; and for extended discussions of his scientific work see Oppenheimer, 1966, 1971; Haraway, 1976).

Harrison's early career spans the period of biological research in the USA (1890–1920) that has recently become the subject of controversy (Allen, 1978, 1979, 1981; Maienschein, Rainger & Benson, 1981). Harrison plays a key role in this debate as one of the 'Johns Hopkins Four' (together with T. H. Morgan, E. B. Wilson and E. G. Conklin) and his changing research program has been analysed by Maienschein (1981, 1983). It could be argued that the discussion of the work and career of only one scientist cannot give a full picture of the changing nature of the research strategies of the time. However, Harrison was undoubtedly one of the major figures in the field of embryology, and I think that there are two special reasons for choosing his work as a case study to illustrate this period at the turn of the century. Firstly, Harrison's success at in vitro culture justifies description as an historic event in its own right, both for its achievements at the time and for its subsequent importance in all research concerned with cells. Tissue culture might be considered to be the 1900s equivalent of the monoclonal antibody and DNA recombinant techniques of the 1970s and 1980s. Secondly, throughout his career spanning 50 years, Harrison extolled and defended the use of experimentation in embryology and in biology in general. I believe his discussions of the role of experimentation illustrate clearly the tensions that arose as the principle that only observation could provide valid data on normal embryonic development was supplanted by the belief and later the tacit assumption that experimental analysis was equally valid.

Harrison's use of experimentation and enthusiasm for an experimental approach can be traced through a series of progressively more radical experiments on the embryo beginning in 1898 to the publication of his definitive paper on tissue culture in 1910. These experiments were concerned with the problem of nerve fibre development, then a highly controversial subject. In this paper, I shall describe these controversies and Harrison's work related to them that led to tissue culture. I shall then discuss current arguments on changes in American biology at the turn of the century before returning to Harrison's writings on the use of experimentation in biology.

## Nerve fibre development – the theories

There were three main theories of nerve development current at the time Harrison began his research. (For a full discussion see Billings, 1971). The first theory, that nerve fibres had a *multicellular* origin, was due initially to Schwann (1839) and was subsequently taken up and modified by various workers. Nerve fibres were supposed to arise by fusion of chains of cells, or from the secretions of such cells, or by formation within these cells. The second theory was proposed by Hensen (1864) who believed that cell division did not go to completion during embryonic development and that the daughter cells remained in *protoplasmic continuity*. The reduction of this random network of connections to the specific pattern of the nervous system came about as a result of the atrophy of those pathways that were not used. This was a particularly attractive feature of this theory. The third of the theories was proposed by His (1886) who suggested that the nerve fibre arose as the *outgrowth from a single nerve cell* and that no other cytoplasmic material was involved in its production. Earlier workers had proposed similar theories but this theory became associated particularly with Ramon y Cajal because of his championship of it (Cajal, 1907, 1908; see also Billings, 1971, and Hamburger, 1980). A problem remained to account for the development of specific pathways, and Held attempted to provide a solution by bringing together features of the outgrowth and protoplasmic bridge theories. He suggested that nerve fibres arose as cellular outgrowths but that these moved within the protoplasmic bridges (Held, 1907, 1909).

## Nerve fibre development – the controversies

These then were the three theories that had been proposed over a period of 60 years and they or variations of them were still hotly debated at the turn of the century. It is not clear why this topic was so controversial. Billings has discussed this in relation to the Cell Theory and points out that it was readily accepted that most tissues were composed of discrete cells. Both Schwann and Hensen had suggested that the nervous system was reticular, and this set the nervous system aside as something quite exceptional. Its development might require the operation of factors not involved in tissues in general. Hamburger (1980, p. 600) referred to this as '. . . a fallacy that hindered all progress in neurology'. It also constituted a challenge to the Cell Theory that had proved so valuable; Wilson (1901) described it as one of the 'great lines of progress', and Harrison (1937, p. 372) went so far as to say that 'The reference of developmental processes to the cell was the most important step ever taken in embryology'. From this viewpoint, the reticular theory of the nervous system would be regarded as a challenge. It also seems from their writings that the participants in the controversy relished the debate, although by and large it seems to have been conducted amicably. (Rivals inspected each others' preparations, Cajal and Marinesco visiting Held and Golgi respectively. The notorious exception was Golgi's speeches on the occasion of the award of the Nobel Prize to Golgi and Cajal.)

Nevertheless, the problem was precisely that the participants in the controversy were obliged to *debate* them. Despite the tremendous improvements in histological techniques and microscope performance in the nineteenth century, data obtained by observation of fixed, sectioned and stained preparations were inadequate; they could not decide in favour of one of the theories just described. An idea of the vigour of this debate can be gained from Harrison's listing of some of the participants between 1890 and 1920; there were 15 supporters of the outgrowth theory who published 27 papers in this period and 16 advocates of the multicellular theory who contributed 22 papers to the debate. In addition, there were two workers (Braus and Dohrn) who modified their original stand and moved at least partly to the outgrowth camp (Harrison, 1924). (Detailed contemporary reviews will be found in Marinesco (1909) and Neal (1914); Billings (1971) gives a more considered analysis of the controversy; an interesting sociological analysis of the use scientists make of lists of the kind given by Harrison is provided by Gilbert & Mulkay, 1984). Ramon y Cajal was the central figure throughout this debate and in his preface to Marinesco's book he wrote '. . . qu'on assiste aux peripipéties intéressants du combat renouvele neuronistes et antineuronistes'. Cajal pointed out that these '. . . controverses interminable . . .' had in fact stimulated '. . . l'invention de méthodes nouvelles de plus en plus en plus exactes' but the results of each new advance in histological technique became in turn the subject of '. . . polémiques violentes . . .'. (Cajal in Marinesco, 1909, p. xii). Harrison was well aware of these problems: '. . . it seems that we are really very far from a satisfactory solution of the question, which even the invention of new and marvelously refined histological methods has failed to bring to a final settlement' (Harrison, 1910*b*, p. 788). W. H. Lewis remarked that 'The controversy over the origin and development of the peripheral nerves has of late years assumed considerable importance, in that vigorous attacks have been made on the neurone doctrine as formulated by His . . .'. (Lewis, 1907, p. 461). This controversy reached a climax in the debate between Ramon y Cajal and Hans Held in the period 1900–9.

Cajal was, of course, the outstanding microscopist in this field and he applied his modification of Golgi's silver staining method to studies of the developing nervous system of the chick embryo. (Cajal's papers are available in translations by Guth, 1960.) Cajal's preparations showed a complex structure that he called the growth cone at the tips of the fibres (Cajal, 1890) and he was struck by the similarity between the growth cone and the terminal differentiation of a nerve at its end organ. The demonstration of the growth cone was taken as strong evidence in favour of the outgrowth theory.

However, supporters of the other theories were able to obtain histological preparations that were convincing evidence to them of the
truth of their own views. For example, Held had used Cajal's
modified Golgi stain technique so that their results were strictly
comparable, and had published photomicrographs instead of
engravings, yet he believed that these new preparations supported
his previous views (Held, 1907). Cajal was able to examine the
preparations themselves on a visit to Germany in 1907. In a paper
published that year, he wrote that he was astonished to find that
Held's preparations were to all intents and purposes identical with
his own! (Cajal, 1908, p. 3.) Marinesco had a similar experience
when he examined preparations made by Golgi (Marinesco, 1909,
pp. 15–16). It was clear, Harrison wrote, that the evidence rested on
'. . . such minute histological details that a decision to which all
would subscribe was impossible of attainment' (Harrison, 1910*b*,
p. 789).

It was apparent that the problem was one of interpretation of what
was seen down the microscope and it demonstrates the extent to
which what are taken as facts can be determined by the hopes and
expectations of the observers. As Cajal pointed out in 1909: 'Dans
les autres sciences, on discute seulement les théories et les
hypothèses; en histologie, on discute aussi bien les faits que les
théories'. Cajal continued, perhaps with a touch of bitterness: 'c'est
pour cette raison que, dans notre domaine, il est si difficile de triom-
pher'. (Cajal in Marinesco, 1909, p. xiii). Harrison also recognised
that the problem was one of interpretation: 'When one compares the
careful analyses of their observations, as given by various authors,
one cannot but be convinced of the futility of trying by this method
to satisfy everyone that any particular view is correct' (Harrison,
1906, p. 121). He had no doubts about the way out of this impasse:
'The only hope of settling these problems definitely lies, therefore,
in experimentation' (Harrison, 1906, p. 121) because 'It is generally
recognized in science that the experimental method . . . is a vastly
more efficient means of analysis than the method of merely observing
phenomena as nature presents them to us' (Harrison, 1908, p. 385).

## Heteroplastic grafting

In 1892, Harrison went to Bonn to begin work for an M.D. degree in
the laboratory of Moritz Nussbaum. Both actions – studying in

Germany and taking a medical degree – set Harrison apart from his contemporaries. The effect of this background on Harrison has been discussed by Maienschein (1983) and Oppenheimer (1966) has referred to Nussbaum's influence on Harrison's views on polarity. Harrison had a high regard for Nussbaum, and Nussbaum's research interests are reflected in Harrison's own work.

Harrison's first major experimental paper was published in 1898 in Roux's *Archiv fur Entwicklungsmechanik der Organismen* (an English translation was published in 1899 in the *Bulletin of the Johns Hopkins Hospital*). In 1897 Born had made the serendipitous discovery that fragments of young amphibian larvae could fuse together (Born, 1897). Harrison took up this new technique and developed it further by fusing portions of larvae of species differing in pigmentation. In this way he was able to determine the contributions of host and donor cells to various tissues and the method became known as heteroplastic grafting (Harrison, 1935). As Maienschein (1983) says, at first sight this paper seems to be unexciting – simply the application of a technique in a variety of seemingly unrelated ways. But I believe it is a very important guide to Harrison's subsequent work, and I see it as a paper in which Harrison, as a young research scientist, demonstrated both his skills as an experimenter and the versatility of a new and remarkable experimental technique. It is possible to discern in this paper the embryological problems that were to concern him throughout his life. These were: (1) the problem of nerve fibre development (Harrison, 1910 *a*, *b*, 1924); (2) the question of the origins of polarity and symmetry in the developing embryo (Harrison, 1936); and (3) the study of growth (Harrison, 1935).

Harrison devoted a large part of this paper to observations on the distributions of nerve fibres in tadpoles in which the tails had been transplanted. He observed that the epidermis of the head portion moved back over the tail in a manner that '. . . corresponds in direction and relative amount in the different regions with the displacement of the sensory area as compared with the motor belt of the same segmental nerve' (Harrison, 1899, p. 178). This observation suggested that the connection between '. . . each ganglion cell and its end organs in the integument is established early in development' (Harrison, 1899, p. 178) and that the nerve fibre was drawn out as the epidermis moved over the tail. In a footnote, Harrison wrote that his results supported some of Hensen's views, but they were not to be taken as evidence against the outgrowth theory '. . . at present almost universally accepted' (Harrison, 1899, p. 178).

It does not seem profitable to try to determine why Harrison turned first to studies of nerve fibre development and only later took up research on the origins of polarity in the developing embryo. Nussbaum had published studies both on polarity in *Hydra* and nerve and muscle development, and Harrison cited these papers in his own work. It is probable that the controversies surrounding research on the nerve fibre stimulated Harrison to begin with that, and it was perhaps more directly amenable to experimental attack. (The most authoritative accounts and analyses of Harrison's early work are to be found in essays published by Oppenheimer in 1966 and 1971.)

# Development of the musculature

In 1904, Harrison published a paper describing the role of innervation in muscle development. This paper is important because it illustrates clearly Harrison's experimental approach to controversial problems. The subject was in similar state to that of nerve development: '. . . the data are as yet of such a varied and conflicting nature as to preclude the possibility of satisfactory generalizations' (Harrison, 1904, p. 197) but Harrison hoped that '. . . through the study of comparatively simple particular problems we may advance towards some general conclusion' (Harrison, 1904, p. 197). He went on to review critically previous investigations and concluded that studies such as those of acephalic foetuses were unsatisfactory. The only way to proceed was by '. . . direct experimentation, but in devising experiments for this purpose it is necessary to formulate clearly just what is to be determined . . .' (Harrison, 1904, p. 199).

Harrison decided that there were two questions to answer: firstly was a stimulus from the nervous system necessary to initiate differentiation? Harrison examined this by removing a strip of tissue containing the cord with neural crest and a dorsal portion of the myotomes, and the dorsal fin fold. The embryos developed quite well despite this drastic operation and although the spinal cord was entirely absent, in the best specimens it was quite clear that the muscle had differentiated quite normally.

The second question was whether muscles had to be exercised to differentiate properly. Harrison left embryos intact but prevented muscle contraction by raising the tadpoles in water containing chloroform. Tadpoles could be maintained totally immobilised for up to seven days but again normal muscle differentiation took place. These experiments did not contribute directly to the question of nerve fibre growth except that in cases where the dorsal strip was not removed entirely, nerves were found in the body only at the head end. This suggested that nerves could not arise without the cord.

## Development of peripheral nerves *in vivo*

Harrison realised the importance of the step he had taken in using experimental techniques to unravel the complexities of the development of the neuromuscular system and in 1906 he dealt directly with the problem of the development of nerve fibres. I believe that this paper is a classic example of the application of the experimental approach to an embryological problem, and that it illustrates clearly Harrison's genius at recognising crucial questions and designing simple experiments to answer them. He began the paper by setting out two such critical questions.

Firstly, is the nerve fibre the process of a single cell or derived from a chain of cells? Harrison cut off a thin strip from the dorsal part of the cord, and so removed the sources of Schwann cells and the spinal ganglia. In these embryos, only motor nerves developed and these were without Schwann cells. The converse experiment was more difficult, requiring the removal of the ventral half of the cord and then allowing the remaining dorsal strip to heal back again. In the successful cases, sensory nerves with Schwann cells developed but motor nerves were entirely absent.

The second question dealt with how the connection between neuroblast and end organ arose: what were the relative contributions of nerve fibre outgrowth and protoplasmic bridges? Harrison proposed tackling this question by performing two types of experiment. In the first, he would remove the source of the neuroblasts to determine whether protoplasmic bridges alone could give rise to nerve fibres. In the second set of experiments, the tissue in which nerve fibres were to develop would be altered to determine whether neuroblasts could give rise to nerve fibres even where the normal pattern of protoplasmic bridges had been disturbed. The experiments gave strong support to the outgrowth theory. In the absence of cord no fibres developed indicating that protoplasmic bridges (if present) could not by themselves form nerve fibres. However nerve fibres could be found growing out of the brain into the mesenchymal tissue that filled space left by the cord, '. . . a tissue as unlike that forming the normal path [of nerve fibres] as it could possibly be' (Harrison, 1906, p. 129), indicating that nerve fibres could develop even where protoplasmic bridges had been disrupted. And finally, fragments of spinal cord transplanted to the abdominal wall also gave rise to nerve fibres, in even more unlikely surroundings. Harrison concluded that these results could be interpreted in only one way: 'The nerve center (ganglion cells) is shown to be the one necessary factor in the formation of the peripheral nerve' (Harrison, 1906, p. 129) and '. . . the nerve fiber is the outgrowth of a single ganglion cell, with which it remains in continuity throughout life' (Harrison, 1906, p. 131).

## Unsolved problems

By this stage then, Harrison had established that:

(1) Schwann cells were unnecessary for nerve fibre development.
(2) The role of protoplasmic bridges was in doubt – nerve fibres could develop in abnormal situations.
(3) Neuroblasts in the spinal cord appeared to be essential.

However, several problems remained. An objection could be raised against Harrison's experiments demonstrating, as he saw it, nerve fibre outgrowth into abnormal surroundings. Protoplasmic bridges were supposed to be ubiquitous and so nerve fibres might be formed from such bridges wherever they were. Furthermore, Harrison's experiments had not suggested any mechanism by which a nerve fibre could grow out from the cord, nor how the nerve fibre was guided to its destination. In 1907, Harrison published a long paper dealing with the first of these problems (Harrison, 1907a), and also a short note summarising various experiments dealing with the latter problems (Harrison, 1907b). The work described in this preliminary note of 1907 was reported in detail in two papers in 1910 and it is to these papers (Harrison, 1910a,b) that I shall refer.

## Limb transplantation

The first of these problems arose in an acute form as a consequence of a paper published by Braus in 1905. Braus performed a series of ingenious transplantation experiments using limbs of amphibian larvae that he believed supported the protoplasmic bridge theory (Braus, 1905). For example, Braus had transplanted forelimbs to various regions of the host and had found a normal distribution of nerves within the limb, although the source of the nerves might be quite abnormal for a limb. Braus believed that it was the presence of protoplasmic bridges within the transplanted limb that gave rise to the correct forelimb nerve pattern no matter where the limb was transplanted.

In analysing Braus' paper, Harrison found (1907a) that the rigorousness that he expected of his own work was absent from that of Braus. Ten pages of Harrison's paper were devoted to a critical analysis of Braus' results and he concluded that Braus' arguments did '. . . not constitute a logical proof of the continuity theory . . .'. Furthermore, 'The experiments do not approach the problem directly enough . . . and there are too many loopholes left to permit of a rigid proof' (Harrison, 1907a, p. 241). Harrison pointed out that the result of the experiment I have just quoted had another equally plausible explanation. On the basis of the outgrowth theory, there could be structures within the forelimb that guided any nerve fibres entering the limb into the appropriate pattern. (It should be noted that in his discussion of the guidance of nerve fibres growing into the transplanted limb, Harrison again referred to Nussbaum's work on the courses taken by nerves in muscle.)

I want to describe in this paper only one of the experiments that Harrison reported, an experiment designed to test the dependence of the already formed nerve fibre on its continuity with its cell body. Harrison first produced nerve-free tadpoles by extirpating the cord from embryos and then because these were unable to feed, they were joined in parabiosis with normal tadpoles. An already innervated limb was then transplanted to the nerve-free tadpole and the survival of nerves in the limb followed. Not surprisingly, at least to us, the nerves degenerated; protoplasmic bridges, if present, and Schwann cells were unable to sustain even a fully formed nerve fibre once its connection with the cell body had been severed.

## *In vivo* culture

In his paper 'The development of peripheral nerve fibers in altered surroundings', Harrison reviewed his previous work, for example the transplantation experiments, and reported one new experiment (Harrison, 1910*a*). The importance of this experiment is that Harrison himself referred to it as forming '. . . a link, so to speak, between those performed on embryos and those in which living fibers were observed growing in clotted lymph' (Harrison, 1910*a*, p. 16). Harrison had concluded that the best way to deal with the omnipresent protoplasmic bridges was to remove all traces of them. To do this Harrison not only removed the medullary cord from an embryo, but he replaced it with a cylindrical clot of adult frog blood that had been formed in a capillary tube. In six out of eight cases, he found that nerve fibres had grown from the brain and penetrated the blood clot, for as far as 170 $\mu$m. It was clear that nerve fibres could appear in cell-free and therefore protoplasmic bridge-free surroundings and Harrison concluded that '. . . these experiments must be regarded as decisive against the protoplasmic bridge theory' (Harrison, 1910*a*, p. 30). It should be pointed out that these experiments resemble those performed earlier by Leo Loeb. Loeb had originally observed that cells in regenerating wounds would penetrate a blood clot, and, in 1901, he embedded a piece of tissue in a blood clot that he then implanted in an animal. Using this animal 'incubator', he found that the epithelium continued to grow and that there were even single cells detached from the main mass of tissue (Loeb, 1901). There is, however, no evidence to suggest that Harrison followed Loeb's example (Witkowski, 1983).

## Tissue culture

The second problem – that of the mechanism of nerve outgrowth – was tackled in Harrison's great paper 'The outgrowth of the nerve fiber as a mode of protoplasmic movement' (Harrison, 1910*b*). As Harrison pointed out in his introduction, one of the major advances made by Cajal in relation to the outgrowth theory was his discovery of the growth cone. In a famous passage, Cajal described it as '. . . a sort of battering club or battering ram, endowed with exquisite chemical sensitiveness, with rapid amoeboid movements, and with a certain impulsive force, thanks to which it is able to press forward and overcome obstacles met in its way . . .' (Cajal, 1909, p. 599). It should be remembered that despite the eloquence of Cajal's description of the activities of the growth cone, it was only surmise based on the observation of fixed and stained material. For Harrison's work to be complete, he had to examine this crucial part of the outgrowth theory and this was declared in the first sentence of his 1907 note: 'The immediate object of the following experiments was to obtain a method by which the end of a growing nerve could be brought under direct observation while alive...' (Harrison, 1907*b*, p. 116).

Harrison's first attempts were failures. Tissues did not survive in simple saline solutions and did not grow when transplanted to cavities within the body such as the ventricles of the brain. However, Harrison noticed that if fragments of neural tissue adhered to the walls of the ventricles, nerve fibres grew out and penetrated the walls. Acting on the assumption that cell movement could not occur in a liquid medium, Harrison embedded fragments of cord in gelatin or clots of adult frog lymph. The former failed, but the latter succeeded in what can only be described as a spectacular fashion. Harrison succeeded because he recognised the need for a more natural medium than physiological saline and because he went to great lengths to work aseptically.

I think that this paper can be best discussed by reference to just a part of one of the plates illustrating it. The relevant part is reproduced here (Fig. 1) and consists of a series of beautiful drawings of a group of cells made over a period of 34 h. Harrison drew three important conclusions from specimens such as these. Firstly, and most importantly for the further use of tissue culture, Harrison concluded that '... the mode of procedure employed in the experiments permits the characteristic differentiation of various tissues to take place' (Harrison, 1910*b*, p. 812). He observed fibroblasts and pigment, epithelial and muscle cells, and he was able to observe the general morphology of the nerve fibre and its elongation. Secondly, he was able to show clearly that the nerve fibre arose from a single cell and did not require chains of cells or protoplasmic bridges. Thirdly, he was able to study the tip of the nerve fibre in detail and demonstrate its complete identity with Cajal's growth cone. He was also able to observe those amoeboid movements of the tip that Cajal had inferred from his histological preparations.

It seems clear that Harrison was particularly impressed that the outgrowth of the nerve fibre was '... but a specific form of that general type of movement common to all primitive protoplasm' (Harrison, 1910*b*, p. 840). As he remarked on another occasion, there was 'nothing mysterious, nothing hypothetical' about the formation of the nerve fibre; there was no need to '... postulate the

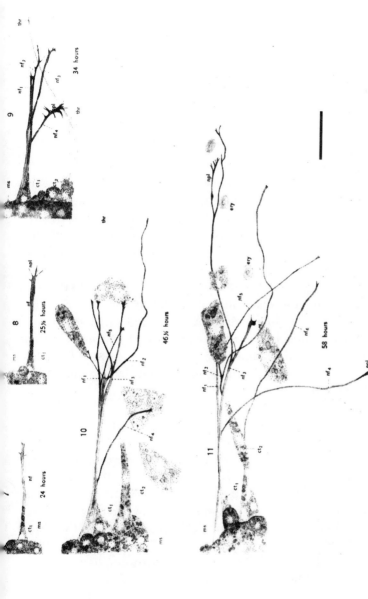

Fig. 1. This figure is part of plate 2 of Harrison (1910b) and shows the development of a small fragment of medullary cord tissue from a *Rana palustris* embryo, 3·3 mm long, growing in a clot of lymph from an adult *Rana pipiens*. The scale marking is 50 μm. The figure shows (1) that nerve cells differentiate in vitro, (2) the similarities between the tips of these fibres and the growth cones described by Cajal, and (3) the amoeboid-like changes of the nerve fibre ends. (Reproduced by permission of Alan R. Liss, Inc.)

existence of invisible and otherwise unknown structures' like protoplasmic bridges (Harrison, 1907a, p. 277). Development of the nervous system was 'no less capable of rational analysis than is development in general' (Harrison, 1910b, p. 841).

I think that there can be little doubt that the dramatic nature of the cell differentiation that Harrison obtained, demonstrated the power of tissue culture as an experimental technique and established his position as its 'founder'.

## Cellular stereotropism

Harrison had discussed the third problem – that of how nerve fibres found their way to the appropriate end organs – at some length in his 1910 paper (Harrison, 1910b, pp. 835–40). It was clear that the directions taken by nerves were profoundly influenced by the surfaces of and the spaces between the tissues surrounding them. In many cases the nerve pathways were laid down very early in development when the distances to be travelled by the nerve tips were small. He suggested that one or two fibres – 'pathfinders' as Harrison called them – made the initial contact and then acted as guides for later fibres to follow. The final connection between nerve tip and end organ must be a 'sort of specific reaction' and Harrison drew an analogy with the sperm–egg interaction. It was only at this stage that he believed that the chemotactic mechanism postulated by Cajal became important.

Leo Loeb had recognised stereotropism as early as 1898 but had not studied the phenomenon systematically (Loeb, 1898). Harrison investigated the mechanical influences on cell movement in his last experimental paper on tissue culture published in 1914 (Harrison, 1914). He performed a series of experiments in which he varied the three factors in tissue culture: the tissue, the medium and the solid support. He demonstrated that a solid support was required for cell movement and in a remarkable tour-de-force showed how the fibres of spiders' webs guided cell movements. He compared this with the way in which cells in the embryo were closely applied to blood vessels and nerve fibres. However, cultures using nervous tissue were largely unsuccessful and Harrison was unable to comment specifically on factors guiding the nerve fibre tip.

# Harrison's later work

With the exception of a paper published in 1924 that was mainly a review of evidence against the multicellular theory (Harrison, 1924), this paper on stereotropism completed the series of experiments on the outgrowth of the nerve fibre. Harrison did not pursue the problem of how the nerve pathways arose; he may have recognised the difficulties involved and in the 1910 paper he wrote of the failure of the few experiments he had attempted, referring to the 'crudities' of the methods he had used (Harrison, 1910b, p. 840).

It should not be surprising that Harrison was not inclined to continue using the technique of tissue culture. Techniques played a very important role in Harrison's work (heteroplastic grafting, tissue culture), but only in relation to particular problems that interested him. Harrison was first and foremost an embryologist and the mere novelty of a technique did not deflect him from that; the application of tissue culture to other fields did not interest him. He recognised the importance of tissue culture in studies of cell differentiation– his own work on the nerve fibre had demonstrated that – but this was not the problem that concerned him. Instead, he turned again to the whole embryo and followed up another of the interests shown in his 1898 paper. He began work on the nature and development of polarity and symmetry during growth (Harrison, 1921, 1936; Haraway, 1976; Witkowski, 1980a), and as tissue culture was inapplicable to this type of work, Harrison quite simply stopped using it. He had noted in 1913 that 'While work with these methods [tissue culture] has been successful in respect to the study of cell differentiation, cell movement and cell division, it has failed to be of use in the study of the gross form of organs' (Harrison, 1913b, p. 71).

# Impact of Harrison's tissue culture

Harrison never published experimental work on tissue culture again, but his success stimulated further development of his methods and their application in almost every field of medical and biological research. I have tried to quantify this by counting the numbers of references to tissue culture that appeared in the *Index-Catalogue of the US Surgeon-General's Office* for 1913 and in Albert Fischer's book, *Tissue Culture*, published in 1925 (Fig. 2). The former is the forerunner of *Index Medicus* and the latter was the first comprehensive text on tissue culture. I have included all the papers in these bibliographies and these include a proportion dealing with topics unconcerned with tissue culture, for example with biochemical methods. I do not think that these distort the figures significantly except for the pre-1910 period in Fischer's bibliography. Of the 31 papers cited in this period, only three are cited in the *Bibliography of Tissue Culture Research* (Murray & Kopech, 1953). One of these is the 1907 report by Harrison of his tissue culture findings and the other two are by Leo Loeb. Of the latter, one does not deal with culture in vitro (Loeb, 1901) and while the other (Loeb, 1897) has been cited as a forerunner of tissue culture (Rubin, 1977), I have shown that this claim cannot be maintained (Witkowski, 1983).

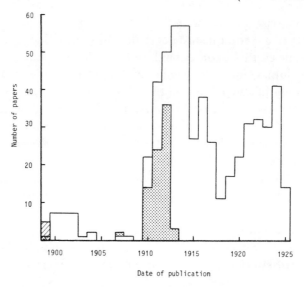

Fig. 2. Histograms of the distributions by year of tissue culture papers cited in Fischer (1925) and in the Index Catalogue of the US Surgeon-General's Office (1913) are shown open and stippled respectively. Papers published prior to 1900 are cross-hatched.

The results of these counts show clearly that tissue culture was adopted rapidly following publication of Harrison's work. An event that was particularly important in bringing tissue culture to the attention of other research workers was Harrison's Harvey Lecture given in New York in 1908. It was probably at this meeting that Alexis Carrel became aware of the potential of tissue culture and adopted and promoted its use. Of the 21 papers published in 1910, 9 were from Carrel's laboratory at the Rockefeller Institute, and of the 542 papers cited in Fischer's book, some 140 were by Carrel and his collaborators. I have discussed elsewhere the influences – both baneful and beneficial – of Carrel's work (Witkowski, 1979, 1980*b*). The rate at which the publication of tissue culture papers reached its pre-First World War peak following cessation of hostilities in 1918 demonstrates that it was already a well established technique for the study of cells.

This rapid acceptance of tissue culture seems to conflict with my discussion of the need Harrison felt to defend the use of tissue culture and experimental methods in general in biology (see below). This apparent conflict can be resolved by considering the purposes to which tissue culture was put, for it seems to me that from a very early stage two quite distinct groups of workers were involved. One group, the larger and more influential, was concerned with what might be called pathological problems and studied the growth and behaviour of cells. This group seems to have welcomed tissue culture whole-heartedly so that by 1914, Smyth was able to review its application in areas as diverse as neurobiology, cancer, bacteriology, virology, immunology, studies of cell degeneration and metabolism and the effects of agents such as snake venoms, gases and ultraviolet and X-ray irradiation. The second group followed Harrison's example more closely and applied tissue culture methods to studies more embryological in nature, in particular to cell morphology and differentiation. The work of this group is exemplified by that of Margaret and Warren Lewis (Corner, 1967) who reviewed work in this field in their 1924 article.

The existence of these groups may be inferred from addresses such as the one Harrison gave at the Congress of American Physicians and Surgeons in 1913. Harrison wrote: 'To you who are participants in the triumphs of modern medicine, which has not hesitated to avail itself of effective experimental methods, it may seem superfluous that this plea [to use such methods] is made. Unfortunately, however, experimentation does not enjoy the confidence of all embryologists and anatomists' (Harrison, 1913*b* p. 75). The contrast Harrison draws between the acceptance of experimental methods by medical research workers and their cool reception by those scientists who might have been expected to welcome it most readily is striking. Similar sentiments were expressed at about the

same time by the Lewises who wrote: 'The sooner anatomists come
to realize the limitations of the histological method and come to
utilize more and more the experimental method the better for the
science of anatomy' (Lewis & Lewis, 1912, p. 8). I suggest that
Harrison was concerned with promoting acceptance by such reac-
tionary anatomists, histologists and entomologists of this type of
experimental work when he defended the use of tissue culture.

I want now to refer briefly to current views on the changing attitudes to the use of experimental methods in embryology and biology
in general, before going on to discuss Harrison's role in promoting
their use.

## American biology at the turn of the century

Between the start of Harrison's research career (1893) and the
completion of his research on nerve development (1914), there was a
significant change in the way in which biological, and especially
embryological, research was carried out. Harrison noted this change
in his address to the Zoological section at the AAAS meeting in 1936.
Referring to the corresponding meeting of 1897, Harrison pointed
out that of the 35 papers presented at that meeting, only one was an
experimental study: 'It is safe to say that had a Rip Van Winkle gone
to sleep after that meeting and not awakened until now, he would
have scarcely understood any of the papers on this year's program'
(Harrison, 1937, p. 369).

This change to experimentally based research has been characterised as a revolt from morphological methods (Allen, 1978) or as a
contrast between naturalist and experimentalist approaches to biological research (Allen, 1979). However, this view has been challenged by others who see not an abrupt change but rather '... more
complex changes that cannot be stated in oversimplified terms as
dichotomies' (Maienschein et al. 1981, p. 86). Maienschein has supported this latter view by studies of the work of T. H. Morgan, E. B.
Wilson, E. G. Conklin and R. G. Harrison that she believes demonstrate that 'It is precisely the gradual and continual reworking of
assumptions and the concern with developing new research efforts
based on the old that characterizes turn-of-the-century embryology'
(Maienschein, 1981, p. 111). Benson in the same symposium
examined the work of W. K. Brooks, the teacher of these four, and

demonstrated how Brooks' morphological studies changed gradually and exhibited an 'increasingly sophisticated approach to embryological events' (Benson, 1981, p. 120; but see McCullough, 1969, for a different view of Brooks). As a result of these criticisms, Allen has modified his position somewhat, but still regards the changes that occurred between 1890 and 1910 as marking '... an historical discontinuity in biological research.' (Allen, 1981). Recently Maienschein (1983) has discussed Harrison's work in detail in relation to this topic, but in this article I want only to comment briefly on some general aspects of the discussion.

I do not intend to enter into a long debate on the relative merits of the points of view expressed by Allen, Maienschein, Benson and Churchill (1981) in his commentary on the formers' papers. There is complete agreement between the participants in the debate that between about 1890 and 1920 there was a major shift in what was thought to be an acceptable way of doing biological research, an 'extraordinary shift' as Churchill put it, '... towards the experimental and narrow-focus ends of the relevant continua' (Churchill, 1981, p. 182). The disagreements centre in part around the rate at which this occurred. Is it possible to decide on how radical and rapidly changing scientific attitudes and behaviour must be to be considered revolutionary rather than evolutionary? A difficulty is that any such generalisation must to some extent represent a compromise between the common factors that unite, and the idiosyncracies that distinguish, the individuals on whom the generalisation is based. Churchill made the point that it would be possible to find passages in the writings of the biologists of this period that would support revolutionary, evolutionary and revolutionary–evolutionary perspectives. I do not mean to suggest that studies of change in science are futile but only to emphasise that grand generalisations can be made only with the greatest reserve and caution. We may come to believe so strongly in our generalisations that they prejudice our outlook. Maienschein appears to recognise this (Maienschein 1981, p. 112), but in her insistence in the early parts of this paper on the evolutionary nature of change, she may be subject to the same criticism that has been made of Allen's dichotomies. Instead, detailed studies of individual case histories should be made to help in understanding overall change. Maienschein makes this point when she remarks that '... it will be useful to draw accurate descriptive pictures and rigorously illus-

trated analyses of shifting individual and then group assumptions.'
(Maienschein, 1983, p. 124).

Unfortunately, the task may be difficult to achieve for it is evident
from these papers that it is not easy to reach agreement on what
constitutes accurate description and rigorous illustration. Allen's
response to his critics is an attempt to demonstrate that the dichotomy
he observes was perceived by and influenced the behaviour of the
biologists at that time. To do this he quotes extensively from contemporary sources and comments that his critics '. . . need to look more
carefully at what people *at that time* saw the situation to be . . .' (Allen,
1981, p. 170, his italics). But Maienschein, Benson and Churchill see
dangers in this. Maienschein emphasises the need to distinguish
between the gradual changes in outlook of the pioneers in a field and
the later rapid acceptance of that field by outsiders (Maienschein
1981, p. 112). Churchill makes a related point when he refers to the
possibility of self-delusion creeping in when the pioneers make post
hoc justifications for what they did: 'Retrospective claims about a
scientist's accomplishments must be clearly distinguished from the
actual events' (Churchill, 1981, p. 178). The difficulties arise when
there are not independent sources available to enable this to be done.
In such cases, I endorse Allen's approach of using whatever contemporary documentation is available. If Harrison or Morgan or whoever
made some comment, then we have no reason to reject it if there is no
evidence to support such action.

The importance of Harrison in this discussion is underlined by
the recent paper by Maienschein (1983) dealing with various aspects
of Harrison's background and work. Maienschein is particularly
concerned with the period 1895–1910 but I believe much of interest
in Harrison's views on experimentation is to be found in his later
writings and it is these that I want now to discuss.

## Harrison's approach to experimentation

In 1904, an event of considerable significance for the status of experimental biology in the USA took place with the publication of the first
volume of the *Journal of Experimental Zoology*. The leading roles in
the formation of the journal appear to have been taken by T. H.
Morgan and E. B. Wilson, and the editorial board finally included,
in addition to Morgan and Wilson, such notables as E. G. Conklin,
Jaques Loeb, W. E. Castle and F. R. Lillie; Harrison was invited to

become managing editor. (W. K. Brooks was also on the editorial board but, according to Harrison, while Brooks was sympathetic to aims of the journal despite never having worked in the experimental field, he was not active in the affairs of the journal. Harrison, 1945, p. xxvii; see also Maienschein, this volume.) This group saw the use of experimentation together with observational studies as the only way forward for embryology. Indeed, they saw it as the only way forward for the biological sciences in general. A fascinating insight of the views of this most influential group of American biologists and of the way in which they set about exercising that influence is given in the essay that Harrison wrote on the occasion of the publication of the one-hundredth volume of the journal in 1945. Churchill's caution about the retrospective writings of eminent scientists applies here in some force, but there is no reason to believe that there is any serious distortion in Harrison's recollections.

Harrison wrote that by 1903 when Morgan and Wilson initiated events leading to the formation of the journal, an '. . . interest in experimental zoology had taken root and was rapidly growing'. He singled out experimental embryology, genetics, cytology and general and comparative physiology as areas of especial significance, and commented that these were '. . . attracting men from the more conventional fields of systematics and descriptive morphology' (Harrison, 1945, p. xiii). Those interested in these diverse areas were united in the '. . . common conviction that the application of radical experimental methods was feasible in zoology and other descriptive sciences, and that only through this course could the biological sciences expect to approach the exactness and systematic consistency of the physical sciences' (Harrison, 1945, p. xiii). The importance attached by the members of the editorial board to the experimental method can be seen in the reason given by Harrison for their decision to incorporate the word 'experimental' in the journal's title: '. . . it was recognised that the experimental method is par excellence the method of analysis and it was one of the prime objects of the Journal to go beyond purely descriptive science' (Harrison, 1945, p. xvii).

I think it is important to show that these views Harrison expressed in 1945 were in fact held by him from very early in his scientific career and are not simply the result of hindsight. I have already in the course of describing Harrison's early work quoted his remarks on experimentation that show he was well aware of the significance of his approach. Indeed, the opening paragraph of his paper first published in 1899 shows his immediate acceptance and advocacy of the experimental analysis of developmental processes:

In the method of grafting we have a means of experimentation for which no substitute is offered. Born's discovery that certain amphibian embryos lend themselves with readiness to such operations, is of especial importance in that it renders the method applicable to the study of developmental problems. How important this form of experiment is, may be well appreciated from a careful study of the paper in which Born records the results of his work in this field.

Harrison did not regard experimentation as an end in itself, or believe that the experimental method conferred a special value on the results so obtained. In his 1913 address to the American Association of Anatomists, he remarked that 'The plea for the general use of experimental methods in anatomy is not based on the contention that a fact discovered by experiment is of more importance than one based on observation. . . . There are many footless uncritical experimental studies that lead nowhere, while carefully weighed observations very often lead to important generalizations' (Harrison, 1913*a*, p. 407). It is noteworthy that Harrison's 1910 tissue culture paper, usually regarded as the supreme example of his early experimental work, is almost entirely observational in its methodology. The paper is distinguished by the great rigour Harrison employed in making those observations, 'carefully weighed' as he put it in the quotation just given. The experimental method was straightforward, on a par with Roux's needle or Driesch's shaking, but the design of the experiments was precisely tailored to the problem.

Harrison did not advocate the use of experimental methods in biology in imitation of the physical sciences (Maienschein, 1983, p. 123). He did not at this time seek to reduce biological phenomena to physics and chemistry in the way that Jaques Loeb might have done. Nevertheless, he seems to have been acutely aware of the strides made in the physical sciences by the use of experimentation. In his Harvey lecture of 1908, Harrison in a discussion of the use of experimental methods remarked: 'The sciences that have used this means of advancement [experimental method] have achieved a much higher degree of perfection than those that have not,' and in the following sentence referred to the physical sciences (Harrison, 1908, p. 385). This theme becomes more strongly stated in Harrison's later writings and towards the end of his research life he did indeed endorse a physicochemical approach. For example, in 1933 he wrote that studies of the changes during differentiation would require '. . . all the ingenuity of the embryologist, using the most refined methods of physics, chemistry and general physiology, not only those of the present but many others still to be invented' (Harrison, 1937, p. 373). Between 1916 and 1940, Harrison studied the development of polarity and symmetry in the embryo, using the transplantation method of Born that he had first employed in 1898 (see for example Harrison, 1921, 1935). Harrison attempted to account for these phenomena in biophysical terms (the orientation of protein molecules) and in a 1936 essay (his only major speculative paper), he wrote of the 'quality of "wholeness"' exhibited in the parts of the embryo: 'It is the capital problem of embryology to find the physicochemical basis for it' (Harrison, 1936, p. 220, reprinted in Wilens, 1969, p. 17; see also Haraway, 1976, and Witkowski, 1980a).

Nevertheless, despite the admitted virtues of observation as a scientific tool, 'The purely morphological conception of anatomy . . . has led us to a barren field in which we are still plodding' (Harrison, 1913a, p. 404). (Here Harrison may be contrasting 'morphological' with 'functional' rather than with 'experimental'.) Similar sentiments had been expressed by the Lewises (quoted above), and it raises the question of why, if it was so clear to scientists like Harrison and the Lewises that experimentation was as essential for the biological sciences as for the physical sciences, it was necessary for Harrison to have to plead repeatedly the cause of the experimental approach. The reason was, he wrote, that 'There are many who would deny to this type of experiment validity in elucidating the phenomena of normal development, maintaining that the experimental conditions are too radically different from normal to be of value in interpreting the latter' (Harrison, 1912, p. 183). Sceptics applied this argument to all experiments where any manipulation might induce 'abnormal' regulatory responses, from the simplest procedures on embryos to the most extreme like tissue culture. However, as Harrison pointed out, the physical sciences were able to cope with this problem: 'If chemistry had stopped when it became necessary to go beyond the study of the "normal" substances found on the earth's surface, it would have been little but descriptive mineralogy' (Harrison, 1913b, p. 74). 'Neither the physicist or chemist allow themselves, however, to be deterred from experimenting by difficulties of this kind' (Harrison, 1912, p. 184). If the chemist and the physicist were prepared to use analytical methods to go beyond the study of normal phenomena as nature presents them to us, 'Why, then', Harrison asked, 'should we, in morphology, be still so dominated by the conception of the object as it occurs in nature, the organism as a whole, which to many seems to be a sort of fetish not to be touched lest it show its displeasure by leading the offender astray?' (Harrison, 1912, p. 184).

Harrison could not respond easily to this attitude because the appropriate controls for showing that the investigator was not being led astray were often not available. He wrote of the problem in general that 'In fact we have almost stated that there are no criteria by which we may be absolutely certain that the same combinations of circumstances are operative in producing an effect with the un-touched organism and in the one experimented upon.' (Harrison, 1912, p. 185). Harrison attempted to justify the use of tissue culture for the study of nerve development by comparison or analogy with the behaviour of other cell types in vitro (Haraway, 1976). That is, he asked the question: are the conditions in culture sufficiently like those in the embryo to justify any comparison at all between cell behaviour in vitro and in vivo? By comparing the phenomena [in vitro] with those observable in the embryo, we may with a varying degree of probability draw conclusions regarding the causal nexus of the factors acting within the normal body' (Harrison, 1912, p. 187). 'Such an empirical determination must have more weight than any amount of a priori argumentation upon the subject' (Harrison, 1910*b*, p. 823).

Harrison compared what could be observed of cell movement and cell differentiation in vitro with what was known of these phenomena in the embryo. Among the examples of cell movement he cited were those of mesenchymal and neural crest cells, the cells of the lateral line rudiment, Schwann cells along nerve fibres and the movement of cells during wound healing. Harrison concluded that the move-ments of cells in vitro '... must be considered as manifestations of activity similar in kind to those shown by cells within the normal embryo' (Harrison, 1910*b*, p. 825).

The examples of cell differentiation in vitro were even more striking. The occurrence of pigment cells, the development of muscle cells and the presence of neurofibrillae in nerve cells, all showed that 'Each type of cell follows the same course of differenti-ation it would have taken had it not been removed from the embryo'

(Harrison, 1910b, p. 825). Again, it is interesting to note that it was from *observational* considerations such as these that Harrison concluded that experimental techniques even as extreme as tissue culture were valid methods for investigating the developing embryo. The whole embryo was no longer to be left untouched and merely observed, but 'On the contrary we should endeavour to extend our experimental analysis wherever possible... (Harrison, 1912, p. 184).

## Conclusion

I have shown how Harrison's development of tissue culture arose as a logical consequence of his investigation of nerve fibre development, and how this whole approach was founded on the conviction that progress in understanding embryonic development demanded experimentation. Harrison wrote: 'The old fear of getting hold of something that is abnormal must be cast aside as obstructive to progress' (Harrison, 1913b, p. 74). Experiment in itself was not sufficient. Experimental analysis involved an understanding of the problem, the framing of questions that were both interesting and capable of being answered, and the design of experiments that provided answers that were unequivocal if possible. This strategy is apparent throughout Harrison's work, whether on nerve fibre growth or on polarity of the limb.

One of the themes of the symposium has been to examine whether there has been, and if so why, a lack of progress in embryological research. I should like to end with one last quotation from Harrison in which he recommended how such a lack of progress might be remedied: 'Experimental embryology will be placed on a sounder basis if its questions are framed more carefully' (Harrison, 1933, p. 319).

I am very grateful to the Wellcome Trust for a travel grant that enabled me to study the Harrison papers at Yale University. Dr T. Horder made many useful comments during the preparation of this work and Dr G. E. Allen kindly criticised a draft of the manuscript itself.

## References

Abercrombie, M. (1961). Ross Granville Harrison, 1870–1959. *Biographical Memoirs of the Royal Society of London* **7**, 111–26.
Allen, G. E. (1978). *Thomas Morgan Hunt*. Princeton: Princeton University Press.

Allen, G. E. (1979). Naturalists and experimentalists: the genotype and the phenotype. *Studies in the History of Biology* **3**, 179–209.

Allen, G. E. (1981). Morphology and twentieth-century biology: a response. *Journal of the History of Biology* **14**, 159–76.

Benson, K. R. (1981). Problems of individual development: descriptive embryological morphology in America at the turn of the century. *Journal of the History of Biology* **14**, 115–28.

Billings, S. M. (1971). Concepts of nerve fiber development, 1839–1930. *Journal of the History of Biology* **4**, 275–305.

Born, G. (1897). Ueber Verwachsungsversuche mit Amphibienlarven. *Archiv fur Entwicklungsmechanik der Organismen* **4**, 349–465 and 517–623.

Braus, H. (1905). Experimentelle Beitrage zur Frage nach der Entwickelung peripherer Nerven. *Anatomischer Anzeiger* **26**, 433–79.

Cajal, S. R. y (1890). A quelle époque apparaissent les expansions des cellules nerveuses de la moelle épinière du poulet? *Anatomischer Anzeiger* **5**, 609–13 and 631–9.

Cajal, S. R. y (1907). Die histogenetischen Beweise der Neuronentheorie von His und Forel. *Anatomischer Anzeiger* **30**, 113–44.

Cajal, S. R. y (1908). Nouvelles observations sur l'évolution des neuroblastes, avec quelques remarques sur l'hypothèse neurogénétique de Hensen-Held. *Anatomischer Anzeiger* **32**, 1–25.

Cajal, S. R. y (1909). *Histologie du Système Nerveux de l'Homme et des Vertébrés.* Madrid: Instituto Ramon y Cajal.

Churchill, F. B. (1981). In search of the New Biology: An Epilogue. *Journal of the History of Biology* **14**, 177–91.

Corner, G. W. (1967). Warren Harmon Lewis, 1870–1964. *Biographical Memoirs of the National Academy of Sciences, USA* **39**, 323–58.

Fischer, A. (1925). *Tissue Culture. Studies in Experimental Morphology and General Physiology of Tissue Cells in vitro.* Copenhagen: Levin & Munksgaard.

Gilbert, G. N. & Mulkay, M. (1984). *Opening Pandora's Box. A Sociological Analysis of Scientists' Discourse.* Cambridge: Cambridge University Press, especially chapter 6.

Guth, L. (1960). *Studies on Vertebrate Neurogenesis.* Springfield: Thomas.

Hamburger, V. (1980). S. Ramón y Cajal, R. G. Harrison and the beginnings of neuroembryology. *Perspectives in Biology & Medicine* **23**, 600–16.

Haraway, D. J. (1976). *Crystals, Fabrics and Fields: Metaphors of Organicism in Twentieth-Century Developmental Biology.* New Haven: Yale University Press.

Harrison, R. G. (1899). The growth and regeneration of the tail of the frog larva. *Bulletin of the Johns Hopkins Hospital* **10**, 173–94.

Harrison, R. G. (1904). An experimental study of the relation of the nervous system to the developing musculature in the embryo of the frog. *American Journal of Anatomy* **3**, 197–220.

Harrison, R. G. (1906). Further experiments on the development of peripheral nerves. *American Journal of Anatomy* **5**, 121–31.

Harrison, R. G. (1907a). Experiments in transplanting limbs and their bearing upon the problems of the development of nerves. *Journal of Experimental Zoology* **4**, 239–81.

Harrison, R. G. (1907*b*). Observations on the living developing nerve fiber. *The Anatomical Record* **1**, 116–18; also *Proceedings of the Society for Experimental Biology & Medicine* **4**, 140–3.

Harrison, R. G. (1908). Embryonic transplantation and development of the nervous system. *The Anatomical Record* **2**, 385–410.

Harrison, R. G. (1910*a*). The development of peripheral nerve fibers in altered surroundings. *Archiv für Entwicklungsmechanik der Organismen* **30**, 15–33.

Harrison, R. G. (1910*b*). The outgrowth of the nerve fiber as a mode of protoplasmic movement. *Journal of Experimental Zoology* **9**, 787–846.

Harrison, R. G. (1912). The cultivation of tissues in extraneous media as a method of morphogenetic study. *The Anatomical Record* **6**, 181–93.

Harrison, R. G. (1913*a*). Anatomy: its scope, methods and relations to other biological sciences. *The Anatomical Record* **7**, 402–10.

Harrison, R. G. (1913*b*). The life of tissues outside the organism from the embryological standpoint. *Transactions of the Congress of American Physicians & Surgeons* **9**, 63–76.

Harrison, R. G. (1914). The reaction of embryonic cells to solid structures. *Journal of Experimental Zoology* **17**, 521–44.

Harrison, R. G. (1921). On relations of symmetry in transplanted limbs. *Journal of Experimental Zoology* **32**, 1–136.

Harrison, R. G. (1924). Neuroblast versus sheath cell in the development of peripheral nerves. *Journal of Comparative Neurology* **37**, 123–205.

Harrison, R. G. (1933). Some difficulties of the determination problem. *American Naturalist* **67**, 306–21.

Harrison, R. G. (1935). Heteroplastic grafting in embryology. *The Harvey Lectures, 1933–1934*, 116–57.

Harrison, R. G. (1936). Relations of symmetry in the developing embryo. *The Collecting Net* **11**, 217–26. [A revised version was printed in *Transactions of the Connecticut Academy of Arts & Sciences* **36**, 277–330, 1945. It was reprinted, with illustrations, in Wilens, S., 1969].

Harrison, R. G. (1937). Embryology and its relations. *Science* **85**, 369–74.

Harrison, R. G. (1945). Retrospect, 1903–1945. *Journal of Experimental Zoology* **100**, xi–xxxi.

Held, H. (1907). Kritische Bemerkungen zu der Verteidigung der Neuroblasten- und Neuronentheorie durch R. Cajal. *Anatomische Anzeiger* **30**, 369–91.

Held, H. (1909). *Die Entwicklung des Nervengewebes bei den Wirbeltieren.* Leipzig: Barth.

Hensen, V. (1864). Ueber die Entwickelung des Gewebes und der Nerven im Schwanze der Froschlarve. *Virchows Archiv für pathologische Anatomie und Physiologie und für blinische Medizin* **31**, 51–73.

His, W. (1886). Zur Geschichte des menschlichen Ruckenmarks und der Nervenwurzeln. *Abhandlungen der Sachsischen Akademie der Wissenschaften Leipzig* **13**, 479–513.

Lewis, W. H. (1907). Experimental evidence in support of the theory of the outgrowth of the axis cylinder. *American Journal of Anatomy* **6**, 461–71.

Lewis, W. H. & Lewis, M. R. (1912). The cultivation of sympathetic nerves from the intestine of chick embryos in saline solutions. *The Anatomical Record* **6**, 7–31.

Lewis, W. H. & Lewis, M. R. (1924). Behavior of cells in tissue culture. In *General Cytology*, ed. E. V. Cowdry, pp. 383–447. Chicago: University of Chicago Press.

Loeb, L. (1897). *Ueber die Entstehung von Bindegewebe, Leucocyten und roten Blutkorperchen aus Epithel und uber eine Methode, isolierte Gewebsteile zu zuchten*. Chicago: M. Stern.

Loeb, L. (1898). On certain activities of the epithelial tissue of the skin of the guinea-pig, and similar occurrences in tumors. *Bulletin of the Johns Hopkins Hospital* **9**, 1–5.

Loeb, L. (1901). On the growth of epithelium. *Journal of the American Medical Association* **37**, 1024–25.

Maienschein, J. (1981). Shifting assumptions in American biology: embryology, 1890–1910. *Journal of the History of Biology* **14**, 89–113.

Maienschein, J., Rainger, R. Benson, K. R. (1981). Were American morphologists in revolt? *Journal of the History of Biology* **14**, 83–7.

Maienschein, J. (1983). Experimental biology in transition: Harrison's embryology, 1895–1910. *Studies in the History of Biology* **6**, 107–27.

Marinesco, G. (1909). *La Cellule Nerveuse*, vol. 1. Paris: Octave Doin et fils.

McCullough, D. M. (1969). W. K. Brooks's role in the history of American biology. *Journal of the History of Biology* **2**, 411–38.

Murray, M. R. & Kopech, G. (1953). *Bibliography of the Research in Tissue Culture, 1884–1950*. New York: Academic Press.

Neal, H. V. (1914). The morphology of the eye muscle nerves. *Journal of Morphology* **25**, 1–188.

Nicholas, J. S. (1961). Ross Granville Harrison. *Biographical Memoirs of the National Academy of Sciences USA* **35**, 130–62.

Oppenheimer, J. M. (1957). Embryological concepts in the twentieth century. *Survey of Biological Progress* **3**, 1–46.

Oppenheimer, J. M. (1966). Ross Harrison's contributions to experimental embryology. *Bulletin of the History of Medicine* **40**, 525–43.

Oppenheimer, J. M. (1971). Historical relationships between tissue culture and transplantation experiments. *Transactions & Studies of the College of Physicians, Philadelphia* **39**, 26–33.

Rubin, L. (1977). Leo Loeb's role in the development of tissue culture. *Clio Medica* **12**, 33–56.

Schwann, T. (1839). *Mikroskopische Untersuchungen uber die Uebereinstimmung in der Struktur und dem Wachsthum der Thiere und Pflanzen*. Berlin: Sanders.

Smyth, H. F. (1914). The cultivation of tissues in vitro and its practical application. *Journal of the American Medical Association* **62**, 1377–81.

Wilens, S. ed. (1969). *Organization and Development of the Embryo*. New Haven: Yale University Press.

Wilson, E. B. (1901). Aims and methods of study in natural history. *Science* **13**, 14–23.

Witkowski, J. A. (1979). Alexis Carrel and the mysticism of tissue culture. *Medical History* **23**, 279–96.

Witkowski, J. A. (1980a). W. T. Astbury and Ross G. Harrison: The search for the molecular determination of form in the developing embryo. *Notes & Records of the Royal Society of London* **35,** 195–219.

Witkowski, J. A. (1980*b*). Dr Carrel's immortal cells. *Medical History* **24,** 129–42.

Witkowski, J. A. (1983). Experimental pathology and the origins of tissue culture: Leo Loeb's contribution. *Medical History* **27,** 269–88.

# The emergence of experimental embryology in Germany

If one were to judge by textbook accounts, experimental embryology emerged in its most easily identifiable form in Germany. The experiments by Roux and Driesch on mosaic and regulative development, described earlier in this volume by Maienschein, provide clearly definable landmarks, and the approach to which they contributed can be readily followed through the achievements of Hans Spemann. By comparison with his contemporaries elsewhere, Spemann was surely unrivalled in the originality and thorough-going nature of the conceptual and experimental framework he established during the first three decades of the twentieth century. However, to concentrate on Roux and Driesch would be to disregard the great complexity of the German biology community. Table 1 gives some indication of the numbers of, and complex interconnections between, centres of biological research contributing to embryological issues. In the circumstances, as Russell says, 'It is hard to say whether Roux's work was cause or consequence' (*Form and Function*, 1916, p. 330). The elements combined in Roux's experiment each had long antecedents – Haeckel had divided blastomeres twenty years earlier, His had argued for a physiological approach to embryos, the idea of qualitative differences between chromosomes was implicit in the work of Rabl, Boveri and others at the time, and so on – and it was only differences in context which meant that those earlier contributions were relatively neglected. The situation in Germany differed markedly from that in Britain or the United States of America not only in respect of scale but also in the wealth of expert and mature viewpoints which, it may be reasonable to suppose, makes possible the balanced integration of information that underpins the ultimate insights of figures such as Weismann or Boveri.

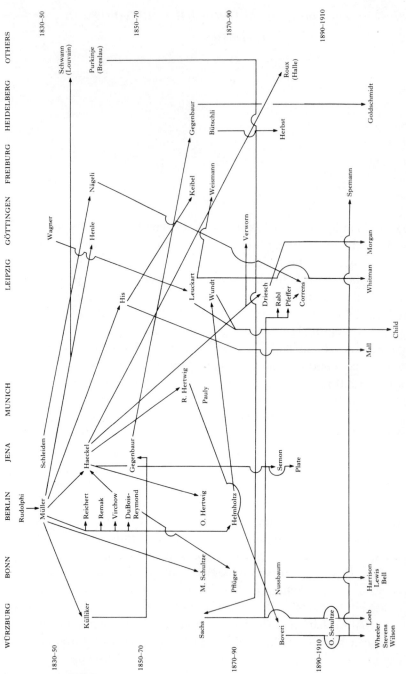

Table 1. *Figure showing principle nineteenth century German embryologists and their American links*

Arrows indicate teacher-student or collaborator relationships; associated American embryologists are listed below.

The continuity and breadth that comes from well-established scientific organisation such as this may help to explain the relative absence of any split between genetics and embryology in Germany. German biologists were no less capable as experimentalists than the Americans and the discipline of genetics was recognised early in Germany. Indeed the world's first genetics journal was *Zeitschrift für induktive Abstammungs- und Vererbungslehre* founded in 1908. Yet German biologists managed to maintain a tradition that saw genetics through its physiological, and therefore also embryological, effects: 'the hereditary factor . . . as a determiner for a given mass of certain ferments' (Loeb & Chamberlain, 1915, *J. Exp. Zool.* 19). How different this approach was to that being followed in the United States can perhaps be judged by the hostility that attended Goldschmidt's arrival there. Major contributors to genetics such as Boveri, Goldschmidt, Woltereck, Kühn and even Correns and deVries retained an awareness of, and interest in, cytoplasmic inheritance. On this question too it is perhaps impossible to say what was cause and what was effect, but we can recall how so many of the German biologists seeking solutions to the problems of evolution and inheritance had for so long turned to microscopic material and especially embryos, from Haeckel, His and Weismann, culminating in Roux, whose use of the frog embryo had as great an influence on the emergence of genetics as a discipline as it had on embryology.

In comparing this situation with that in the United States, one simple but easily overlooked eventuality can be considered to have been peculiarly decisive in determining the direction of primary interests. When Driesch, in such rapid, unexpected and stark contrast to Roux's finding, stumbled on the phenomenon of embryonic regulation – the phenomenon that posed once and for all the defining problem of embryology, namely how, given their initial equivalence in terms of developmental capacities, different parts of the embryo end up forming different adult structures – he was using the favourite marine invertebrate of the Europeans. This was the sea-urchin, whose embryo happened not to behave in a mosaic fashion. Now both a clearly defined problem and a suitable choice of species were available and were to be combined in the work of Spemann. Embryology in Germany up to the turn of the century has been relatively well served by historians of biology. The paper which follows examines the main line of advance thereafter.

# Hans Spemann and the organiser

T. J. HORDER

Department of Human Anatomy, South Parks Road, Oxford, OX1 3QX, UK

and P. J. WEINDLING

Wellcome Unit for the History of Medicine, 45–47 Banbury Road, Oxford OX2 6PE, UK

Spemann is a figure of considerable potential interest for the historian of biology but, given the continuing uncertainties of our understanding of the subject of embryology, a particularly difficult one to assess. Though a Nobel Prize winner – indeed the only embryologist to have been so honoured – and voted the second best known biologist by European zoologists[1], the work that he initiated ended in disillusionment and is unlikely to have been read by the average developmental biologist today. He is widely thought of as having introduced what amounted to anachronistic, 'vitalistic' concepts, particularly that of the 'organiser'. But much as his work does not fit into the style of modern biology, it is not the case, as any current textbook treatment of the phenomenon of embryonic induction will show, that the concepts and techniques that he pioneered have been superseded or invalidated. He has been relatively neglected by historians, particularly in the English language. Yet there are rich sources of information available in an unfinished autobiography covering mainly the period up to 1914 and in a remarkable document summarizing all Spemann's scientific output, his *Embryonic Development and Induction* (1938).[2] Many interesting issues emerge when one begins to consider a man who, in his time, was acknowledged to be a shining example of the best of early twentieth century biology. Is it significant, for instance, that Spemann was working in Germany in a period marked by a sense of moral and economic catastrophe following the Versailles Treaty and culminating in the rise of Nazism? To what extent is the style and direction of his work the product of the German biological tradition and to what extent might this have made his work inaccessible in the eyes of embryologists in other countries and of different generations?

# I. The development of Spemann's scientific ideas

## Background and scientific training

Hans Spemann was born in 1869 in Stuttgart, the first child of a publisher and bookseller.[3] His mother, from a medical family, died when he was one and a half years old but his father remarried. Spemann was brought up in prosperous, intellectual and artistic surroundings. He received a classical *Gymnasium* education; of the little science that was offered, Spemann was worst at zoology. His career began as an apprentice in his father's business (1888–9). After a year's military service he returned to bookselling and it was then that his reading of Goethe and Haeckel decided him in the direction of biology. At this time, on a visit to the Lausanne natural history museum, he was especially intrigued by a specimen of a calf with two heads!

Spemann began his training as a medical student in Heidelberg and Munich (1891–3). He was most strongly attracted to comparative anatomy as taught by Gegenbaur and became friendly with Gustav Wolff (re-discoverer in 1894 of regeneration of the lens in the newt's eye) and August Pauly (a charismatic Munich zoologist with strong psycho-Lamarckian beliefs). The latter's influence was such that Spemann gave up medicine in favour of research into 'general biological problems'. Accordingly, from 1894, Spemann chose to study under Theodor Boveri in Würzburg, a close friend of Pauly's and already well known, though only seven years older than Spemann.[4] His other teachers in Würzburg included Sachs and Röntgen. In spite of Spemann's immediate enthusiasm to embark on the experimental approach already being pursued in Würzburg by Boveri and Oskar Schultze, Boveri set him to work on descriptive embryology of fixed specimens of the parasitic nematode *Strongylus paradoxus*. This does not put Spemann's earliest work into the mould of the Haeckelian subjugation of embryology to phylogenetic considerations; Haeckel's influence was now on the wane; Boveri was sceptical and had never been much involved. The resulting paper (1895) does not refer to the biogenetic approach. We know that Boveri regarded it as essential training for his students to have acquired skill in observation, description, reconstruction and

drawing of biological specimens. The choice of species presumably reflected the popularity of nematodes for studies of chromosomes and cleavage,
and the comparable work on *Ascaris* by Boveri, published in 1892.[5] Boveri
had a deep interest in comparative anatomy based on 'a desire to
focus, above all else, on the gradual alteration (in evolution) of the
vertebrate body organization as a whole'.[6]

After receiving a doctorate in 1895 Spemann was again actively
looking for a project; 'My strongest inclination was for a problem of
general biological interest requiring technical invention'.[7] He was
particularly keen to use living material and, having been interested
in chick embryos for some time, he suggested a physiological or
chemical study on their yolk and blood supply. However he was
advised by Boveri to do a project more likely to produce results that
would relate to the widest possible scientific issues. Boveri suggested
the development of the middle ear in *Rana*, and this was the work
Spemann undertook to qualify (*Habilitation*) as a lecturer (*Privat-
dozent*). Here the concern was homologies underlying head evolution, another of Boveri's interests. Spemann published this work in
a painstaking paper in 1898 and would return to similar themes later
(1915). This was Spemann's introduction to the amphibian embryo,
to which he remained faithful henceforward, but he maintained
wide interests in comparative anatomy.

Boveri had firmly held views on how science should be conducted
and, judging by the close friendship between the two men, it seems
likely that Spemann shared similar attitudes. Baltzer describes how,
virtually every day, Boveri would follow his students' progress and
encourage theoretical discussion. A clue as to what Spemann admired in a scientist comes from the qualities in Boveri that he
picked out in his obituary of his teacher: powers of observation;
self-criticism; reliance on facts. Boveri wrote to Spemann's father
about this time saying that 'I know many zoologists and many
students have passed through my hands, but I know none for whom
such good prospects can be envisaged in the future.'[8]

During six months of the winter of 1896–7 Spemann was forced to
convalesce in the Alps following a chest infection. He relates in some
detail how the only scientific book he took with him was Weismann's
*Das Keimplasma* and the impact it had on him. He later said that 'all
my work has its roots in Weismann'.[9] He seems to have been struck
particularly by Weismann's demonstration that bold theories, if sufficiently clearly stated, become testable experimentally. His choice

of theme for his *Habilitation* lecture later in 1897 reveals that his interests centred on the much-discussed interpretation of the conflicting results of Roux and Driesch concerning the development of complete or partial embryos from single blastomeres (see Maienschein, this volume) – it was entitled 'Critical consideration of experiments on the effects on development of removal or killing of single blastomeres' – while his disputation covered a wide range of issues concerning evolution, heredity and the aims of science.[10]

# First experiments

As was the custom in Germany at that time, Boveri encouraged independence in his students. In 1897 Spemann started experiments on salamander eggs arising directly out of his reading of Weismann and concern with the possibility that Roux's failure to obtain Drieschian regulation (which was prime evidence in support of the Roux/Weismann theory of qualitative division of nuclear factors) could be connected with the way he had performed his experiments, by killing rather than removing one blastomere. Spemann 'thought this experiment could be extended for further analysis, if one cell was not killed, but was kept static in its development in relation to the other'.[11] In order to achieve this aim he planned to subject the blastomeres to different temperatures, a procedure already shown, in 1888 by Roux, to change rates of development and to produce deformities during gastrulation. It was perhaps for his need to orientate the embryo for insertion into a cooling tube that he started using a hair loop.[12] The resulting double-headed embryos, first obtained in June 1897, were therefore fortuitous. It was the aspect of duplication that he chose to emphasise in the title of his first paper on the work in 1900, 'Experimental production of double-headed embryos'.

As Spemann later suggested[13], his thinking at this time was close to Boveri's. It is well to remember, as Wilson put it[14], that Boveri has

sometimes been placed with those who ascribe a 'monopoly of heredity' to the nucleus. He it was, nevertheless, who first gave an experimental demonstration of the determinative activity of the protoplasm in ontogeny ... As early as 1892 he concluded that the form of cleavage is determined wholly by the protoplasm of the egg ... In later years he gave increasing attention to the role of the protoplasm in development, returning to this problem again and again in successive works. Some of these are among his best.

By 1888 Boveri had been in opposition to Weismann's theory of qualitative nuclear division because his own studies showed how individual chromosomes retained their identities in different cells. In 1887 he had described chromosome diminution in *Ascaris* – which enabled him to provide early support for the distinction between germ-line cells and somatic cells – but he returned to the topic repeatedly over the following twenty years and by 1899 believed that the phenomenon was due to cytoplasmic factors. His first successful proof of this was in 1904. In 1900 he had started work on the egg of the sea-urchin *Paracentrotus*, attracted by the fact that its ring of cytoplasmic pigment might mark the region giving rise to specific cells in the embryo and thus directly suggesting cytoplasmic determinants of future structure. By 1901 he had shown regional differences in egg potentialities by the different patterns of development of mechanically separated dorsal and ventral parts. Only parts containing the ventral pole developed, usually into whole *plutei*, while dorsal parts failed to differentiate at all. At this point he introduced the concept of a 'privileged region', a forerunner of the notion of a cytoplasmic gradient which he was the first to describe in 1910.[15] Boveri stated his views on development most clearly in 1902 as follows:

It appears to me that the quite peculiar interaction of the cytoplasm with its simple structure and differential division and the nucleus with its complex structure and manifold total multiplication may still achieve what Weismann and Roux attempted to explain with the help of differential nuclear division . . . a continually increasing specification of the originally totipotent complex nuclear structure, and consequently, indirectly, of the cytoplasm of the individual cells, appears conceivable on the basis of physico-chemical events once the machine has been set in motion by the simple cytoplasmic differentiation of the egg.[16]

Spemann published the results of all the hair-loop experiments which were completed in 1900, in three parts between 1901 and 1903, each concentrating on a different aspect of the findings, determined, as it transpired, by the orientation of the loop. Whereas a loop placed parallel to the future longitudinal axis of the embryo produced varying degrees of head and body duplication, with loops placed transversely, one half produced one whole embryo while the ventrally derived fragment (*Bauchstück*) frequently failed to

differentiate in any way. Spemann recognised that the single underlying explanation was that only embryo parts containing the dorsal
lip of the blastopore continued development. He also made the connection between the role of this particular region in the amphibian
embryo and Boveri's 'privileged regions', but by 1903 it was clear
that his primary concern was with questions of 'polarity' and
bilateral symmetry. These were the terms in which many of the
principal embryological issues of the day were discussed, such as the
half and double embryos of Roux and Driesch, the nature of neurulation as a 'concrescence' of two initially separate half-embryo rudiments, the 'promorphology' of the organisation of the egg cytoplasm
that anticipates the fundamental plan of the future adult organism.[17]
Furthermore, as the discussion in the 1903 paper makes clear,
Spemann was by now fully immersed in problems arising from the
development of the eye, which raised obviously analogous issues in
phenomena such as cyclopia.

## Lens induction

In 1898 Spemann had happened to notice, while looking at a frog
(*Rana fusca*) embryo at the neurula stage, structures which he took
to be rudiments giving rise to the lens of the eye. These were sufficiently far removed from the site of origin of the eye-cup itself,
which he knew to arise from the neural plate, that, as he now
realised, it might be possible to destroy the eye while leaving the lens
rudiment unaffected, and in this way test the interdependence of the
development of the two structures. Though he soon recognised that
he had been wrong and that what he had taken to be the *anlage* of the
lens was in fact the neural crest or cranial ganglionic placodes, the
idea could still be put into action the following year. Using a hot
needle or an electrocautery he destroyed the rudiment of the eye in
the neural plate and, having allowed time for further development
and then prepared microscopic sections, subsequently found that on
the side of the head on which the eye was now missing, the lens had
failed to form.

This apparently quite distinct departure in Spemann's interests
was the product of a prepared mind. It is conceivable that Spemann
had had his attention drawn to the eye by Boveri since this was a
subject he wrote on in 1904, but one link of which we can be sure is
from Wolff. Spemann indicates[18] that he was first alerted to causal

interactions in eye development, a subject of some interest to a
number of workers at the time, by discussion with Wolff, although a
more prominent issue than the role of the eye-cup in lens development was the ability of the eye-cup itself to develop normally in the
absence of the lens. Spemann also saw significance in the double-
origin of the lens, shown in Wolff's work, from ectoderm in development and from the iris in Wolffian regeneration; he refers to this as
'one of the main reasons that I undertook these experiments.'[19]
Given the way in which development and regeneration were equated
at the time and given his already established interest in doubling (he
had already noted the varying numbers of eyes in his double-headed
embryos and had been concerned with whether cyclopian eyes were
the result of fusion of rudiments from two separate head rudiments
or failure to divide of a single eye rudiment[20]) it is not surprising that
in Spemann's mind the problems raised by his work with hair-loops
and the lens were closely related and centred on the development of
organised structure as revealed by 'symmetry' and doubling.[21]
    We can be fairly confident about the way in which Spemann
viewed lens induction at this time. As presented in his own papers,
and in other reviews at the time, it was just one more example of the
loose and wide category of phenomena known as 'correlative' interactions, and based on concepts which Spemann traces back to
Cuvier's ideas on the integration of morphological structure.[22] We
can sense what this meant by considering the examples given by
Herbst in his review of 1901. As examples of correlative phenomena
he includes the dependence of taste buds on nerve supply, denerva-
tion atrophy of muscle, functional modelling of bone structure and
blood vessels, determination of sexual differentiation by environmental factors and the effects of lithium salts and other external
conditions on embryonic development. Viewed according to a
modern definition of induction – as applying to embryos rather than
to differentiated organisms, and to interactions between specific
tissues within the embryo – only two examples are quoted; firstly
eye/lens interactions and secondly the dependence of arm formation on the skeleton in sea-urchins. Spemann included within his
understanding of the phenomenon both the mechanical influence of
the eye on the separation of the lens from the ectoderm and a possible
role in the actual histological differentiation of the lens. Whereas
Herbst goes further, Spemann is inclined to limit the interaction to
one of release, while attributing the localisation and specific

differentiation of the lens to pre-existing mechanisms, presumably built into the egg itself. Nonetheless, in 1901, he set out the experimental strategies that would be required to analyse other possibilities.[23]

## Technical innovations and 'the remarkable fact of "double assurance"'

Spemann's work on the lens appears to have aroused some interest. Mangold records the considerable stir caused by Spemann's first lecture and demonstration on the subject at the Bonn meeting of the *Anatomische Gesellschaft* in 1901. Further afield King was beginning experiments on lens development in 1900 at Bryn Mawr, presumably at the instigation of Morgan. Apparently under the influence of Spemann's work, Lewis started work on the subject at Johns Hopkins University in 1903. Lewis was able to exploit Born's method of heteroplastic transplantation of embryo fragments, the method already being used by Harrison in the same laboratory, and the first arrival of the recently introduced dissecting microscope in 1903. By recombining a head portion of the embryo of one amphibian species with the belly portion of another species, he became the first to demonstrate induction of a lens out of ectoderm far removed from the normal site of formation of the structure, and the fact that induction operates normally between tissues of different species.[24]

However in 1903 Mencl announced the counter-example of a case of autonomous self-differentatiaion of a lens in the absence of an eye in a cyclopian teleost. In 1905 this was followed by further inconsistencies in amphibian species reported by King. In a paper in 1907 Lewis was still defending the conclusion that lens development was solely due to the influence of the eye, in the light of his own and LeCron's data in *Amblystoma*.[25] He was quite clear about the implications of the issues at stake; induction offered an alternative to what he considered the preformationist positions adopted by Weismann and Conklin. Spemann, meanwhile, was increasingly aware of the need to avoid any possibility that the techniques he had used to ablate the eye might also have affected the prospective lens rudiment. The techniques used so far were crude; even Lewis's technique was limited to the manipulation of essentially half-embryo fragments. Spemann's increasing awareness of the necessity to re-test his conclusions in species showing autonomous lens development with more refined techniques, coincided with the acquisition of a binocular dissecting microscope in Würzburg in 1905. In order to achieve the required manipulation of small embryo fragments he at the same time developed a new range of surgical instruments, made from glass. With the newly available techniques Spemann proceeded to confirm formation of a lens in the absence of the eye in *Rana esculenta*, even where the eye had been removed by surgical dissection rather than by the original methods, but he also confirmed, by substitution of the lens-forming ectoderm by ectoderm transplanted from other regions of the embryo, that lens potentiality was initially shared more widely in the ectoderm and that the eye alone had the capacity to elicit it.[26]

In 1907 Spemann found a way to reconcile the conflicts in the evidence from different species and indeed to capitalise on them. He invoked the phenomenon of 'double assurance', introduced one year previously by his friend Hermann Braus, as a result of his discovery that the opening of the operculum covering the limb-bud of the frog larva, which is normally associated with the outgrowth of the limb itself, still occurs after the early removal of the limb rudiment. Spemann found this 'remarkable'[27] because he, like Boveri, saw it as a beautiful example of the purposefulness of developmental processes which was a central theme of Pauly's psycho-Lamarckism. It remained an important aspect of his understanding of lens induction, even though he subsequently presented only one further example of it. The idea that Nature had provided two routes to lens formation, one due to the influence of the eye and another established earlier in development, presumably as part of the initial laying out of polarity and symmetry, could explain the variation in the data from different species; the balance of the two mechanisms was just different among them.

Although lens issues dominated Spemann's writings until 1912, when he wrote his last review paper on the subject, his interests had already begun to return to the more fundamental matter of polarity in the egg. Although his work on the lens received passing reference internationally in many textbooks of the period, the complexities of the data and implications meant that it remained a peripheral concern for most embryologists.

## Diversification of interests

By the time when, in 1908 at the age of 39, Spemann was appointed as a Professor to the Zoological Institute at Rostock, his new techniques made possible a number of new approaches to old problems. A paper of 1906 describes a battery of techniques and methods of analysis which were to recur frequently over the next twenty five years. By 1904 Spemann had started to study the extent to which ectoderm surrounding the normal site of lens formation possessed the potentiality to form lens. By ablating ectoderm at the normal site he caused neighbouring tissue to heal in its place and this he found initially capable of lens formation. From 1906 he could extend the analysis by substituting ectoderm transplanted from the flank and by rotating a flap of head ectoderm so that new cells came to occupy the lens site. He discovered that the spatial distribution of lens potentiality varied in different species and correlated with variations in capacity form lens in the absence of the eye. From 1905/6 he also rotated grafts of neural plate with the objective of establishing the time at which the rudiments of the eyes themselves become self-differentiating. A byproduct of this experiment was the occurrence of *situs inversus*, a phenomenon that he followed up later. In embryos with double heads Spemann had earlier noted how, like the eye, the numbers and locations of the rudiments of the inner ear was closely associated with the arrangements of the neural plate. As with the eye, he now also investigated the time and position of potentiality for the formation of the inner ear by rotating ectoderm in the appropriate part of the head region. As a result of conflicting results subsequently obtained by Streeter on the polarity of structures *within* the inner ear itself, Spemann later addressed this issue specifically.[28]

Despite the apparent variety of the specific themes that Spemann's papers now increasingly appeared to address, his work is marked by a considerable internal consistency, which can be traced back to the concerns of his first experiments on the salamander egg with hair-loops. The continuity of his thought is indicated by the frequency with which newly applied experimental approaches had been explicitly suggested by him long before.[29] Although a particular paper may be dealing with the inner ear or *situs inversus*, Spemann used each publication to rework the central theme of the principles underlying morphological organisation. Even when, in later experiments, he investigates the effects of delayed nucleation of parts of the egg cytoplasm or turns to the technique of fertilising anucleate cytoplasm with sperm of different species, he is recalling issues and techniques (in these cases, Weismann's theory and the technique of merogony pioneered by Boveri) which derive from his first research interests.

## The discovery of the organiser

Spemann's growing reputation was demonstrated when, in 1914, he was appointed head of one division of the newly founded Kaiser-Wilhelm Institute for Biology in Berlin. In 1919 he moved to the even more prestigious Department of Zoology at Freiburg and to the chair earlier occupied by Weismann himself. From this point Spemann acquired an increasing number of research students – twenty three graduated under his supervision in Freiburg between 1920 and 1940 – and a 'Spemann school' can be said to have been created. Certainly, in part, he owed his high repute to the originality of the techniques he had introduced and the way in which he had shown they could be applied. There is no doubt that he himself enjoyed the practicalities of designing and making new instruments, and that he rated surgical skills highly. While still in Würzburg he had introduced, in the face of disapproval from Boveri, photomicrography and later founded a photographic archive.[30] It is worth recalling some of the practical difficulties faced by experimental embryologists at that time which would have contributed to concern with technique. Amphibian embryos were only available between April and early July. Survival of operated embryos was poor; in the experiments which led up to the discovery of the organiser by Pröscholdt in 1921–2, for the six specimens that could be finally used as material, several hundred had had to be operated upon. The pressure to continue operating throughout waking hours during the season cost the Nature-loving Spemann dearly, in foregoing spring over some forty years. Mangold implies that it was this pressure that accounted for an obscure wasting of Spemann's left arm that started to occur in Würzburg.[31]

What exactly was the question that Spemann hoped to answer by way of the succession of transplantation and rotation experiments in the period from 1905? As a consequence of the results he had obtained when he applied hair-loops to embryos at different stages of development, he had already deduced that the stage of gastrulation is associated with a loss in the capacity of morphological organisation to compensate for experimental disturbances. Now Spemann's aim was to define in precise detail the time at which the capacity for self-differentiation of embryonic parts replaced the capacity to reveal their initial wider potentialities. Using the rotation technique he had, by 1912, shown that the potentiality to form lens diminished gradually at greater distances from the site of lens formation, in what he described as 'circles of diffusion'.[32] During his time in Berlin he started to examine the nature and operation of the responsible mechanisms directly, by rotating halved gastrulae with respect to one another. Always implicit in the study of timing was the idea that the loss of potentialities might occur at different times at different places in the embryo.. 'By the fact that, for the purpose of testing potencies, the embryo can be divided up into almost unlimited small particles, one is enabled to push the causal analysis of its development much further'; the aim being to find a 'centre of differentiation and then, by finer and finer transplants, further localise it'.[33] With these aims in view Spemann was led to one further technical advance in the period 1912–-15. In order to improve on the limited spatial resolution that he could achieve by rotating ectodermal transplants, he came across the idea of *reciprocal* exchange of smaller ectodermal fragments between different sites on the embryo. The speed with which such exchanges had to be achieved to get good healing of the grafts also required the design of glass micropipettes.[34] To facilitate location of the grafts after healing, Spemann exchanged tissue between embryos (of the same species) distinguished by differing levels of pigmentation. However this method of marking transplanted tissue had the limitation that the pigment differences disappeared after about two days and by 1918 he was using the technique of 'heteroplastic' exchange of permanently distinguishable tissue of two different amphibian species.[35] However, although in the experiments he described in 1916 he had obtained, for the first time, the result which would later be recognised as the 'organizer' effect, the significance was not yet apparent to him.

At some time between 1919 and 1921 Spemann received a letter from Hans Petersen, an embryologist from Braus's department in Heidelberg with interests in lens development and regeneration, concerning particular grafts described in the 1916 paper which resulted in host embryos having duplicate axial structures including secondary neural plates. He pointed out that these grafts 'completely or in the greater part, were taken from the invaginating region of the gastrula'[36]. Hitherto Spemann had assumed that this finding indicated the earlier onset of determination nearer the blastopore, shared equally by prospective neural plate and underlying mesendoderm and reflecting a hypothesised caudo-rostral determining signal. But Petersen appears to have drawn his attention to the fact that neural determination might have been secondary to mesendoderm invagination, so that these particular grafts might have consisted only of tissue with invaginative capacity and the secondary neural plates might have been of host origin.[37]

So it was that in 1921 he gave his student, Hilde Pröscholdt, the task of repeating his transplantations of the dorsal lip of the blastopore into the ventral region of a second gastrula using heteroplastic exchange. The first successful result came on 8 May 1921.

The experiment is described in detail by Saxen and Toivonen elsewhere in this volume. It happened that Spemann was at the time correcting the proofs of a paper and he took this opportunity to append a note describing the new result. He also coined the term 'organiser' here as a contraction of 'Organisationszentrum'.[38] In 1922 Pröscholdt obtained the five additional specimens that were used as material for the definitive paper published in 1924. By the time that paper was published Hilde Pröscholdt, who had in the meantime married Otto Mangold, had died as the result of an accident.

## The initial presentation of Spemann's concept of the 'organizer'

In analysing the sources and steps underlying Spemann's thinking in 1921 we can make full use of his own account as presented in *Embryonic Development and Induction*. He appears to have intended this book as a historical record of his work. To a remarkable extent it is indeed a scientific epitaph and a virtually complete description of his scientific output. That it was not meant as a review of the whole field at the time of its publication is evident: the English translation, published in 1938, was unmodified – except for several cuts including omission of a Preface and of a review of organisers in non-amphibian species – from the German version of 1936. References stopped at 1935 and were noticeably incomplete for the period after 1933, which was the time at which the text was mapped out for delivery as the Silliman lectures at Yale in that year. Thus the book reflects

Spemann's position in 1933 and, effectively, in 1921 because, despite certain changes in emphasis that we will describe, the mechanisms and concepts that he deals with remain unchanged on the evidence of his writings between 1921 and 1938, as we might expect given that the organiser concept was merely the culmination of considerably older ideas.

The book is complex and its layout by no means self-evident. The sequence of chapters follows the historical sequence of Spemann's work, if one allows for the occasional explanatory interjections (chapters 9, 16) and some historical transpositions which make the arguments more intelligible. (Much of chapters 1 and 5 describe fate maps which were not known until *after* the discovery of the organiser, the same being true of the account of the work of Vogt and Goerttler on the mechanics of gastrulation in chapter 5.) Thus after a preliminary survey of the early hair-loop experiments (chapter 2) Spemann summarises his work on lens induction (chapter 3), and then discusses his final interpretations (chapter 4). Chapter 5 returns to the gastrula, chapters 6 and 7 giving a more or less historical account of the discovery of the organiser.

It is typical of Spemann that in the paper of 1924 he emphasises his uncertainties regarding the interpretation of the organiser effect: 'The causal relationships in the origin of the secondary ... anlage are still completely in the dark'. However he also indicates his own inclinations: 'If wishful thinking were permissible in questions of research, then we might hope in this case that the second of the previously discussed assumptions would prove to be the correct one'.[39] Spemann is here referring to the two very different interpretations of the mechanisms underlying the organiser effect, which he presented in this first paper on the subject and whose relative contributions he would never resolve. The interpretation that Spemann actually preferred in 1924 can be seen as a direct extension of the views he had already formed regarding the establishment of basic morphological organisation. Running throughout his previous work is the theme that organisation is established in the form of axes of bilateral symmetry by mechanisms operating from the earliest stages of development, and probably built in to the egg cytoplasm. The way in which he actually envisaged such a mechanism is suggested by his introduction of the term 'organisation field' at about this time, a term which he had 'taken over from physics'. He had previously arrived at the concept of 'a region of the embryo that has preceded the other parts in determination and thereupon emanates determination effects of a certain quantity in certain directions'.[40]

What had perhaps most struck Spemann about his first organiser results was the way in which the transplanted cells became perfectly integrated as parts of the complex pattern of organs that they had induced in the surrounding cells of the host embryo. Unlike the eye in relation to the lens, no anatomical distinctions allowed the inductor tissues to be separated off, or in any way characterised as different, from the tissues they had undoubtedly caused to be induced. 'It would appear', Spemann wrote, 'that such a secondary embryonic anlage is induced by some superior power which uses the material at hand without any consideration as to its origin or its species'.[41] Here he is referring to two further aspects of the data which seemed to demand that the organiser be considered an entity operating quite independently of the cells actually generating it or being controlled in their development by its influence. The fact that the influence could operate across tissues of different species indicated an independence from any particular characteristics of cell differentiation. Furthermore, comparing different specimens it became evident that there was variation in the precise point of origin of the grafts at the dorsal blastopore, and yet quite equivalent and fully integrated organiser effects had been produced by them all.

The alternative mechanism for the organiser effect was one with which Spemann was thoroughly familiar. This was the possibility that the secondary neural plate, about which the secondary embryo was gradually formed, was the product of direct induction by the underlying archenteron roof, which itself was formed by gastrulation movement by way of the blastopore[42] A prime reason why he considered that this could not provide a complete explanation for the results was the fact that inductor and induced tissues were not confined to separate germ layers; already existing knowledge of gastrulation would have required that the neural plate, which is derived from the ectoderm, would be of host tissue only, whereas the mesoderm, along with the archenteron, being products of gastrulation movement, should have been graft-derived. Spemann had not found this; parts of the neural tube were in some cases graft-derived whereas the mesodermal structures were invariably made up of complex, integrated combinations of cells from both sources. Combining phrases from several sources we can see that he thought of the inductive explanation as a 'common chemical reaction', mediated by 'direct contact' and passing 'from cell layer to cell layer', in contrast to the organiser field which operates 'from cell to cell within the same layer'.[43]

It is clear that, from the start, Spemann associated the problems and alternative mechanisms of the organiser with the concepts he had already arrived at in his study of the lens. He associates the field concept with 'circles of diffusion' defining regions of lens potentiality, and he associates the alternative mechanism of induction by the archenteron with the ill-defined role of the eye itself. Indeed he was soon describing lens induction in terms of the new phraseology: 'The eyeball may ... be called the organiser of the lens'.[44] There is therefore no reason to suspect that Spemann was uncomfortable with the position that he finally adopted in the 1924 paper, and adhered to subsequently, namely that _both_ mechanisms play a part collaboratively in normal development. This, after all, would be another example of double assurance. The 1924 paper ends on a confident note; 'The designation "organiser" (rather than, perhaps "determiner") is supposed to express the idea that the effect emanating from these preferential regions is not only determinative in a definite restricted direction, but that it possesses all those enigmatic peculiarities which are known to us only from living organisms'.[45]

## 'Chaos out of order'

By 1928 Spemann was a powerful figure in German biology. In that year he was mentioned as a possible President of the _Kaiser-Wilhelm Gesellschaft_. Through contacts with friends and ex-students the influence of embryological ideas derived from Spemann was felt in zoology departments throughout Germany. His sixtieth birthday was regarded as a sufficiently important event to merit five celebratory volumes of Roux's _Archiv_. Among German zoologists his influence could now probably only be compared with that of Richard Hertwig, Kühn or von Frisch. Increasingly travel and lecturing occupied his time. A first visit to the United States in 1931 was followed by the preparation of the Silliman lectures, which he delivered at Yale in 1933. He received the Nobel Prize in 1935. From around 1931 Spemann appears to have ceased active research and his last research paper was published in 1933.[46]

In the very first papers on the organiser, Spemann set out his strategy for the further analysis of the phenomena. His priorities reflect his picture of how the field properties of the organiser might work; they direct attention most specifically to the idea of fields. [47] 'At the beginning (the two mechanisms) seemed to be of fairly equal value, the actual facts urge us more and more toward the exclusive adoption' of the possibility that induction of the secondary medullary plate spreads 'after invagination, from the deeper layer toward the superficial one'.[48] The period from 1924 to about 1933, when Spemann effectively summed up his final position, shows a clear shift in emphasis in the weight to be assigned to the two mechanisms, but Spemann did not finally abandon either or fundamentally change his perceptions of them. Events in this period are, however, made considerably more complex due to two other developments. Whereas in 1924 Spemann had directed attention primarily to the organiser as the source of developing morphological organisation, the evidence increasingly revealed the great importance of the potentialities of the target cells that express organisation through their differentiation. As Spemann put it: 'the share of the system of reaction turned out greater and greater so great at last that the very conception of the organiser became problematical'.[49] The second development which only served to reinforce the same conclusion was the chemical analysis of the organiser, which started in earnest in 1932.

Spemann's analysis of the organiser began with Bautzmann's mapping out of the distribution of organiser action in the gastrula, by implanting grafts directly into the blastocoele cavity of the host embryo, a technique suggested by Spemann.[50] Using the same method Geinitz extended heteroplastic transplantation between more widely different species, by transplanting the dorsal lip of the blastopore from anurans into urodeles. At the same time, in 1925, Marx showed that archenteron from a later gastrula itself had organiser properties, direct evidence that the archenteric inductor might be an important mechanism. One more direct attack on the role of this mechanism in normal development was Lehmann's demonstration of the effects of deficiencies in the mesoderm and archenteron resulting from lesions placed in the dorsal lip. The results failed to provide any basis for distinguishing the alternative organiser mechanisms; although corresponding deficiencies of the neural plate occurred, potential inductor tissue was not sufficiently completely removed to prove that other mechanisms must be operating to explain the neural tissues that were nonetheless formed. However, Lehmann's experiments were important because they introduced the issue of whether the prospective neural ectoderm had some prior disposition towards neural development, before the arrival of the archenteron. The same issue had first arisen in connection with the work of Vogt who, in 1923, had argued that, before gastrulation, cells already acquired independent and differing mechanical potentialities and that this itself might be the basis of their later differences in differentiation.[51] From this point Spemann became increasingly concerned with the role of the responding cells rather than with the nature of the organiser itself.

But how could these characteristics themselves be tested? In 1925 Durken had transplanted ectoderm into the eye of an adult frog on the assumption that this would be a neutral environment in which only features characteristic of the ectoderm would be revealed in its later differentiation. In 1931, Holtfreter and Bautzmann first explanted embryonic tissue into physiological culture solutions with the same end in view. Only after considerable confusion was it realised that neutrality was a mirage, and that there was no way of proving that *any* such conditions did not affect embryonic tissue behaviour.[52] In 1927 Vogt introduced the more explicit notion of '*Bahnung*', referring to an inbuilt, sequentially changing restriction of cell potentialities. *Bahnung* 'may be the consequence of a definitely planned order, established in earlier stages, as far back as the ovum, according to which the partial processes take their course, independently of each other, according to a fixed rhythm' and 'may be of a mosaic nature'.[53]

By 1931 doubts about the nature of the organiser stimulus as such acquired sudden new prominence. Following Marx's demonstration that anaesthetised organisers do not lose their activity, Spemann attempted to destroy activity by squashing. The following year Spemann, in a paper with Holtfreter, Mangold and Bautzmann, showed that a killed organiser retained its properties. There followed a massive outpouring of work directed, particularly by Holtfreter, towards discovering the distribution of organiser activity in different tissues and organisms. Particularly influential was his finding that ventral ectoderm, the very tissue that responds to the organiser, acquired organiser activity when killed. This led to the concept of the unmasking of organiser factors latent in the responding cells; 'indirect induction'. The discovery of the dead organiser, coming as it did at a time of increasing interest in endocrinology, enzymes, and macromolecules (see papers by Olby, Saxén & Toivonen, and Brachet in this volume) drew many new workers to the field, attracted by the possibility that the molecules responsible could be extracted and identified. Particularly disconcerting was the discovery in 1936, by Needham and Waddington, that non-biological molecules such as methylene blue also had organiser actions. By 1937, with the realisation that different laboratories in Germany, Britain and the United States had come to equally firm, but entirely different, conclusions about the identity of the organiser substance on the basis of purified embryo extracts, it was finally evident that there was nothing in common between agents having organiser action and that there were no clear criteria for deciding which was the naturally occurring agent itself.

Spemann was less and less directly involved in these developments. His final concerns, as indicated by the research of his students Schotté and Rotmann at this time, and as shown in the concluding chapters of the Silliman lectures, represented a return to the powers of the responding tissues as explanations for 'wholeness'. As he approached retirement he left behind him, in many people's view, a situation of 'chaos'.[54]

While most attention was concentrated on the *identification* of the organiser in chemical terms, the very idea that the organiser might be simple and chemical in nature served to confirm the overriding importance of '*Bahnung*' and to raise the question of whether specific inductions involved several distinct agents or, alternatively, were due to *quantitative* differences in the effects of only one chemical factor. Spemann's work on 'regional induction' (chapter 13) lead him to lean towards the former view. Overall, the evidence supporting the archenteric inductor was, by 1935, substantial. But as for the alternative mechanism, for which Spemann had seen such remarkable justification in Hilde Pröscholdt's specimens, this had become increasingly unnecesary and hard to detect or measure.[55]

# Retirement

In 1937, two years after getting the Nobel Prize, Spemann was retired, apparently one semester earlier than he had expected. Until his death in 1941, at the age of 72, he spent most of his time in the home, which had always been a most important element in his life, with his wife and sons, and the books of Goethe. He was, however, able to continue some research in the laboratory; his last work, carried out mainly in 1938 and published posthumously, concerned the persistence into adult life of the 'determination fields' with which most of his life's work had been involved. Until the end he held to his belief in the dual mechanisms of organiser action. In the closing paragraph of the Silliman lectures, one that Spemann was particularly insistent on keeping intact during translation into English[56], he left what is probably the most bold and memorable statements of his position.

Again and again terms have been used which point not to physical but to psychical analogies. This was meant to be more than a poetical metaphor. It was meant to express my conviction that the suitable reaction of a germ fragment, endowed with the most diverse potencies, in an embryonic 'field', its behavior in a definite 'situation', is not a common chemical reaction, but that these processes of development, like all vital processes, are comparable, in the way they are connected, to nothing we know in such a degree as to those vital processes of which we have the most intimate knowledge, viz., the psychical ones'.[57]

# II. Spemann's embryological thinking in context

When Spemann began research in the 1890s, embryology occupied a crucial position in the sciences. The embryo was regarded as containing not only vital clues to the secrets of heredity, but it was also a simple model for more complex developmental processes in biology, psychology and the social sciences. Embryologists employed an impressive range of analogies for the interpretation and communication of their empirical evidence. On the one hand there were those, like Roux, who used primarily physical and chemical analogies. On the other hand, many – if not most – embryologists preferred to use historical, psychological and social analogies and explanatory models. Psychologists and philosophers keenly followed the results of the rigorous application of experimental methods by Driesch, Oscar and Richard Hertwig, Pflüger and Roux. During the 1880s Nietzsche was an early critic of Roux's mechanistic theories of organising forces. Freud found inspiration in Roux's injury experiments that showed the importance of the early stages of development. That the historical sciences were pre-eminent in Germany, was not detrimental to embryology; it could claim a strategically important place in the historical chain of cause and effect, which ran from the simplest of organisms, to the highest organism of all – Germany.[58]

This section of the paper will assess Spemann's science in its cultural context by showing the place of non-empirical factors in Spemann's science. We will extend Part One's analysis of Spemann's scientific papers by suggesting that the theoretical structure and scientific values of German biology from the 1890s until the 1930s meant many German embryologists could incorporate a different range of explanatory factors in their science from those embryologists, particularly in the USA, for whom fate-mapping and biochemical approaches became major concerns. It has been tempting to dismiss German biologists after 1914 as exhausted volcanoes. After pioneering German research on the cell and its components like the chromosomes, US researchers have been seen as taking over the torch of knowledge with research on Mendelism and biochemistry. Historians like Lukács and Ringer have seen German professors after 1914, as falling prey to irrational *Lebensphilosophie*, culminating in fascist totalitarianism and racism.[59] Not only have the achievements of distinguished biologists like Spemann been disregarded, but there has been a lack of analysis of the distinctive aims and ideals of 'German biology'. Spemann's concern with 'the forces that govern development' can be better understood if his deep intellectual concerns with psychic factors, Lamarckism and vital organisation are recognised as major preoccupations. Moreover, if his institutional sense of responsibilities as a leading German professor and concern to defend German national values are included in the analysis, then further ideological issues confront any historical assessment of Spemann. A shift occurred away from explanation of development in terms of the formative powers of individual cells, as in Virchow's liberal and democratic concept of the organism as a 'cell state'. The new concern with dominant regions like the organiser coincided with German nationalism breaking with its liberal traditions and becoming increasingly authoritarian. Spemann warned that 'the organiser' was only a provisional term to interpret some new facts, rather than a comprehensive theory. Others were not so cautious.[60]

Spemann compared his publications to an archaeologist's fragments, being parts of a greater whole. Although he cautioned that it would be premature to over-generalise from exact experimental knowledge, he insisted on the scientist's need for qualities of artistic inspiration and vision of a greater theoretical unity.[61] Concentration on the published scientific papers cannot, because of Spemann's rigorous definition of criteria for publication, tell the whole story of Spemann's science. To recreate the mentality that lay behind such experimentalism, conceptual issues must also be taken into account. It therefore seems appropriate to supplement Spemann's scientific papers with other publications and additional sources. These indicate how cell theory prepared and conditioned the scientist's perception of his evidence, and how very different contemporary criteria of rational explanation operated at the time in embryology. Spemann confessed in moments of humility that he was often prone to forget relevant literature which had earlier impressed him, so adding to the problems of the search for appropriate sources. While not able conclusively to 'prove' Spemann's motives, we can provide what is on balance a plausible and convincing historical explanation. In order to understand the rational processes of Spemann's scientific work it is possible to look at a range of more personal sources relating to Spemann's *Weltanschauung*, like speeches on the nature of science, the autobiography, letters, and the reminiscences of his students. These provide material to recreate the thought of a past generation of embryologists with aims and ideas very different from those of today.

# The cultural value of German biology

In order to clarify the status of the occasional political and psychological references that intrude in Spemann's writings, and to gain a sense of the aims implicit in his research it is necessary to clarify his views on the cultural value of biology. It was common among German professors not to engage actively in politics, which they disdained as divisive and partial. Instead they considered themselves guardians of national values, and were convinced of German cultural supremacy.

Boveri had a major role not only in directing Spemann's early research and career but also in inculcating a range of cultural values in addition to those which the enthusiastic young doctoral student brought to his research. This role of Boveri accords with his reputation as an outstanding teacher. Boveri's own teacher, Richard Hertwig, had recommended him in glowing terms to the Prussian ministerial director, Friedrich Althoff, the mastermind dictating development and staffing of the Prussian universities.

My judgement on Boveri is based on intimate knowledge of his research and personal acquaintance lasting for years. He is one of the quite *exceptionally* gifted men, who are met with only seldom not only in zoology, but in all the academic disciplines. In any university, even the largest, he would fill with honour any position, not only as a researcher, but also as a teacher. His lectures are exceptionally clear, and the vividness of his explanatory powers rivets his listeners.[62]

It is not surprising that armed with these powers, Boveri should have made a deep imprint on Spemann's scientific, cultural and social opinions. As colleagues and friends, and with a sense of shared scientific aims, Boveri and Spemann made a powerful combination. Horder's biographical introduction has shown how Spemann came to Boveri through Pauly as a mutual friend, with whom both had ardent discussions on psychic and Lamarckian factors in development. Not only did Spemann learn from Boveri the techniques of how to approach scientific research and to present research conclusions, but they shared an interest in adding to cell theory, as itself inadequate to explain many aspects of development.

Given their common scientific interests, and common social background in wealthy middle class families with commercial interests, it is not surprising to find shared social opinions. Spemann's strong sense of nationalism was shown when he accompanied his father on what amounted to a communal pilgrimage in 1892 to pay respects to the recently dismissed Bismarck. At that time Haeckel proclaimed Bismarck 'Doctor of Phylogeny', as recognition for his work in creating Germany as among the highest forms of cultural organisms. Boveri and Spemann shared similar nationalist views. For example, during the 1914–18 war when many academics supported what they regarded as a defence of their cultural values, Boveri wrote expressing his hopes from annexation of new territories, despite some doubts as to the fate of non-German peoples. Boveri had disclosed antisemitic prejudices in letters to Spemann. He expressed dislike of the face of the geneticist Goldschmidt, and concern to block the appointment of Herbst, as possibly Jewish, so that a third Jew should not become Director of a Kaiser Wilhelm Institute. On 4 May 1915, Boveri wrote that in the current struggle for existence, he was prepared to accept the Aryan racist views of H. S. Chamberlain, on Germanic supremacy. This shows how heightened nationalism meant that what had hitherto been written off as unscientific could now be recognised as containing new truth.[63]

Holtfreter has, in his address in honour of Hamburger, vividly described the psychological impact of the collapse of the old imperial order. In the political and cultural turmoil, there was a craving for prophetic philosophies of men such as Nietzsche and Spengler, and particularly at Freiburg the student prophets were Husserl and Heidegger. Hamburger felt Spemann was impressive as a prophet of analytical biology. During the Weimar period, Spemann was looked to not only for his leadership in embryology, but as a guardian of standards and values in science at a time when Germany's economic difficulties jeopardised future research. There was a new emphasis on biologists' duty to use their science in the service of the nation. Erwin Baur, the geneticist, turned to Spemann for advice on the centralised funding and coordination of science under a scientific 'Führer'. Biologists attained a new professional recognition and prestige during the 1920s owing to their patriotic stance for applied biology.[64] Biological laws of development were regarded as means of guiding the nation towards unity, harmony and social progress. In his speech on development as Rector of the University of Freiburg he proclaimed: 'Pure science with the aim of knowledge is one of the strongest roots of the German spirit, and to serve this spirit is to serve the Vaterland.' He took a staunchly nationalistic stance in university affairs. He refused to lecture in England in 1925 in protest against foreign occupation of the Rhineland. When the nationalist Schlageters was executed by the French in 1923 during the occupation of the Rhineland, Spemann led the Freiburg professors in a cortège of hommage. Mangold's obituary of 1941 depicts Spemann as approving of the new order. Visitors could be subjected to a monologue on the benefits of the new regime. Mangold's view was undoubtedly coloured by his own Nazism.[65] But it seems that at first Spemann like so many of the Germans who voted for Hitler, sympathised with the Nazi movement because it promised to liberate the fatherland from the moral and economic catastrophe precipitated by the ignominy of the Versailles treaty. Spemann later became sensitive to Nazi interference in Science. He was retired one semester earlier than he might have wished. The new order clearly resulted in a complex inner struggle for Spemann, and one must be cautious in delivering any categorical judgement on his views 1933–41. That he was a staunch nationalist is beyond dispute.[66] How these social views might have been imprinted on Spemann's embryology is a difficult and complex question. One might object that there is no logical connection between Spemann's politics and his science. But this interaction is to be linked to the gradual development of his move away from cellular analysis, and to how his views on dominant regions were infused by his broader cultural and political commitments. Spemann regarded embryology as having a special cultural mission to the *Volk*. Common root causes could be seen as underlying social and biological problems at a time when organicist social thought was the norm.

Spemann valued biology as having an important cultural role in the education of youth. He was actively interested in educational experiments furthering handicrafts and outdoor activities, and he took a leading role in extramural studies when at Freiburg. Spemann admired the nationalist educationalists, Hermann Lietz and Eduard Spranger, and wrote of his reflections as to whether it would not be preferable to become a school teacher in order to fulfill his sense of mission to the German People. That in the German edition of the Silliman lectures he introduced his work as an offering to the *Volk* illustrates his cultural view of national role of biology.[67] In the 1920s, Uexküll proclaimed 'Spemannism' as a new type of biology. By this he meant a new experimental approach to research on the living organism. During the Third Reich, Uexküll's work was much admired, and his positive attitude to the racial theorist, Houston Stewart Chamberlain, was an added inducement to recommend his work. That the German Zoological Society also provided a forum for the Nazi party member, Konrad Lorenz, indicates the type of dynamic new behavioural biology that was favoured.[68] Spemann's position here is problematic. Politically he was a nationalist of the old school, formed in a very different cultural environment of Imperial Germany. It is important to distinguish this from any implied sympathy specifically with Nazism. Mangold was investigated for Nazi connections after the war and there were 'Brown shirts' in the laboratory in Spemann's time. Eakin suggests (1963) Spemann was probably most uncomfortable having to say 'Heil, Hitler'![69]

The award of the Nobel Prize in 1935 was an accolade not only personally for Spemann, but was regarded in Germany as recognition for the new order. In Spemann's speech of acceptance, he depicted himself as the 'Führer' of a comradely band of researchers, and as the son of his *Volk*, which aspired to respect from other cultured peoples for its work in the building of a better future for humanity. Yet Spemann's address to the German Zoological Society in 1936 when it met in Freiburg conveyed a sense of distance from his immediate social environment. He regarded biology as enshrining permanent truths that the disruption of the last war and ensuing social change could not disturb. He addressed a younger generation of German biologists, whom he characterised as keen for revolutionary change, with cautionary words: that the young and old alike had to abide by the rock hard criteria of scientific truth.[70]

Although Spemann's views on the national importance of pure science reveal consistency between the 1920s and 1930s his views on pure science were challenged by the Nazi control of universities and stress on the national value of applied science. On 30 January 1938 Hitler instituted a national prize for Arts and Sciences and forbade German scientists to accept the Nobel Prize. In a remarkable pair of speeches, published in 1938, Spemann considered the international and the national character of science. Spemann addressed a problem that lay at the heart of his own research: to what extent was scientific work universally rational, and to what extent was science a product of national psychology. Spemann pleaded for pure science to be accepted as valid on both the international and the national levels. Every people has its own culture and history, yet there are transcendent universals valid for the whole of humanity in Man's common confrontation with Nature. Spemann separated universal reason from the volkisch spirit, although he regarded these as in continual interaction. On 10–12 June 1938 Spemann addressed the Nazi Student *Bund* in Freiburg on science in the service of the nation. While acknowledging the relevance of applied science, Spemann pleaded for the higher spiritual value of pure science. He emphasised the value of pure research to the *Volk*. He said that just as the Führer favoured art for its racial value, so the scientist also had a vocation to serve his people. He emphasised the higher value of certain truth, 'die Forderung unbedingter Wahrhaftigkeit', necessitating that research be free from the constraints of time and place. Science provides man with spiritual nourishment and has great educative value. It provides orientation and boosts psychological confidence. Science showed how man was part of nature, and the *Volk* was obliged to serve the highest of values. He concluded: 'Alles Segen über unseren Führer und über unseren Volk!'.

These speeches reveal how Spemann was a defender of scientific truth, while at the same time showing his deeply nationalist commitments. These were ideals which put Spemann at odds with certain Nazi values, like the priority of applied science, and the dictates of the Nazi policy of *Gleichschaltung*. This may have been a factor in Spemann's premature retirement in 1937. From this it seems that Spemann could not condone that the state should interfere in science, as this was a matter for the individual in experimental dialogue with Nature. It was clearly at variance with the Nazi view of applied biology as eugenics or racial hygiene. Spemann had a view of science that had unyielding standards, and was independent of the ebb and flow of social change. While his embryology has retained much scientific validity, it should nonetheless be recognised as containing cultural presuppositions and as according to the very different criteria of scientific rationality that prevailed in the Germany of Spemann's day.[71]

## Methodology

Hamburger has emphasised Spemann's interest in philosophy, and it is worthwhile trying to see how philosophical concerns permeated scientific thought. Spemann was typical in his broader cultural interests of German professors who were reluctant to be just narrow specialists, and felt compelled to relate their research to generalities. Despite Spemann's rigorous specialisation in vertebrate embryology, he was responsive to philosophical issues agitating the academic community. For professors considered themselves torch bearers of German national cultural traditions, as well as being dependent on the state for their appointments and research funds. There was widespread concern with the disintegration of society owing to mechanisation, urban life and an associated rationalism. Spemann's emphasis on living Nature as symbolised by the whole-embryo experiments rather than dissection of dead material can be seen in the context of a rising tide of *Lebensphilosophie* between the 1890s and 1930s. In a classical article on the growth of a causality and organicist configurational thought among Weimar physicists, Paul Forman has referred to Spemann's dilemma between the dominance of *Entwicklungsmechanik* and the organicist views which he favoured. It is important to recognise that judged by the contemporary standards of the most progressive physicists, the use of analogies referring to *Gestalt* psychology and other branches of organicist thought was acceptable scientific practice.[72]

Spemann wrote to Driesch that pure science as an end in itself had no interest to him[73], although he never ceased to believe in objective, empirical and rigorously logical science. Yet Spemann regarded his biology as a particularly German achievement, and believed that the enduring values of biology should be an inspiration to future generations. Indeed, he felt his type of embryological studies of the living organism as not so much a personal achievement, but as a product of the special affinity of the German for Nature. While intellectually Spemann was avowedly dispassionate and objective, he also recognised the need for intuitive gifts in scientific research. He considered his experiments as a dialogue with living Nature, and that the German success in biology was a racial achievement. Spemann made great show of technical virtuosity and of the need to develop appropriate manual skills in laboratory work, being as much intuitive craft as science. It is well symbolised by his insistence on a baby's blond hair (obtained from his first child who happened to be blond), as opposed to a less Aryan hair colour, which was more flexible for separating the two halves of the embryo at the cleavage stage.[74]

Spemann occupied an intermediate position of great significance between Roux's *Entwicklungsmechanik* and Driesch's vitalism. Methodologically he was indebted to Roux, but intellectually his inspiration was Driesch. In 1939 he wrote to Driesch that Roux was a mere technician and if there was nothing to embryology but mechanistic factors, Spemann expressed a preference for being an engineer. It was Driesch that he regarded as having raised the fundamental problem for his research.[75]

The border we accept today between the humanities and the sciences had not yet appeared. Institutionally zoology remained for a considerable period part of the philosophical faculty. Scientists were not shy of discussing their method and ultimate aims and assumptions by identifying themselves with total philosophies like monism, historicism, vitalism. Figures like Haeckel, O. and R. Hertwig, Driesch, and Pauly held and voiced philosophical views. It was a long and beguiling tradition (especially in Germany) to look for unifying principles such as monism, sustained after all by the giants of German culture, Leibniz, Kant, Goethe, and evident in the works of founding fathers of embryology, C. F. Wolff,

Pander, von Baer and Johannes Müller. There was a common regard for organic form, for the organism and for developmental growth. Physics provided the scientists' model for the design of causal and explanatory experiments, but in the tradition of Johannes Müller it was a common belief that vital phenomena would not be reducible to the forces and entities described by physicists. Pure reductionism in cell theory was suspect; one of the underlying reasons for the strong reactions to Weismann's theory was (apart from its preformationism) widespread suspicion about the arbitrariness and improbability of the inferred atomistic determinants. Along with Morgan, Thompson, Hertwig and Driesch, both Boveri and Spemann were very suspicious of such a scientific line of reasoning. History provided a much more attractive model especially given that embryology was known as *Entwicklungsgeschichte* (literally, history of development). In the context of comparative anatomy, embryology was meant to generate evidence for reconstruction of the history of life. The concepts of division of labour, inheritance, organisation and of the organism as a 'cell state' provided a thoroughly acceptable alternative explanatory currency. Although biologists may have been careful to guard themselves by stating these concepts as *analogies*, it is clear that they were used in powerful ways to structure their conceptual frameworks. What emerges is that Spemann followed well established methodological rules in (1) confining himself to the interpretation and evaluation of empirical evidence in scientific papers, and (2) reserving opinions on broader issues like vitalism and Lamarckism, to which he was sympathetic, but which were as yet unproven.[76]

Embryology had passed through a series of causal frameworks: Haeckel's evolutionary recapitulation, His's physical reductionism, Roux's *Entwicklungsmechanik*, and Driesch's and Hertwig's variants of organicism. Spemann's sense that vital processes required distinctive causal entities drove him to attribute a range of causal powers to 'the organiser'. That politically, the organised *Reich* of Bismarck had collapsed, and that the Weimar experiment in democracy was disdained by Spemann, suggests a wish to assert Germany's need for an 'Organiser' or 'Führer'.

# The conceptual origins of the 'organiser'

## (i) Cell theory

A major feature in Spemann's work concerns an apparent shift away from a conventional cellular analysis of developmental processes, that culminated in what Oppenheimer has called a 'supracellular' approach. Oppenheimer has interpreted this shift in terms of Spemann opening the way to biochemical explanations of embryological form; but it can alternatively be seen as a shift to a higher level of organisation, as represented by the organiser, having its methodological counterpart in a switch from histology to whole embryo experimentalism.[77]

In the 1890s the German biological establishment had its attention focused on major discoveries as to the cellular mechanisms of heredity and development. These related to reduction division, and to experimental investigations of the formative powers of cells by injury experiments. The dominance of the cellular approach is indicated by Spemann's teachers – Gegenbaur, Richard Hertwig, and Boveri – having made fundamental innovations to cell biology. Gegenbaur had been among the first of the anatomical biologists to recognise in the 1850s the universality of cells in all organisms, and the continuity of cells ensured by cell division. He applied cell theory to evolutionary comparative anatomy, which had many similarities with the evolutionary synthesis of Ernst Haeckel that so inspired Spemann. Both Haeckel and Gegenbaur sought to reconstruct the historical sequences of organic forms, attributing to these a causal value as lower forms recapitulated higher stages. Both sought a synthesis between Darwinism and the new protoplasmic theory of the cell of the 1860s that emphasised the formative powers contained in the substance of the cell, rather than in the previous view of the cell as a hollow chamber filled by a structureless fluid (Schleim). The new synthesis stimulated the development of embryology in conjunction with analysis of the cellular structure and substance to ascertain their formative powers. Richard Hertwig and his student Boveri were leading exponents of the precise investigation of the nucleus and chromosomes with an eye to ascertaining their controlling functions in heredity and development. In the 1870s Richard Hertwig conducted pioneering research on the chromosomes during fertilisation, and in the 1880s he conducted innovative cellular and embryological experiments. Study of cellular properties under experimentally modified environmental conditions contributed to knowledge of the role of the nucleus during fertilisation and cleavage. Richard and his brother Oscar maintained a view of the cell as an elementary organism. This was challenged by Boveri's shift of emphasis to the chromosomes, by Roux's attempts to derive causal forces that shook the organisational presuppositions of the Hertwigs, and by Weismann who gave explanations of heredity that went beyond what was empirically observable in the cell. Around 1888 major debates flared up regarding the reduction division of chromosomes during fertilisation, and with regard to whether the formative forces at cleavage and gastrulation were cellular in nature or based on chromosomes as subcellular 'organisms'. Despite the fact that most of our present-day concepts of cell organisation and function had been identified by 1900 it is most important to recognise that their *understanding* was still open to differing interpretations. There was still room for heated debate about fundamental issues and for serious misapprehensions which were only resolved much later – e.g. status of organelles, universality of meiosis.

Spemann thus began his career when the orthodoxy of the cell as an elementary organic unit was being challenged by a variety of perspectives and discoveries. The previous emphasis on a bipartite view of the cell – that of a dominant nucleus that contained all hereditary powers and information – and a nutritive protoplasm responsive to environmental change, began to seem oversimplified. Oscar Hertwig had abandoned evolutionary phylogenies as causal mechanisms for the organising powers of the cell as the elementary organism, and Spemann was to take this process a step further.[78] It was possible that the organising powers attributable to the cell might instead be located in controlling regions of the organism, without the rigorous reliance on cellular regularities. During the 1890s Spemann learnt not only the newly refined techniques of microscopy and experimental investigation, but he also was initiated by Boveri into the technicalities and broader interpretative issues in cell theory and embryology. On 15 February 1897 Boveri advised Spemann as to the value of Weismann's germ-plasm theory and attempted to wean Spemann away from Gegenbaur's, Haeckel's and Pauly's variants of Lamarckian evolutionary theory, and onto experimentalism. Boveri considered that Weismann's importance lay in his introducing principles in areas requiring theoretical clarification, and in his criticisms of Lamarckism. Boveri warned in 1901 that one experiment was better than 100 pages of the comparative morphologist Gegenbaur. Yet Boveri was not enthusiastic over Weismann and was critical of the particulate views of Roux. He wrote from Naples to Pauly on 14 January 1886 that the poor impression gained from Weismann's writings was confirmed by personal contact. Spemann's doctoral thesis on *Strongylus paradoxus* provided additional points of comparison to research on another parasite over which controversy was raging – *Ascaris*.[79] On the basis of his cell studies in *Ascaris*, Boveri had challenged both Weismann and Oscar Hertwig. If the chromosomes could be proved to be persistent structures during interphase, it would refute not only Hertwig's view of a controlling totality of the nucleus, in which the chromosomes simply disintegrated, but also Weismann's presupposition of invisible hereditary germ-plasm contained only in special germ cells. Hertwig's view of the organism as a democratic association of cells would thereby be challenged as an absurdity. Instead the controlling elements were chromosomes that – to use another contemporary metaphor – manoeuvred like soldiers, and exercised a controlling function over development. Boveri's 'infracellular' views contributed to the making of Spemann's 'supracellular' approach to development.[80]

Boveri's work was for a time to take him in the direction of a synthesis of chromosomal theory with Mendelism but later back towards an emphasis on the cytoplasm and the whole embryo in sea-urchins: Spemann proceeded to questions of the experimental investigation of later stages of embryological development in amphibians. Their research can be regarded as complementary. Spemann always kept up with advances in Mendelism as in a series of general review articles and Boveri admired a view of the organism as a historically developing entity, as his major speech on that topic in 1906 showed when he accepted elements of Pauly's psycho-Lamarckistic doctrine of the cell psyche. Boveri's stress on the 'individuality' of chromosomes went with a view of the chromosome as a *'Zentralorgan'* consisting itself of elemental chromatin particles, which 'manoeuvred' like an infantry regiment. Although Boveri rejected the concept of the nucleus as a *'Zentralorgan'*, his analytical studies of the constitution of the cell retained hierarchical social analogies to express concepts of organisation and integration.[81]

Boveri's criticisms of the cellular approach to morphology consequently raised broader theoretical questions as to controlling forces and regions. The concept of dominant regions (*Vorzugsbereiche*) was suggested by Boveri in 1901 but its prehistory dates from the debates on cleavage of 1882. Rotation experiments clarified the respective roles of the animal and vegetal segments of the embryo at cleavage. Debate on the formative powers of the blastopore occurred between Oscar Hertwig and Wilhelm His in 1892. His's *Concresenztheorie*, postulating the germinal mound (*Keimwülste*) as the key to embryological growth, was rejected by Hertwig, who observed that many processes like the development of axial organs were traceable to the blastopore. Hertwig raised the issue of how processes of cellular growth in the blastopore region determined gastrulation. His research on spina bifida in amphibians in 1893 included the attempt to separate blastomeres of the newt.[82] That Spemann later successfully executed this operation raises the question of whether not just a transfer of techniques had taken place from the first generation of experimentalists, but also a transformation of concepts of dominant regions. Whereas Hertwig had interpreted the blastopore's powers in terms of its cellular constituents, Spemann excluded the cellular element, and concentrated on the totality of the organising region.

Thus Spemann was trained at a time when the status of the cell or its organelles as 'individuals' or fundamental 'living units' was a serious and controversial issue. This may have served as a precedent for him in his drift towards treating, not the cell, but the whole organism as his unit of analysis. Certainly the example of Boveri's 'dismantling' of the contemporary concept of the cell as an organism could have set a precedent for Spemann leading to his avoiding strictly cellular issues and giving him freedom to move to other levels of analysis.

*(ii) The rise of genetics*
The German response to Mendelism was cautious and slow. It has been suggested that Boveri unified cytology with genetics, and finally broke with morphology in 1907.[83] It was, however, an American researcher, Sutton, who pointed out in 1902 that Boveri's proof of the individuality of chromosomes supported Mendelian theory. Boveri himself was more cautious, as he only acknowledged this confirmation as a possibility in 1907. His research continued on lines independent of Mendelism. Boveri wrote to Driesch on 31 January 1903 regarding the role of chromosomes in heredity that he

did not consider that each chromosome represented a single character, but that there were perhaps 10 000 qualities on each chromosome. Moreover, perhaps each chromosome was only a container in which those particles which were the true characters could recombine during mitosis. He remained unimpressed by any single-minded reliance on Mendelian factors.[84]

Boveri's plans for a new research institute, the *Kaiser Wilhelm Institut für Biologie* indicates the cautious German response to Mendelism. It is appropriate to consider the planning of this major new institute, as Spemann became its Assistant Director. That the Institute was established after consultation with twenty-seven leading German biologists, justifies its use as an indicator of the German response to Mendelism. The Institute was hailed as a 'German Oxford', in that it would be a colony for pure research; but it was to draw on the funds of industrialists like Gustav Krupp von Bohlen, so emulating USA successes with foundations like Carnegie and Rockefeller. Certain industrialists regarded Boveri as representing a staid and sterile type of research, and preferred an approach based on the study of living organisms and plants, as represented by von Uexkull. On this issue American referees were sought. Loeb approved of Uexküll's imaginative approach as did Morgan, but Uexküll was also opposed as 'wild and unsound'. That Boveri was offered the Directorship showed that the opinion of the German biological community was decisive. However, that Spemann actually became employed in an environment where he would have been aware that a new direction for research on living organisms was called for, was indicative of the direction that his research was to take.[85]

The planning of the new Institute was revealing of the diversity of approaches to biology. For some advisers biology included anthropology, psychology and sociology, and others wanted little more than a research station, offering resources for specific projects. That the Institute ultimately included two geneticists – Correns as its Director and Goldschmidt – was far from inevitable. Boveri was not keen on the appointment of Goldschmidt and preferred Frisch, the zoologist noted for his study of bees. Spemann recommended the Lamarckian, Paul Kammerer. Boveri ultimately had to withdraw owing to the psychological strain of the negotiations, ill-health and reluctance to leave a university post. Boveri's concern with his professorial status, salary and pension caused serious tensions with the Institute's administrators. Spemann, appointed in 1913, was glad to return to the university environment and accepted the offer of a Chair in Freiburg in 1919.[86]

The evidence provided by the Kaiser Wilhelm Institute discussions is revealing of the strong alliance of Boveri and Spemann, and their sense of the complementarity of their approaches. Their antagonisms to Roux's mechanistic proposals and reserved attitude to Mendelism should be taken as a cautionary warning of an overly reductionist interpretation of the direction that they proposed for German biology. Although Boveri is usually portrayed as one of the pioneers of Mendelism and chromosome theory (and has indeed been promoted as an unfairly neglected and truly model scientist by Oppenheimer, Baltzer, Fankhauser and Wilson) this is only half the picture. As we have attempted to demonstrate, he openly expressed his reservations about Mendelism and indeed followed up other lines of interest, increasingly embryological, after 1905. Although he was relatively uninfluential in terms of the size of his laboratory and in the production of textbooks (which was the norm at the time), he was involved in a number of heated, if not openly hostile, debates suggesting that his work was valued for analytical and theoretical contributions to developmental biology.[87]

*(iii) Psychic analogies*

We would be distorting the truth if we did not take quite seriously the sorts of analogies and distinctions that were common currency at the turn of the century though so alien to us now. The use of the psychic 'analogy' by Spemann was, as he himself quite clearly insists, no mere metaphor. There is every reason to think that it represented something essential to Spemann's overall thinking. The fact that he still insisted on it in 1938 (evidently aware of how it was likely to be taken; hence his apologetic, explanatory tone) bears this out. If we remember the tradition of monism, unifying the physical with the psychic, and bear in mind the sorts of scientific problems that it seemed to bear on when Spemann was being trained we can see it was 'reasonable'. As Spemann mentions more than once, the psychic realm is a reasonable starting point in an overall view, as Leibniz had said; it is what we have the 'most intimate knowledge' of. Moreover it could serve as a model to solve the problems of heredity; hence ancestral memory and the theories of Semon and Butler. This linked very directly to Lamarckism which remained a clearly identified alternative to natural selection as late as 1883 (when Weismann first showed that natural selection could potentially explain evolution without any element of inheritance of acquired characters which Darwin and many of the evolutionary theorists had assumed to be one component). Haeckel's concept of cell souls was therefore a perfectly plausible position, fully in line with the German preoccupation with chains of being and monism. This conception of psychic matters as central and legitimate attributes of any account of living systems explains the ease with which workers like Loeb and Child could discuss simultaneously problems of behaviour, neurophysiology and development or heredity. The easy links between psychology and development were exemplified by the fact that Child was a student of Wundt; that Sachs analogised developmental and regenerative responses to behavioural stimulus and response (tropism); that one of the few and most potent examples of correlative development was the dependence of limb regeneration on nerves.[88]

The crisis of Darwinism during the 1890s – provoked in part by the discovery of the self-organising properties of embryos – resulted in increasing interest in alternative psychic and environmental explanations of evolution rather than strictly reductionist explanations. In response to Weismann's criticisms of Lamarckism, a number of psychic theories suggesting adaptive willpower of organisms were propagated.

The anti-reductionist position of Spemann contributed to an interest in psychic factors in the explanation of organic processes. From Spemann's friendship with the Lamarckian Pauly while a student in Munich, there grew a concern with psychic factors in adaptation. Indeed, it is possible that Spemann had been impressed by Haeckel's view that protozoan cells were 'cell souls', from which the nervous system and mental faculties evolved in higher organisms. Cells (as Boveri conceded) could be regarded as having the capacity to 'learn' in their responses to environmental and physiological changes. Pauly's psycho-Lamarckian theories remained a perennial enigma to Spemann as Hamburger has stressed. Pauly stressed that although an organism functioned according to mechanical laws, it was also composed of irreducible levels of organisation that had regulative powers. He emphasised psychic and teleological factors in evolution, and their applicability to understanding human culture and society. After Richard Semon's mneme theory (1904) had suggested that all cells had innate qualities of memory and Pauly completed his Lamarckian work in 1905, Spemann added his comments on correlation and evolution in 1907. Rather than seeing the evolution of the lens as the result of $10^5$ generations of cellular growth, he preferred an explanation in terms of the imprinting of *Engrammes* of hereditary information. His emphasis on cellular processes as intelligible only with regard to the totality of the organism indicated a search for explanatory factors of a higher scale than the cell. Uexküll developed environmental explanations based on interactive patterns, a concept approvingly cited by Spemann in the Silliman lectures. During the 1920s and 1930s interest in Uexküll's *Umwelt* theories increased.[89]

Psychic factors were more than just analogies to Spemann; they were factors inextricably involved in heredity and development. He remained convinced that all matter was endowed with psychic properties *(beseelt)*. He wrote in his autobiography that his experiments had shown the basic kinship of all life processes, vividly illustrated by the example that: 'we are standing and walking with parts of our body which could have been used for thinking if they had developed in another part of the embryo'. He drew interesting analogies between two-headed malformations and the development of schizophrenia. This approach makes Spemann's concluding remarks in the Silliman lectures (1938), that his concepts and experimental results pointed towards psychic analogies, less surprising than it must have seemed to those approaching embryos from reductionist perspectives. The gap between the psychic and the organic was considerably narrowed by the rise of biologistic psychology and eugenics. The prevalence of anti-reductionism was vividly demonstrated when the US researcher, Ernest Just visited the Kaiser Wilhelm Institute in the early 1930s, and found holistic and philosophical concerns with organic development to be normal among German zoologists, whereas Americans scorned philosophy for a purely empirical analysis of cell lineage.[90]

*(iv) The leadership analogy*
A number of the issues that were central for biologists at the turn of the century appear utterly foreign to us now. Prominent among these are their concerns with 'individuality', 'symmetry', 'twinning' and 'sexual differentiation'. The very property of wholeness, integrity and the maintenance of the anatomy characteristic of a species and an individual, which was so vividly demonstrated in the phenomenon

of regeneration (as well as heredity and embryonic development) seemed to biologists the essence of the problem, and they tended therefore to assume that this represented directly a property characteristic to life itself. Spemann several times states how he regarded the explanation of this wholeness and individual differentiation as the starting point of his research and its explanation as his ultimate aim.

Psychic analogies and authoritarian social thought emphasising the primacy of the whole in terms of *Ganzheit* or *Gestalt* theory did not necessarily go together. Driesch progressed from vitalist philosophy to democratic and pacificist politics during the 1920s. Monistic philosophy was as much a breeding ground for socialist zoologists like Schaxel, as for those who ultimately were sympathetic to the Nazis like Plate. We must stress that it is necessary to look carefully at each individual synthesis of biology and philosophy, and that no particular biological theory had a necessary political or social corollary. Two biological concepts are of interest in Spemann's case: inductive potency and the organiser. It might be argued that the theory of inductive potency neglected the responsiveness and movements of the future lens cells in favour of a virtual dictatorship of the optic cup. The concept of the 'organiser' represented a similarly dictatorial explanation in supracellular terms. The significance of gastrulation had been a persistent interest of Spemann; Haeckel had interpreted gastrulation as a primary causal process in the development of all multicellular organisms, and Spemann was also deeply impressed by Boveri's concept of 'privileged regions' of 1901. On the basis of his research on severing blastomeres he elaborated a view of the dorsal blastopore lip as a formative centre (*Bildungszentrum* or *Organisationszentrum*); and recognised that only certain cells in the blastopore lip could be regarded as anlagen. He used a military analogy to interpret the results of xenoplastic transplantations.[91]

Some clues as to what Spemann might have meant when he used the term *Organisator*, can be found when in 1920 he used the term in an anthropomorphic sense. He spoke of Roux as 'Organisator' and 'Führer' of experimental biology. He later referred to the educationalist Lietz in these terms, when describing the relevance of his rural schools for the Weimar State's need to train leaders. At the same time Spemann was becoming increasingly critical of cellular theories. He approvingly cited Vogt's comment on gastrulation:

'regarding this as not a "wandering of cells" but rather passive obedience to a superior force'. This may be why Spemann chose the term 'organiser' rather than 'determiner' in order to convey the sense of the distinctive qualities of living organisms. In the Silliman lectures, of which Professor Oppenheimer was one of four translators, Spemann further revealed that he wished to imply a psychic as opposed to a physical analogy. In many passages in the Silliman lectures one can detect hostility to physicochemical reductionist causal models. For example on gastrulation: 'The movements are regulated not in a coarse mechanical model, through pressure and pull of the single parts, but they are ordered according to a definite plan'. He saw formative tendencies as 'imprinted', and Spemann adopted the term 'gastrulation experience'. Holtfreter's transplanting of a 'dead inductor' dismayed Spemann. Spemann's comments can be seen as fragments of a comprehensive *Weltanschauung* that was personally sympathetic to Lamarckism and a modified form of vitalism, even though these were scientifically unproven. But since they were unproven, his scientific papers remained largely silent on these issues.[92]

Spemann's work cannot be understood if divorced from the cultural context, which gave his concepts their meaning, and which established his importance as the leading zoologist of the 1920s. To see 'the organiser' as the product of purely empirical research is to trivialise the meaning of a complex and strikingly bold concept, by which he sought to break away from atomistic, cellular explanations and to arrive at an explanation in terms of 'fields' and controlling hierarchies. The 'organiser' concept must be rooted in the early 1920s, when political chaos seemed to threaten German's unity and cultural traditions. The 'organiser' arose from *Sehnsucht* for order, as a means of national survival. Spemann attempted to combine two very different sets of methodological precepts: the experimental and the intuitive. At some future point in time, it would be possible to integrate the two realms and use science to 'prove' psychic and organisational factors. Until then, the scientist was constrained by his fragmentary evidence. The 'organiser' was a point at which such integration seemed – at least in the early 1920s – to have been achieved. By the 1930s two conflicting trends grew in opposition. On the one hand, biochemical and genetic reductionism secured new empirical foundations. On the other hand, Nazi totalitarianism forced German biology into racial, mystic vitalist moulds, which were in contradiction to empirical fact. Spemann's unifying scientific values like the organiser to provide cultural inspiration were no longer viable. The historian is left with a duality of experimentalism and idealism in Spemann's case. The conclusion must be that Spemann worked simultaneously with two methodologies, the experimental and the organicist–psychic.

# III. A half century later

The scientist's perception of the causes of scientific progress inevitably tends to differ from the account that is likely to be offered by outside observers and commentators such as historians or sociologists of science. The first two parts of this paper reflect these alternative perspectives; the first written with the aim of tracing the conceptual and experimental path underlying Spemann's scientific contributions; the second, a necessarily selective consideration of certain 'externalist' viewpoints. In presenting these separate views we hope to have established our first most important conclusion; that any adequate account of a scientist must include both types of analysis. Spemann is a particularly interesting case in point because the published material available offers a rich source of information on the scientist's working assumptions and methods, which can be set against the equally rich cultural context in which he was working. Following Collingwood's precept that 'history is the reconstruction of past thought', we have attempted to reconstruct Spemann's 'conceptual model' and to identify input into this model by social and historical factors together with their impact on the course of his scientific work. Our overriding obligation has been to try to see the problems of embryology as Spemann himself would have seen them. This therefore means that we must give as much weight to his unsubstantiated assumptions, prejudices and 'mistakes' as to the things he got 'right'. One objective of this paper has been to show that by taking into account *all* the influences impinging on him (social, cultural, educational, personal etc.) we can, at the very least, get useful extra clues about the origins of his scientific output.

Furthermore we have divided our account into two parts in acknowledgement of the fact that, although the line of separation is wholly artificial, it is no easy matter to establish relations between the two sources of evidence. Perhaps it will always remain a meaningless question to ask how much a given scientific discovery was the

Fig. 1. At left, Hans Spemann; right, Ross Harrison.

result of scientific reasoning and to what extent it was the result of social influences and beliefs. Even allowing for the complexities of the unconscious derivations of scientific ideas and of the effects of retrospective rationalisations, the 'rational' basis of scientific arguments and evidence is obvious enough. On the other hand, the fact that it is unrealistic to hope for any 'scientific proof' that a given social factor explains a given scientific development, does not mean that we do not require such explanations.

In this section we want to concentrate our attention on one of the more important and accessible points of interaction between a scientific advance and its context; this is the question of the effects of the way in which perceptions of the discovery of the organiser affected the history of the concept after Spemann's time.

## Reactions to the organiser by Spemann's contemporaries

Phases in work on the organiser have been aptly summed up by Saxén and Toivonen as the 'hopeful' 20s, 'rather confused' 30s, 'depressed and inactive' 40s and the 'new optimism and activity' of

Fig. 2. 'Die Institutskutsche', reprinted with the permission of Professor Johannes Holtfreter (University of Rochester) who as a student sketched it in 1920.

The Zoologisches Institut der Universität Freiburg i.Br. is shown as a stage coach with an animal cage on top. The horses bridled by the coach driver Spemann are the 'Großpraktikanten', third- or fourth-year students doing full-time laboratory work supervised by 3 'Assistenten' who are shown as the horse-driver brandishing a whip (Fritz Baltzer) and the two barking dogs (Bruno Geinitz and Otto Mangold). Seated in the rear is the enduring lab factotum Johann Mayer who served four professors in succession (Weismann, Doflein, Spemann, Mangold). In the distance the lab servant Mr Hund is seen in full action. Of the three persons running after the coach or clinging to it, at least one (Frau von Liebenstein) must also have been a student (see label). Among the horses, the following are of interest to the historian of embryology: Hilde Pröscholdt (later Hilde Mangold) (second in the left row, eyelid bashfully lowered); Victor Hamburger (running nose up beside Hilde Pröscholdt); Johannes Holtfreter (covered by Baltzer's right arm and whip), and Spemann's future son-in law, the geologist Ernst Cloos (the bucking horse at the left rear).

The puns in the frames read as follows (clockwise, starting from top left):

Vorne trabt auch Litzelmann, allen anderen voran

Seht unsern Junker, den bekannten Sonnengott der Praktikanten.

Eier suchen diese beiden und Buketts - in Laubheuschrecken. Hamburger läuft rechts von Beiden, Pröscholden nicht zu verdecken.

Die Biene sticht, Herr Legewie wohl mit dem Eier-Leger wie?

Herr Snethlage, sie haben recht Gott schütze uns vor dem Geschlecht.

Dieses ungleich große Paar glaub ich hat die Würmer gar Die stattliche Figur ist Koether die kleine Dicke da Holtfreter.

Arbeitslust ohne Erlahmen, schätz ein jeder an der Stute. Und so sind auch diese Damen Aller Stolz im Institute Rechts die beiden heißen Schmidt Fräulein Mense ist die Dritt.

Ach Herrje, ist das der Cloos? Bei dem ist der Deubel los!

Meine Stentoren fressen üppig dabei lachte Luxburg pfiffig.

DochFräulein Krutmann kann nicht hindern, daß ihre Viecher sich stets vermindern (Spemann):
Diese Fahrt scheint problematisch darum ist sie mir sympathisch.

Der ganzen Bande Mißgeschick fällt stets auf Maiers Haupt zurück.

Bescheiden wirkt im Hintergrund der vielbeschäftigte Herr Hund.

Last not least kommt auch Herr Klähm der Ammonit scheint unbequem.

Blumenfreund ist diese Dame, Schuhmacher so ist ihr Name.

Fest hält und treu trotz Ehepein am Studium Frau von Liebenstein.

Es ist auch Mangolds Hauptbes- treben die Biester etwas anzuregen.

Du faules Aas, willst Du wohl laufen!
Bei Baltzer gibt es kein Ver- schnaufen!

Ist das nicht der Geinitz hier? Voller Streit- und Mordbegier?

The word 'Buketts' refers to the respective stage of meiosis which Hamburger and Pröscholdt were supposed to study in grasshoppers; indeed some drawings of meiotic chromosomes by Hilde Pröscholdt are still preserved. However, a Bukett (ornamental bunch of flowers) is also an inevitable attribute of wedding ceremonies! The 'Würmer' (worms) of Koether and Holtfreter refer to their task of studying *Ascaris* embryos, and also to the fact that Koether was nicknamed 'Würmle' on account of his slim and bent figure (this information was kindly furnished by Professor Holtfreter).

The original drawing was in the possession of Fritz Baltzer; after his death it was given to an Italian scientist (according to a rumour that could not be verified). (*Legend kindly supplied by Professor K. Sander.*).

the 50s.[93] Their paper in this volume, together with Brachet's, explains the development of the field subsequent to Spemann's contributions. The discovery of the organiser evidently created immediate responses. In 1925 Harrison referred to 'the opening up of this new Yukon to which eager miners were now rushing to dig for gold around the blastopore'.[94] Possibly the first reference in English to the work was by Huxley[95] and English readers soon had the opportunity to study Spemann's own account of the work in the *British Journal of Experimental Biology* of 1925. Spemann delivered a Croonian lecture in London in 1927. In Cambridge it was news of these exciting discoveries that attracted Waddington from geology and population genetics and to the initiation of organiser work on the chick embryo by 1929. He visited Spemann's laboratory in 1931 as well as Mangold and Holtfreter in Berlin; he spent six months there in 1931–2 and Needham joined him at that time. In 1929 Huxley, Harrison, Child and Lillie were contributors to volumes of *Roux's Archiv* dedicated to Spemann on his 60th birthday. Huxley and DeBeer's influential book published in 1934, coming at the high point of optimism with regard to the prospects for a chemical identification of the organiser, presented Spemann's results in what was to prove to be their most coherent and positive form.

The responses in the USA were, at least at this time, obviously enthusiastic. In 1929 Conklin spoke of Spemann as someone 'whose experimental analysis of the development of the amphibian egg is the admiration of the scientific world'.[96] Harrison in 1933 wrote that 'The most important advance in embryology of late years has been Spemann's discovery of the organization center and organizer in the amphibian egg'. Harrison was also concerned with the terminology:[97] 'The use of the term "organizer" is likely to be attended by some confusion, for the word may be readily taken to imply more than we are really justified in attributing to the thing itself.' Typically Morgan was quick to make an incisive and critical appraisal of Spemann's concept in the following terms;[98]

Spemann has called this dorsal lip region the 'organizator', meaning that it acts as a center of influence for the rest of the egg. It might appear, without further explanation, that the organizator has a mysterious influence on the neighbouring parts, but the evidence so far at hand would not seem to justify such a conclusion; for, as Spemann has pointed out, the influence seems to be due to the presence of the cells, underlying the neural plate cells, that are destined to produce the chorda-mesoderm.

Three of the most important American embryologists can be said to have taken up the same issues as Spemann before the Second World War: Lillie, Child and Harrison. From about 1930 Lillie had turned his attention to the study of feather formation. Although his interests covered a wide span (including patterns of pigmentation and their variations in different strains of hens), he described his studies in terms such as 'The nature of organizing action'.[99] Child also explicitly associated his work with the organiser, although, under that term, he included mostly work on regeneration in invertebrate organisms carried out considerably earlier. He expressed his position in the following way: 'The hypothesis of specificity of induction and of specifically different substances at different levels of the primary organizer pose more problems than they endeavor to solve'; 'the burden of proof is on the seeker for specificities'.[100] Thus Child insisted on his gradient theory and attempted to encompass the whole range of developmental and regenerative phenomena in terms of respiratory rates.

Harrison's views on Spemann's work are perhaps of particular interest given the number of remarkable parallels in both the lives, experimental techniques and high reputations of the two men. Of all the American embryologists he might have been expected to have come closest in his work to the issues of induction raised by Spemann, but this was hardly the case. He never worked on the early embryo or primary organiser and the only work that *directly* concerns induction is a two-page paper on self-differentiation of the lens in *Amblystoma* in 1920.[101] Perhaps the strongest clue we have to why this might have been comes from an important paper in 1933,[102] which is one of the rare occasions on which Harrison deals specifically with theoretical problems or explicitly with induction. The paper addresses the problems that had already arisen for the German workers as they realised that the organiser could not be defined until the relative roles of inductor and responding tissue had been defined, namely the 'fact that there is no certain criterion by which (the question of the state of determination of an organ rudiment) can be answered. A number of tests involving different conditions may be applied, but they frequently do not give the same answer to the question'.[103] In the course of a penetrating analysis of the alternative and conflicting criteria, Harrison returns to what he describes as 'One of the most baffling results',[104] his own discovery that although the lens fails to develop in *Amblystoma* after removal of the eye rudiment, which is consistent with induction, it *does* develop if the prospective lens-forming ectoderm is transplanted to a new site in the head, which is not. Irrespective of how influential this most puzzling result might have been in shaping Harrison's views on induction, the 1933 paper makes clear how important clear design of experiments was to Harrison: 'Experimental embryology will be placed on a sounder basis if its questions are framed more carefully'.[105] It seems probable that both Harrison and Spemann realised the differences in emphasis in their work. As Harrison puts it, 'there is probably no one who would maintain that the steps in the determination ... of the lens are the same as those in the case of the limb'.[106] For Spemann, Harrison's work on the limb field lies 'outside the narrower frame of inductive development'.[107]

Thus in the United States responses to Spemann's work were disparate and not a little confusing. It is not the case that the issues had not attracted attention; Mangold in his review of the problem of lens induction in 1931 lists twenty-seven Americans who had by then become involved in the subject in some way. Yet when Morgan came to review embryology in the 766 pages of his *Experimental Embryology* in 1927, he devoted one paragraph to the lens.

# World War II and after

The devastating consequences of the Third Reich and World War II for the Spemann school are not hard to document. Hamburger, Schotté, Waelsch and Holtfreter migrated to the United States. For those remaining in Germany work on the chemistry of the organiser did not get going again until the late 1950s under Tiedemann. However, in view of the increasing domination of America in science after the war, we have to consider the effects of significant changes in emphasis continuing there. A transformation can be detected if one compares the contents of the *Growth Symposia* over the period. The first meeting of this series in 1939 was in many respects dominated by the Europeans and had a high input from ex-members of the Theoretical Biology Club, including Berrill, Needham, Woodger, and Waddington. If one compares the issues discussed in the sixth symposium held in 1946 the contrast is dramatic; the issues are now based on techniques and the fragmentation of thinking is indicated by the diversity of the narrow theoretical issues covered. There were only two Europeans at this meeting. Induction was essentially covered in only one paper. The pattern of research interests of the four members of the Spemann school who had settled in the United States also reflected the prevailing trends. None continued to work directly on induction or the organiser; Waelsch entered genetics, Holtfreter cell biology, Hamburger neurobiology and Schotté regeneration.

What had happened, to put it in its broadest terms, was a swing from the perspective of the whole organism to a growing realisation that the potentialities of individual cells encompassed, and might well explain, the phenomena previously studied in the organism. Of the innumerable factors contributing, the development of the electron microscope, increasing appreciation of the properties of biomolecules and membranes, and improvements in cell culture perhaps stand out.[108] In the latter field embryologists themselves played a major part; Holtfreter demonstrated selective reaggregation of amphibian embryo cells in 1939 and in 1943, and with his concept of the 'surface coat' contributed to the growth of interest in the extracellular matrix. As far as the pre-war embryological issues are concerned we can see two sorts of consequences. On the one hand there was a turning away from the vexed issues of the nature of the inductive stimuli or the nature of determination towards studies of the cellular and molecular routes available for such signals: intercellular junctions, the effect of intercellular filters, and the like. Secondly there was a fragmentation of interests among developmental biologists, reflected in the creation of subspecialisations centred on the use of technically increasingly demanding material, such as chick and later mammalian embryos. In another paper in this volume Wallace describes how the study of regeneration, once coextensive with embryology, became totally detached.

The introduction of the term 'developmental biology' in the early 1950s to replace 'embryology' seems to have been not only a clarion call to embryologists to create a discipline to rival 'cell' or 'molecular biology', but also an expression of the broadening of the interests of members of the discipline and the need to maintain links. During the 1960s the same tendencies were evident as molecular biology had increasing impact. As Lederberg put it in the introduction to a new Developmental Biology journal in 1966: 'The field has had enough fancy; more recently its methodology has been under enormous pressure to accommodate the inspirations of molecular biology and the models of development that can be read into microbial genetic systems. But now, as this volume amply shows, it is responding'.[109] Looking back in 1974 Bonner saw it as follows:

During the last 20 years developmental biology has become totally transformed, but the change has been for a curious reason. It has relatively little to do with its own progress, which has been steady but slow and sedate. It is almost entirely due to the staggering success of molecular biology and molecular genetics. They have rushed ahead with lightning speed and as a result we see the problem of development in an entirely new way. The developmentalist looks anxiously at molecular biology to learn how he might gain deeper insights into their problems while the molecular biologist feels that now that he has solved the problem of genetics it is time to annihilate the problem of development with the same vigor.[110]

Although of steadily diminishing interest to most developmental biologists, some of the issues of Spemann's day have been resolved.[111] Barth's discovery in 1941 of 'spontaneous' neurulation of ventral ectoderm in culture seemed finally to take away any role for the inductor. However Holtfreter proceeded to argue that it was the supposedly neutral medium which was acting as an indirect inductor; he argued that damage to the cells ('sublethal cytolysis') was releasing their own masked inductors in the same way as he had earlier shown with killed tissue. Needham pointed out the logical flaw in the concept of masked inductors as an explanation for distinctive patterns of induced response as follows: such 'very different regional effects would require as many different kinds of sublethal cytolysis' for their explanation.[112] The period around the war was dominated by theories that attempted to explain differential actions of inductors despite their lack of chemical specificity. Independently of Child, a number of quantitative models were proposed: Dalcq & Pasteels (1937); Yamada ('Double potency theory', 1939); Nieuwkoop ('Activation/Transformation' theory, 1952); Toivonen ('Two-gradient' theory, 1950). With the discovery of the 'pure mesoderm inductor' by Toivonen in 1953, no identification of the actual physiological inductor can be made, but the demonstration of predictable, chemical selection of specific alternative patterns of differentiation in embryonic tissue has re-established the concept of a qualitative inductive action favoured by Spemann.

What of the situation today? Recently Saxén posed the question: 'Can it be that the pioneering work of one great scientist has led his successors astray or along a pathway that will come to a dead end, despite the superb methods and wealth of information available today?' It is possible to find diametrically opposed answers among leading present-day embryologists. According to Saxén, 'By and large, the results of recent work performed in Japan, in Europe and in the United States are confirmatory rather than revolutionary and show with modern techniques that Hans Spemann was right about many essential points and that he was years ahead of his time',[113] while according to Nieuwkoop, 'Spemann's conclusion that the dorsal blastopore lip of the amphibian embryo represents the "organization center" for the embryonic anlage is still perfectly correct'.[114] On the other hand if one turns to look how induction figures in currently influential models of development, for example Lewis Wolpert's theory of positional information (discussed elsewhere in this volume) one finds it discussed in the following terms;

Induction and its related concepts, which have so dominated embryological thinking, have completely obscured the problems of pattern formation; the failure of inductive theory to consider the problem of spatial organization; Inductive processes will not be discussed, and I regard the misuse of concepts of induction as a major feature preventing progress in understanding pattern formation.[115]

## Some explanations

We suspect that the status of Spemann's work will undergo further reappraisal in the future and that the pattern of events up to the present time may eventually be seen as one of the more bizarre examples in the history of biology of the way timing conspires to influence events. The picture of Spemann that emerges for us is of a supremely gifted experimenter whose discoveries have become significantly inaccessible to us today, simply as a consequence of the way that the passage of time has separated our perspectives and emphases from those of his time. His scientific ideas stem from the pre-Mendelian era when no distinction between embryology and genetics was possible, through a time when organicism expressed people's views of what were the most important features of organisms in need of explanation, and his final position was expressed just prior to the time when cell biology would finally reveal the full extent of the potentialities locked up within every cell. This would explain the difficulties we have with his ideas now, but how was it possible for the progressive neglect of his work to get underway? We have mentioned a number of possible contributory factors in the course of this paper: the break in the tradition of his work due to the war, coming at a time when the field was suffering the disillusionment and confusion that resulted from the chemical studies; the coincidental arrival of the emphasis towards cells and macromolecules, and so on.

One cannot help wondering whether things might not have been different if the precise timing of events in his scientific work had been different. We have seen how, on any number of occasions, the sequence of experiments was largely fortuitous, and there was no reason that Spemann could not have discovered the organiser some fifteen years either side of the time he actually did. Lewis, after all, had made the same observations fourteen years earlier; heteroplastic transplantation had been available since the 1890s. As it was we have seen how, in a number of respects, the way Spemann presented his findings was influenced by circumstances. The discovery coincided with widespread interest in the idea of fields; the emphasis he chose to put on his interpretations was influenced by his earlier work on the lens and yet that work would not have started if he had not happened to select a species in which lens development is dependent on the eye.

However, above all, it was the way that Spemann himself perceived and presented the organiser findings, which bears the unmistakable stamp of a conceptual framework acquired in the 1980s, that explains the subsequent course of events. It was this outdated perspective which caused him to divert attention to a dual mechanism and to stake his life's work on one in particular. If, nonetheless, we can make allowances for our historical distancing from Spemann's language, analogies and assumptions, we may yet rediscover the experimental data, still intact, and of as much relevance now as then.

The authors are greatly indebted to the following for their assistance: J. Brachet, S. Glucksohn-Waelsch, V. Hamburger, J. Harwood, J. Holtfreter, N. Jacobs, E. Mayr, J. M. Oppenheimer, K. Sander and L. Saxen.

## Reference sources on Hans Spemann

Principle sources are designated A–L, and these code letters are used in subsequent references.

(D) Baltzer, F. (1967) Theodor Boveri, life and work of a great scientist. Univ. Calif. Press: San Francisco.

(I) Baltzer, F. (1942) Zum Gedächtnis Hans Spemann, *Naturwissenschaften* **16**. 229–39.
Bandlow, E. (1969) Hans Spemann's Beitrag zur Widerlegung des Neovitalismus. In *Naturphilosophie-von der Spekulation zur Wissenschaft*: Berlin.
Bautzmann, A. (1942) Hans Spemann zum Gedächtnis. *Morphol. Jahr* **87**, 1–26.
Bautzmann, H. (1943) Johannes Müller und unserer Lehre von der organischen Gliederung und Entwicklung. *Anat. Anz.* **94**, 223–56.
Bautzmann, H. (1950) Hans Spemann's Organisatorenlehre im Rahmen des Mechanismus-Vitalismusproblems. *Der mathematische und naturwissenschaftliche Unterricht* **2**, 264.
(J) Bautzmann, H. (1955) Die Problemlage des Spemannschen Organisators. *Naturwissenschaften* **42**, 286–94.
Eakin, R. M. (1975) *Great scientists speak again.* Univ. of Calif. Press: Berkeley.
Fankhauser, G. (1972) Memories of Great Embryologists. *Am. Scientist* **60**, 46–55.
Goldschmidt, R. B. (1959) *The Golden Age of Zoology. Portraits from Memory.* University of Washington Press: Seattle.
Goldschmidt, R. B. (1960) *In and Out of the Ivory Tower.* University of Washington Press: Seattle.
(F) Hamburger, V. (1969) Hans Spemann and the organizer concept. *Experientia* **25**, 1121–5.
(G) Hamburger, V. (1984) Hilde Mangold, co-discoverer of the organizer. *J. Hist. Biol.* **17**, 1–12.
Holtfreter, J. (1968) Address in honor of Viktor Hamburger. *Devel. Biol.*, Suppl. **2**, ix–xx.
Holtfreter, J. (1980) In *Modern Scientists and Engineers*, pp. 83–5. McGraw-Hill: New York.
Mangold, O. (1929) Hans Spemann. *Naturwissenschaften* **25**, 453–61.
(H) Mangold, O. (1942) Hans Spemann als Mensch und Wissenschaftler. *Wilhelm Roux Arch Entwicklungsmechanikorganismen* **141**, 385–423.
Mangold, O. (1957) Hans Spemann 1869–1941. In *Freiburger Professoren der 19 und 20 Jahrhunderts*, pp. 159–82. Freiburg.
(B) Mangold, O. (1982) *Hans Spemann, Ein Meister der Entwicklungsphysiologie, Sein Leben und sein Werk.* Wissenschaftliche Verlagsgesellschaft: Stuttgart.
(L) Oppenheimer, J. M. (1962) *Essays in the History of Embryology and Biology.* M.I.T. Press: Cambridge, Mass.
(K) Oppenheimer, J. M. (1970) Cells and organizers. *Am. Zool.* **10**, 75–88.
Schnetter, M. (1968) Die Ara Spemann – Mangold am Zoologischen Institut der Universität Freiburg i. Br. in den Jahren 1919–1945. *Ber. Naturf. Ges. Freiburg i. Br.* **58** 95–110.
Seidel, F. (1955) Geschichte Linien und Problematik der Entwicklungsphysiologie. *Naturwissenschaften* **42**, 275.
(A) *Spemann, H. (1938) *Embryonic Development and Induction.* Yale University Press: New Haven. (Transl. of: Spemann, H. (1936) *Experimentelle Beitrage zu einer Theorie der Entwicklung.* Springer: Berlin.)
(C) Spemann, H. (1943) *Forschung und Leben.* Engelhorns: Stuttgart.
Twitty, V. C. (1966) *Of Scientists and Salamanders.* Freeman: San Francisco.
Waddington, C. H. (1942) Hans Spemann. *Nature* **149**, 296.

Waddington (1975) Hans Spemann, In *Dictionary of Scientific Biography* **12**, 567–9.

(E)\*Willier, B. H. & Oppenheimer, J. M. (1964) *Foundations of Experimental Embryology.* Englewood Cliffs, NJ: Prentice-Hall. (Transl. of: Spemann, H. & Mangold, H. (1924) Über Induktion von Embryonalanlagen durch Implantation artfremder Organisatoren. *Wilhelm Roux Arch. Entwicklungsmechanik Organismen* **100**, 599–638.)

Woellworth, C. von (1961) Otto Mangold. *Embryologia* **6**, 1–22 (includes list of publications and students).

*Archival Material*

Hubrecht Laboratory, Utrecht: Spemann (1899–1928) protocols, drawings and slides. Also includes material by Rotmann and Otto Mangold (1913–1959; other material in Institute of Human Anatomy, Giessen).

Senckenbergische Bibliothek, Frankfurt: lecture notes, slides, correspondence, photographs, school books, diaries etc.

Bayerisches Staatsbibliothek: letters from Boveri to Spemann.

Zoological Institute, Frieburg i. Br.: lists of students, documents, photographs.

*Spemann's complete bibliography*; see Mangold (1942).

*Works available in English* (in addition to those already indicated\*)

Nobel Lectures, Physiology or Medicine, 1922–1941 (1964). Elsevier: Amsterdam.

Spemann, H. (1925) Some Factors in Animal Development. *Brit. J. Exp. Biol.* **2**, 493.

Spemann, H. (1927) Organizers in Animal Development. *Proc. Roy. Soc. B* **102**, 177.

*Publications by the 'Spemann school'*; see Mangold (1929).
60th Birthday issues of *W. Roux Arch. Ent.* 116–20, 1929.

*Reviews and reference sources on induction and the organiser*

Holtfreter, J. & Hamburger, V. (1955) Embryogenesis: Progressive differentiations: Amphibians. In *Analysis of Development*, ed. H. Willier, P. A. Weiss & V. Hamburger. Saunders: Philadelphia.

Saxen, L. & Toivonen, S. (1962) *Primary Embryonic Induction.* Logos Press: London.

Nakamura, O. & Toivonen, S. eds. (1978) *Organizer – a milestone of the half-century from Spemann.* Elsevier: Amsterdam.

# Endnotes and references

References coded A–L below are given in full in 'Reference sources on Hans Spemann' above.

1. Leclerq, J. & Dagnélie, P. (1966) *Perspectives de la Zoologie Européenne.* Gembloux, Belgium: J. Duculot.

2. In this paper all references by date to Spemann's publications refer to his bibliography as contained in Mangold (1942); Reference H in Reference Sources above.

3. Unless otherwise stated all biographical details in Part I of this paper are taken from References (B), (C) and (D) in Reference sources above.

4. 'I had heard much about him from Pauly.' (C) p. 169. 'In Heidelberg I had often heard (his name). He was held in high esteem, indeed admiration by the Heidelberg anatomists, because he had made a discovery of the first rank in comparative anatomy, the kidney in *Amphioxus*.' 'When I first heard of this man, I . . . conceived the wish eventually to work under his scientific guidance.' (C) pp. 139–40.

5. Dated references to Boveri's publications in this paper refer to his bibliography as contained in (D). 'Investigations on the problem (of chromosome diminution) in the development of intestinal worms had given rise to differences of opinion between Boveri and other authors, which Spemann's introduction of a new species might clarify.' (B) p. 93.

6. (D) p. 129.

7. (D) p. 37. 'I most wanted to repeat and follow up the . . . experiment of Oskar Schultze', who had obtained double-headed embryos by inverting amphibian eggs. (C) p. 173.

8. (B) pp. 21–2.    9. (B) p. 24.    10. (B) p. 23.    11. (C) p. 179.

12. Spemann, 1901*c*, p. 224. The technique had been introduced, using silk thread, by Oskar Hertwig in 1893 with the specific aim of achieving in the frog what Driesch had been able to achieve in sea-urchin embryos, that is cell separation in place of Roux's cell killing. Herlitzka later used hair-loops (fine, sterilised hairs obtained from women). Retrospectively Spemann justified the need for his experiments in terms of the inadequacies of the four previous studies, none of which obtained double-headed embryos ((C) pp. 180–1; Spemann, 1901*c*).

13. (A) pp. 142, 321.

14. E. B. Wilson (1918), In *Erinnerungen an Theodor Boveri*, ed. W. C. Röntgen, p. 80. Tübingen: J. C. B. Mohr.    15. (A) pp. 318–21.

16. (E) pp. 92–3. His views were encapsulated in the title of another paper (Boveri, 1904); 'Protoplasmadifferenzierung als auslösender Faktor für Kernverschiedenheit'.

17. The following quotation (from Boveri, 1902: (E) p. 91) illustrates the use of such terms:

Polarity and bilateral symmetry depend on the cytoplasmic pattern, and all malformations connected with these axial relations, such as duplications of larvae or the perpetual blastulae originating from fragments of the animal half only and incapable of undergoing polar differentiation, are based on disturbances of defects of the cytoplasm. The structure of the egg cytoplasm takes care, if I may say so, of the purely 'promorphological' tasks, that is, it provides the most general basic form, the framework within which all specific details are filled in by the nucleus.

18. Spemann, 1901*a*, p. 63. Early references to work on eye development are reviewed in Mangold, 1931*b*, in (A); Herbst, 1901, in (A); (A) pp. 55–6; Op-

penheimer ((L) p. 300) traces the idea of lens induction back to von Baer. The priority dispute between Spemann and Herbst in 1901 on the subject suggests that the idea might have been new to them.

19. Spemann, 1901*a*, p. 77.    20. Spemann, 1901*a*, p. 64, (A) pp. 47–8.

21. This linking of the work is indicated in extensive references to the lens in his final paper on the hair loop (Spemann, 1903*a*).

22. Other words used in place of correlation include '*Auslösung*', '*Anstoss*' (Spemann, 1901*a*); '*Formative Reize*', '*Induktionserscheinungen*' (Herbst, 1901, in (A)). On Spemann's early views on induction, see Spemann, 1907*a*. On Herbst's views see Herbst, 1901, in (A). For historical background see: L; J. M. Oppenheimer, 1970. Some diverse backgrounds for Curt Herbst's ideas about embryonic induction, *Bull. Hist. Med.* **44**, 241–50. F. B. Churchill, 1969. From machine theory to entelechy – two studies in developmental teleology, *J. Hist. Biol.* **2**, 165–85.

23. Spemann, 1901*a*, pp. 64–5, p. 77.

24. Lewis, 1904*a*, in (A).    25. Lewis, 1907, in (A).

26. Spemann initially thought he could account for the results of Mencl and King, and, such was his confidence that his new experiments would prove him right, he waited six months before looking at the results ((C), p. 194). Prior to this time he had used a magnifying glass (*Lupe*); he acquired the dissecting microscope at Braus's instigation (Spemann, 1906*a*, p. 197).

27. (A) p. 4. Braus himself had adopted the idea from an engineering concept. Spemann thought of it as having wide application ((A) pp. 95–6). Boveri also refers to it (Boveri, 1906, in (D)).

28. Spemann, 1910. It was this work which led Harrison to investigate establishment of polarity in limb-buds (Harrison, 1921*b* in (A) p. 2).

29. (A) pp. 89, 53, 159, 225, 358.

30. Spemann, 1920.    31. (B) pp. 2, 133.    32. Spemann, 1912*a*, p. 90.

33. (A) p. 135.

34. On how he arrived at these ideas, see (A) p. 128, (B) pp. 154–6.

35. (B) p. 159. Spemann had actually attempted to use the technique of heteroplastic transplantation around 1907 (Spemann, 1916*c*, p. 319) and referred to Harrison's earlier use of it as one 'from which I have learned more than from almost any other investigation, not only for technique, but also for methodically advanced analysis' ((A) p. 131).

36. (A) p. 143.

37. 'At first I thought that it had been formed from the material of the implant' ((A) p. 141). The same interpretation had been made by Lewis (1907*c*) when he earlier observed the organiser effect. Spemann had not used the heteroplastic transplants in these first studies (1916*c*, 1918*a*). By 1916 the concept of a fate map was becoming explicit ('*Schicksal*', '*topographische Karte*', 1916*c*, pp. 308, 318): earlier Spemann *inferred* the relation of hair loops to prospective structure on the basis of the later effects of the constrictions.

38. See Spemann, 1921*b*. The German term was '*Organisator*'. Since the note appended to the paper is dated May, 1921, Spemann may not have had much time to consider matters of terminology.

39. (E) pp. 173, 180.    40. (E) p. 147.    41. Spemann, 1925*c*, p. 501.

42. He had considered the matter of possible induction of neural plate by archenteron roof on several previous occasions and as early as 1901 ((A) pp. 8, 53 159).
43. (A) pp. 372, 197–8.
44. Spemann, 1927*f*, p. 185. The title of chapter 4 in the Silliman lectures, on lens induction, is entitled 'First analysis of the process of induction' implying that it was comparable to a second, that of the organiser.
45. (E) pp. 182–3.
46. With the exception of one paper prepared after his retirement.
47. His priorities (Spemann, 1924*a*, 1925*c*) were to investigate (1927*f*) the 'extent of the centre of organization' and its 'origin'; 'intimate structure') (including 'longitudinal structure', 'laterality' and 'regional structure'); 'nature of the inducing agent' (whether 'chemical or by dynamic means').
48. (A) p. 156.       49. (A) p. 369.
50. (E) p. 178. This and related work is reviewed in (A), pp. 148–69.
51. (A) p. 190.
52. Even prior to 1931 the discovery of various subcategories of inductive action had complicated the picture, as described in (A), Chapters 10 and 14.
53. (A) p. 247. Where tissue, at a stage when its fate can still be changed by transplantation to new sites within the embryo, undergoes specific differentiation when in 'neutral' environments, it can be said to possess an early, but reversible, form of determination. *Bahnung* is quite different ((A), p. 197); it refers to a predisposition to react in specific ways to various, potential stimuli. Waddington's term 'competence' lumps both together (*Phil. Trans. Roy. Soc.* (1932), B221, p. 223).
54. V. C. Twitty (1966) *Of Scientists and Salamanders.* San Franciso: Freeman, p. 41.
55. **Summary and index of evidence on the organiser in the Silliman lectures: historical explanations.** *Evidence that the organiser effect is mediated by mesendodermal products of gastrulation via neural induction.* General discussion: Chapters 7,8, pp. 303–5. (a) In the absence of mesendodermal invagination there is no neurulation: i.e. after removal of the blastopore, pp. 173–5; archenteron roof (inconclusive), pp. 172–3; due to exogastrulation, pp. 181–90, 255, 293. (b) Mesendoderm alone is sufficient to bring about neurulation: i.e. following *Einsteckung* of archenteron roof, p. 161 or dorsal blastopore lip, pp. 161–3, 262–6; within ectodermal explants, pp. 227–8. (c) Supporting evidence: ectodermal commitment to neurulation coincides with mesendodermal invagination, pp. 188, 253; vegetal half of embryo controls axis of dorsal, pp. 155, 158, 160; twinning reflects bifurcation of invagination, pp. 159–60; neural orientation reflects gastrulation axis, pp. 148–55, 266, 369; 'organiser' distribution in pre-gastrula corresponds to prospective invaginative tissue, pp. 162–3. *Evidence for Spemann's alternative (non-inductive) organiser mechanism.* General characterisation: pp. 82–5, 95, 147, 156–9, 187, 189, 303, 309, 368–9; 'preparatory process' (pp. 169, 195–6), 'clearing the track', (p. 197), 'smoothing the way' (p. 169), 'meets the inductive action halfway' (p. 195); physical rather than chemical, pp. 158, 221–2, 230–1, 297, 303, 371; proposed experimental tests, pp. 147, 303–5; discussion of dual mechanisms, pp. 92–7, 147, 155–163, 187–9, 197–8, 247,

259, 304, 306, 371; double assurance, pp. 92–7, 169, 189, 197. Developmental parameters potentially independent of induction as possible measures of the alternative mechanism: (a) 'dynamic determination (i.e. inbuilt preconditions of cell motility mediating gastrulation, pp. 101–10, 118–23, 205–8). Primarily due to Holtfreter's (1933) disconfirmation of Goerrtler's (1927) claims, Spemann considered dynamic factors not to contribute to cell fates ('material determination'), pp. 101–10, 118–19, 121, 125, 159, 190–8, 254, 287–8, 344. (b) 'Labile determination. (i.e. reversible early disposition towards specific fate (Note 53), pp. 49, 53–4, 110–18, 136–40, 171–2, 190–8, 206–7, 252, 257, 287) – cannot contribute to cell fate because would reverse in intact embryo; may reflect early inductors. (c) 'Reaction System': timing. Responses to developmental signals are limited by the responding tissue (pp. 89–92, 197, 202, 207, 237, 250, 257, 306–7, 348–66, 369–70' – ascribed to an autonomous 'facilitation of the reaction to the determining influence' (p. 197) ('*Bahnung*'), pp. 197, 277, 341, 345; paradigm cases were lens and homeogenetic induction (pp. 214–15, 250–1, 278, 281, 283–4). Signal and responsiveness (i.e. *Bahnung*, p. 247) show 'time correlation', Chapter 12, pp. 91, 248–59, 280, 283–96, 311, 349, 370–1. (d) Spatial organisation: 'Assimilation' (i.e. coordination of anatomical pattern irrespective of donor or host origins of component parts; within mesoderm (pp. 158, 163–6, 198, 258, 368), neural tissue (pp. 145, 198, 214–15, 281)) was a priority aspect of the organiser effect; it was broadened (p. 278, Chapter 14), after new evidence on the 'wholeness' (Chapter 17, pp. 146, 283, 312) of patterns of which, due to abnormal inductors, responding tissue showed itself capable, as 'complementary' induction if the inductor participates (pp. 279–83), 'autonomous' if not (pp. 283–96). Work on 'regional determination' of the neural plate (Chapter 13, pp. 147, 293–6, 305–11, especially tail, pp. 165–6, 281–3, 305), limb (pp. 215–18, 298–302, 312–3) and eye (pp. 78–85, 313–17) showed such responses to be of a limited number of quite distinct, internally self-organising, initially overlapping (pp. 303, 308–11), patterns (i.e. 'fields' (Chapter 15, pp. 318, 348, 361). *Spemann's final position.* The field concept, used to bring all the phenomena (including long duration (pp. 244–5, 295, 309) signals, pp. 308–17) into a single explanatory framework (pp. 303, 305; Child (Chapter 16) and Weiss (pp. 298–302) were criticised for failure to do so), did not distinguish induction from other mechanisms (p. 303) – *all* signals were inferred indirectly from responses and poorly defined as the limited response options were selectable (Chapters 10, 11) by low-specificity cues (pp. 91, 221, 237). Holtfreter (pp. 187–9) argued that exogastrulation results positively excluded the alternative mechanism; Spemann saw this as 'not yet proved' (p. 188) but the case *for* it was a negative one, a 'possibility ... decidedly undeniable' (p. 158) and 'not yet cogently excluded' (p. 156). *Historical explanations.* Why, with the first heteroplastic organiser results, did Spemann not pursue induction and its ability to encompass the data instead of seeing evidence for a new mechanism, introduced by two new explanatory concepts (i.e. 'organiser', 'field', chosen to contrast with older, more neutral terms – 'determiner' ((E), p. 182) and 'organiser centre' ((E), p. 180), (Goodfield (1969) *Boston Stud. Phil. Sci.* **5,** 421–49))?: (a) The lens work (pp. 170, 189) made dual mechanisms necessary. (b) Assuming early distinctions *between* germ layers (pp. 124, 136, 163 (later modified, pp. 121–2, 136–9); associated with induction), integration *within* layers suggested separate 'cell-to-cell' 'assimilative' signals. (c) Perspectives of the hair-loop experiments (returned to 1916–22), inferring early mechanisms directly from overall final anatomy as due to a wave spreading directly from a centre (i.e. *'Strom'*, *'organisirten Gebilde'*, *'Differenzirungcentren'* (1901–3 papers)), anticipated (pp. 142, 157–8, 193) organisers. (d) Transitional embryonic processes were bypassed and their capacity to explain complex end results underestimated: chains of induction (pp. 166–9), cell movement (pp. 101–110, 148–155) and entrainment (p. 164) explain assimilation; multiple induction stages rather than autonomous change explains *Bahnung* (pp. 207–12, 247); non-identical collaborative inductors explain double assurance (pp. 176–81, 237, 243–4, 253–4, 269–71, 274); fate maps were poorly understood or poorly documented.

56. Personal communication, by Professor R. M. Eakin, who assisted in the translation of the German edition of the lectures.
57. (A) pp. 371–2.
58. W. Müller-Lauter, 'Der Organismus als innerer Kampf. Der Einfluss von Wilhem Roux auf Friedrich Nietzsche', *Nietzsche Studien*, **7** (1978), pp. 189–223.
   F. J. Sulloway, *Freud. Biologist of the Mind*, London, 1979.
59. G. E. Allen, *Life Science in the Twentieth Century*, New York, 1975.
   R. E. Kohler, *From Medical Chemistry to Biochemistry*, Cambridge, 1982.
   F. K. Ringer, *The Decline of the German Mandarins. The German Academic Community 1890–1933*, Cambridge, Mass., 1969.
   G. Lukács, *Die Zerstörung der Vernunft*, vol. **2**, *Irrationalismus und Imperialismus*, Darmstadt and Neuwied, 1974.
60. P. J. Weindling, Theories of the cell state in Imperial Germany. In C. Webster (ed.), *Biology, Medicine and Society 1840–1940*, Cambridge, 1981, pp. 99–155. Spemann, 1938, pp. 367–8.   61. (A).
62. Zentrales Staatsarchiv Dienststelle Merseburg, (henceforth ZSTA Merseburg) Rep. 92 Althoff A1 Nr 80 Bl. 12 Richard Hertwig to Althoff, 29 June 1896. This was not the universal opinion, see Hamburger, 'Hilde Mangold', 1984.
63. F. Baltzer, Theodor Boveri, pp. 30–31. Nachlass Boveri, Bayerisches Staatsbibliothek, Boveri to Spemann 2 May 1914, 4 May 1915. For anti-semitism see also Boveri to Pauly 7 July 1906. M. Boveri, *Verzweigungen. Eine Autobiographie*, Munich and Zürich 1978, pp. 27–9.
64. V. Hamburger, Hans Spemann and the organiser concept, *Experientia* **25** (1969), pp. 1121–5.
   B. Schroeder-Gudehus, 'The Argument for the Self-Government and Public Support of Science in Weimar Germany', *Minerva* vol. **10** (1972), pp. 537–50.
   P. J. Weindling, Weimar eugenics in social context, *Annals of Science* (1985).
65. *Mangold*, 1941.
   *American Scientist*, **60** (1970), p. 46.
   H. Spemann, Zur Theorie der tierischen Entwicklung, in Spemann 1943, pp. 304–22.   66. (H).

67. Spemann, 1943, pp. 230–49. Die Volkshochschule in Freiburg i. Br., In *Spemann, 1943*, pp. 297–303. German edn of Silliman lectures.

68. Friedrich Brock, Die Grundlagen der Umweltforschung Jacob von Uexkülls und seiner Schule, *Verhandlungen der Deutschen zoologischen Gesellschaft* (1939) 16–18.
Theodora J. Kalikow. History of Konrad Lorenz's ethological theory 1927–1939: the role of meta-theory. Theory, anomaly, and new discoveries in a scientific evolution. In *Studies in the History and Philosophy of Science* (1975) **6**, 331–4.

69. J. von Uexküll, *Theoretische Biologie*, Frankfurt a. M., 1973 (1st edn 1928), pp. 240, 250–8. Frankhauser, G. Memories of great embryologists. Reminiscences of F. Baltzer, H. Spemann, F. R. Lillie, R.G. Harrison, and E. G. Conklin. *Amer. Scient.* (1972) **60**, 46–55, *ports.*, refs.
Eakin, 1963.

70. *Frankfurter Zeitung* 12 December 1935.
'Nobelvortrag', in Spemann, 1943, pp. 304–37.
*Verhandlungen*, 1936.

71. H. Spemann, Die übernationale Bedeutung der Wissenschaft, *Jahrbuch der Stadt Freiburg*, vol. **1** (1938), pp. 124–7.
Spemann, 'Die Wissenschaften im Dienste der Nation', *Jahrbuch der Stadt Freiburg*, vol **2** (1938), cited as offprint.

72. Paul Forman, Weimar Culture, causality, and quantum theory 1918–1927. Adaption by German physicsts and mathematicians to a hostile intellectual environment. In *Historical Studies in the Physical Sciences* **3** (1971), 1–115. The financial support and political alignment of physicists in Weimar Germany. In *Minerva* **12** (1974), 39–66.

73. Nachlass Driesch, Universitatsbibliothek. Leipzig, Spemann to Driesch, 11 July 1939.

74. Spemann, 1943, pp. 120, 208, 228.

75. Nachlass Driesch, 3.11.1930, 11.7.1939.

76. P. J. Weindling, *Eine Sonne im Ei*. Oscar. Hertwig (1849–1922), *Darwinismus und Sozialdarwinismus*, 1986.

77. J. M. Oppenheimer, 'Cells and organizers', *American Zoologist*, **10** (1970), pp. 75–88.

78. Weindling *Oscar Hertwig* (19). Also 'Cell Biology and Darwinism in Imperial Germany. The Contribution of Oscar Hertwig (1849–1922), London University Ph.D. Dissertation, 1982.

79. Nachlass Boveri, Munich, Boveri to Spemann, 15 Feb. 1897, 29 Dec. 1901. Boveri to Pauly, 14 Jan. 1886.

80. P. J. Weindling *op. cit.* (3)
T. Boveri, *Ergebnisse über die Konstitution der chromatischen Substanz der Zellkerns*, Jena (1904), pp. 103–5.

81. T. Boveri, *Die Organismen als historische Wesen* (Würzburg, 1906). F. Baltzer, Theodor Boveri, Stuttgart (1962), p. 166.
Hamburger, Organizer Concept, p. 1125.
T. Boveri, *Zellenstudien, VI*, 1907, pp. 230–1.

82. O. Hertwig, Urmund und Spina bifida. Eine vergleichend morphologische,

teratologische Studie an missgebildeten Froscheiern. *Arch. mikroskop. Anat.* **39** (1892), 353–503.
O. Hertwig, Ueber den Werth der ersten Furchungszellen für die Organbildung des Embryo. Experimentelle Studien am Frosch- und Tritonei. *Arch. mikroskop. Anat.* **42** (1893), 662–807.

83. Allen, *Life Science op. cit.* (2)
J. Oppenheimer, *Theodor Boveri.*

84. T. Boveri, *Zellenstudien VI,* Jena (1907).
Nachlass Driesch *loc. cit.,* Boveri to Driesch, 31 Jan. 1903.

85. Archiv und Bibliothek der Max Planck Gesellschaft (MPG). Gen. II 11 adh. 6, Band 1 betr. von Uexküll. Boveri to Goldschmidt, 16 Jan. 1913, 3 Dec. 1912. G. Wendel, *Die Kaiser Wilhelm-Gesellschaft 1911–1914,* Berlin (1975), p. 164.
L. Burchardt, *Wissenschaftspolitik im Wilhelminischen Deutschland, Vorgeschichte, Gründung und Aufbau der Kaiser-Wilhelm-Gesellschaft zur Förderung der Wissenschaften.* Göttingen (1975), pp. 108–16.

86. MPG, KWG Generalverwaltung, KWI für Biologie, Hauptakten. Biologisches Institut Bd 103.
Nachlass Boveri Munich Letters to Walter Boveri, 16 October 1912–2 December 1912. ZSTA Merseburg, Rep. 92 Schmidt Ott Bd 36 Bl 82 on Boveri's proposals of 25 Sept. 1912.

87. J. Harwood, 'The reception of Morgan's chromosome theory in Germany: inter-war debate over cytoplasmic inheritance', *Medizinhistorisches Journal,* **19** (1984), 3–32. J. Oppenheimer, *Theodor Boveri The Cell Biologists' Embryologist.*

88. Spemann, 1938, p. 372.    89. Spemann, 1907, pp. 42–7. Spemann, 1938, p. 107 on Uexküll (who interestingly was not included in the index).

90. Spemann, *Autobiography,* pp. 158–9, 167. K. Manning, *Black Apollo of Science,* New York (1984).

91. On Gestalt psychology and the impact of embryology see M. G. Ash, The Emergence of Gestalt Theory: Experimental Psychology in Germany 1890–1920, Harvard University PhD 1982.
Spemann, Silliman Lectures, pp. 106, 142.
Spemann, *Verhandlungen der physikalischen medicinischen Gesellschaft zu Wurzburg,* vol. **44** (1916)
Waelsch, personal communication.

92. H. Spemann, 'Wilhelm Roux als Experimentator', *Die Naturwissenschaften,* 1920, p. 443.
Spemann, 1938, p. 106.
Oppenheimer, 1970, p. 85.

93. Saxén, L. & S. Toivonen (1962) *Primary Embryonic Induction.* London: Logos.

94. Twitty (1966) *op. cit.* p. 39.

95. J. S. Huxley (1924) Early embryonic differentiation. *Nature* **113**, 276–8.

96. E. G. Conklin (1929). Problems of development. *Amer. Zoologist* **63**, 31.

97. R. G. Harrison (1933) Some difficulties of the determination problem. *Amer. Naturalist* **67**, 316–7.

98. T. H. Morgan (1927) *Experimental Embryology*, p. 239. New York: Columbia University Press.

99. R. S. Lillie (1938) The nature of organizing action. *Amer. Naturalist* **72**, 389.

100. C. M. Child (1946) Organizers in development and the organizer concept. *Physiol. Zool.* **19**, p. 121.

101. R. G. Harrison (1920) Experiments on the lens in *Amblystoma*. *Proc. Soc. Exp. Biol. and Med.* **17**, 199–200. Among the parallels: they were virtually the same age; both trained in medicine in Germany and began research with lineage-tracing studies; subsequently studied amphibian embryos exclusively, with similar techniques and research objects, including neurulation, lens, inner ear, bilateral symmetry, differential growth in heteroplastic recombinants; they shared similar ideas on fields ((A) p. 84; Oppenheimer (1962) p. 15) and the molecular basis of embryonic organisation (Haraway (1976) *Crystals, Fabrics, and Fields*, p. 89) which Spemann called '*Intimstruktur*'. Their theoretical concerns were equivalent (compare Chapter 9 in (A) with Harrison (1933) *Am. Naturalist* **67**, 306–21). They were close friends having first met around 1900 ((B) p. 30).

102. R. G. Harrison (1933) *op. cit.*   103. *op. cit.* pp. 308–9.

104. *op. cit.* p. 315.   105. *op. cit.* p. 319.   106. *op. cit.* p. 319.

107. (A) p. 313.

108. J. M. Oppenheimer (1966) The growth and development of developmental biology. In M. Locke (ed.) *Major Problems in Developmental Biology*, pp. 1–27. New York: Academic Press. Also Oppenheimer (1970) (K).

109. J. Lederberg (1966) *Current Topics in Developmental Biology* **1**, p. ix.

110. J. T. Bonner (1974) *On Development*, pp. 2–3. Cambridge, Mass.: Harvard University Press.

111. Post-war emphases can be readily assessed from the 1978 review edited by Nakamura and Toivonen. Apart from the continuing concern with chemical identification of the inductor, principal themes covered are: certain revisions of the fate map; origin of the organiser region in pre-gastrula stages; revision of Holtfreter's original maps of the capacity of pre-gastrula prospective germ layer tissues to differentiate autonomously as explants and analysis of the interdependence of such early determinations. Given the way in which inductions in individual organs such as lens acquired the terminology of the organiser and, by implication, its problems, there has been little incentive to return to these possibly simpler examples of induction.

112. J. Needham (1955) Developmental physiology. *Ann. Rev. Physiol.* **17**, p. 39.

113. *ibid.* p. 315. L. Saxen, S. Toivonen & O. Nakamura (1978). In *Organizer – a milestone of a half-century from Spemann*, ed. O. Nakamura & S. Toivonen, p. 316. Amsterdam: Elsevier.

114. P. D. Nieuwkoop (1973). *Advances in Morphogenesis* **10**, 35.

115. L. Wolpert (1970) In *Towards a Theoretical Biology*, Vol. 3, ed. C. H. Waddington, pp. 202–3. Edinburgh: Edinburgh University Press. L. Wolpert (1971) Positional information and pattern formation, *Current Topics in Devel. Biol.* **6**, p. 184.

Part 2

# Early interactions between embryology and biochemistry

## J. BRACHET

Department of Molecular Biology, Free University of Brussels, Belgium

Everybody admits today that interdisciplinary research is necessary for scientific progress; this was not realised half a century ago when 'anatomists' and 'physiologists' were fighting each other. I was once accused of being a 'dirty bastard' by a professor of Anatomy, for turning to test tubes, and writing a book entitled *Biochemical Cytology* (Brachet, 1957).

However, in the period 1930–40, which will be our main topic, a few biochemists had discovered that oocytes, eggs and embryos are an ideal material for the solution of biochemical riddles; and an increasing number of embryologists believed that morphogenesis cannot be understood without the help of biochemistry and biophysics. One man did more than anybody else to expound the view that close interactions between embryology and biochemistry are a necessity: this was Joseph Needham. The impact on many young biologists of his monumental treatise *Chemical Embryology* (1930) cannot be overvalued: it paved the way for today's molecular embryology.

In this paper, I shall examine briefly two interactions which took place, in the thirties, between embryology and biochemistry. First, we shall see that much of our early knowledge about nucleic acids stems from work done on eggs. In the second part, we shall look into the attempts, made jointly by biochemists and embryologists, to elucidate the chemical nature of the inducing substance present in Hans Spemann's 'organizer'.

## I
## How eggs have led to the unexpected discovery that animal cells possess a plant nucleic acid

Why did I join, in 1927, the Embryology Laboratory of the Free University of Brussels as a young undergraduate medical student?

Professor P. Gérard, in the first lecture of the cytology course there, had told us that anucleate cytoplasm survived for some time. This struck me as so strange that I wanted to know why. Reluctantly, I decided to work in my father's (Albert Brachet) laboratory, the only place in the University where such studies could be done. Reluctantly, because I had decided never to work in my father's laboratory as he had so much influence in the University as a professor and former Rector. The solution he found was to put me in the hands of his second-in-command, Albert Dalcq (who became his successor after his death in 1930). This was ideal for me, because A. Dalcq was working at that time on nucleocytoplasmic interactions in frog eggs; in addition, this anatomist had some interest in biochemistry: he even spent three months in Cambridge studying cytochromes in D. Keilin's laboratory.

Belgian embryology at that time was still dominated by the memory of Edouard Van Beneden's gigantic figure; my father had been one of his students in Liège and he had the greatest admiration for him. When my father became Professor of Anatomy in Brussels (around 1905), he created an independent course of Embryology and built up a laboratory, where his research dealt with problems of descriptive and especially experimental embryology (which he called 'l'embryologie causale'). Around 1927, he was still doing experiments where the grey crescent of frog eggs was destroyed by pricking eggs and blastulae with a hot needle: the period in which frogs were laying eggs was called 'the season'; it did not last long, because there were no refrigerators in which to keep the frogs; also the experiments were difficult to perform because there were no dissecting microscopes. But 'the season' was a very exciting period for all of us.

There were no other embryological laboratories in Belgium at that time, but my father's laboratory in Brussels had a wide international reputation. There were close contacts with many outstanding French scientists, including Henneguy, Bouin, Prenant, Caullery, Fauré-Frémiet (a chemical embryologist at that time) and many others, because our family had spent the 1914–18 war in Paris. There were also some contacts with J. P. Hill and Barclay Smith in England, but no contacts at all with the German embryologists (H. Spemann and A. Brachet never met, but A. Dalcq later worked with O. Mangold and I paid, at my father's request, a visit to Spemann, in 1929): Belgians had extremely bitter feelings against the Germans during the post-war years. A kind of revolution took place in the laboratory when my father visited the United States in 1929: not only did he meet Conklin, Wilson, Harrison, Morgan and others, but he found out that *Drosophila* genetics is really a serious matter (very few people believed in genes in Belgium and France in those days).

The laboratory was particularly lively in 1927: besides Dalcq and Van Campenhout (who became Professor of Embryology in Louvain), many medical students were doing pieces of research. One of them was my friend Jean Pasteels, who was working under Dalcq's supervision and who became his successor. There were even a few foreign research workers: Slonimski (Piotr Slonimski's father who was killed in Warsaw during the war), Newth (Professor D. R. Newth's father), Rojas from Argentina, and Hall from the USA. This was unusual in those days when travel fellowships were scarce. But it was fully realised that international exchanges are of fundamental importance: everybody in our laboratory hoped to win a 'bourse de voyage' in a contest organised by the Belgian Government.

In 1930, nobody spoke of Molecular Biology which was still at an early embryonic stage. Of course, people working at that time on nucleic acids already were molecular biologists; but they did not know it and were like Molière's M. Jourdain who discovers that he has been talking in prose for forty years without being aware of the fact!

In the thirties, everybody already accepted the existence of two main types of nucleic acids; but they were called *animal* and *plant* nucleic acids. Their prototypes were respectively *thymonucleic acid* (aDNA) for animal cells and *zymonucleic acid* (aRNA) for plant cells. Plant nucleic acids were known to contain a pentose residue; thymonucleic acid had instead a strange sugar, which had been identified as deoxyribose by Levene & Mori in 1929. The erroneous distinction between plant and animal nucleic acids was found in all biochemistry textbooks; it was due to the trend, inherent to the human mind, to generalise hastily. Both types of nucleic acids were believed to be strictly localised in the nucleus. However Feulgen & Rossenbeck (1924) had shown that plant nuclei (and even bacterial 'nucleoids') give a positive cytochemical reaction for thymonucleic acid (DNA). But biochemists did not think much of cytochemistry and did not accept a finding which made DNA a possible candidate for a role in heredity: in those days, DNA was only a tetranucleotide with a molecular weight of about 1300 daltons, much too small to carry any genetic function. For Levene (1931), who was the major authority in nucleic acids, a likely role for DNA was that of a buffer to keep the intranuclear pH constant.

A strong objection to the possible genetic role of DNA was the belief that it disappeared when, during oogenesis, the lampbrush chromosomes are maximally extended. My cytochemical studies under A. Dalcq's leadership (Brachet, 1929, 1940) showed that, if proper fixatives are used, DNA can be detected with the Feulgen reaction in the lampbrush chromosome chromomeres at all stages of oogenesis. One had to wait for the work of Callan (1967) and Gall & Callan (1962) before it was proven that the lampbrush chromosome consists of a continuous DNA fibre and that transcription takes place on the loops and not on the chromomeres (in contradiction to the 'one gene, one chromomere' theory).

Another question puzzled both embryologists and biochemists 50 years ago: is there a *net synthesis* of DNA at the expense of inorganic phosphate, sugar, purines and pyrimidines, during egg cleavage where the number of nuclei greatly increases (J. Loeb, 1910)? Or, as suggested by E. Godlewski (1908), is there a migration of pre-existent cytoplasmic nucleic acids into the nuclei? The fact that the Feulgen reaction for DNA was negative in the cytoplasm of oocytes, spoke against the migration theory; the basophilic cytoplasmic granules called *'chromidia'* by Godlewski (1908) could hardly contain large amounts of DNA if the Feulgen reaction was to be trusted.

On the other hand, the biochemical evidence was in favour of the migration theory: Masing (1910) had shown that sea-urchin eggs have the same content of purine nitrogen whether they have one or 1000 nuclei. This was confirmed by Needham & Needham (1930), who used a delicate method for the determination of nucleoprotein phosphorus in developing eggs of various marine invertebrates.

This led me to a reinvestigation (in Roscoff) of nucleic acid synthesis during sea-urchin egg development (Brachet, 1933). As expected, the Feulgen reaction showed that in sea urchins, as elsewhere, cytoplasmic DNA could not be detected in eggs and embryos. The intensity of this reaction in the nuclei steadily increased during cleavage and embryogenesis. But no biochemist would be convinced by such purely cytochemical observations. Luckily a chemical method for deoxyribose estimation had just been published (Dische, 1930); chemical measurements of the DNA content of developing sea urchin eggs confirmed the cytochemical observations entirely: unfertilised eggs contain very little DNA and there is intensive DNA synthesis during development.

But how could one reconcile these findings with those of Masing (which I confirmed using an improved method) and of Needham & Needham showing that the nucleic acid content of sea urchin eggs remains constant during development? The only way out was to assume that, in contradiction to the then universal belief, unfertilised sea-urchin eggs contain large amounts of a *pentose* nucleic acid, thus a *plant* nucleic acid. I was still a beginner, a young student of medicine. Before making such an unorthodox proposal, I sought advice from Joseph Needham. I awaited his letter with impatience and anxiety; it finally came and said that my problem was so surprising that he had decided to consult his Chief, Sir F. G. Hopkins. 'Hoppy', a recent Nobel Prize winner, was the 'great old man' of British biochemistry. His verdict was typical of his humorous philosophy: 'tell this young man that he should not care too much about what is written in textbooks; they are full of errors. I know it only too well since I wrote one. What he should do is some experiments to prove that he is right'. Since that time, I have written several textbooks; I do not like to read them again after a few years since all of them contain an incredible number of factual errors.

I followed 'Hoppy's' advice and went back to Roscoff where I measured the pentose content of sea-urchin eggs during development despite the ironical comments of my French friends (J. Monod, B. Ephrussi, A. Lwoff): pentoses were believed to exist only in plants and the method I used had been devised for the estimation of these sugars in . . . straw. Nevertheless, these pentose estimations clearly showed that sea-urchin eggs and embryos contain large amounts of a pentose derivative which was later identified as RNA.

At this stage of the work, I made a mistake: since the pentose content of the eggs seemed to decrease during cleavage, I hastily suggested that RNA is the precursor for DNA synthesis. This led me to the 'RNA–DNA conversion' (or 'partial' DNA synthesis hypothesis); use of radioisotopes later showed that RNA cannot be a precursor for DNA synthesis. My error resulted from the fact that egg jelly glycosaminoglycans, which disappear during development, yield after acid hydrolysis the same product (furfural) as the pentoses. However conversion of ribose into deoxyribose does take place in sea-urchin eggs (as everywhere), but at the nucleotide level: inhibition of ribonucleotide reductase, the enzyme responsible for ribonucleotide reduction, results in cleavage arrest at the 8–16 cell stage in sea-urchins (Brachet, 1967, 1968).

But A. Dalcq, an anatomist, did not believe in results obtained with biochemical methods – unless I could show him RNA under the microscope. A simple method for RNA cytochemical detection (Unna staining before and after ribonuclease digestion of the sections) was worked out. This 'Unna–Brachet' test showed that unfertilised sea-urchin eggs contain large amounts of RNA in their cytoplasm; in fact, all cells (animal or plant) possess RNA in their nucleoli and in their cytoplasm, and there is a correlation between the RNA content of a cell and its ability to perform protein synthesis (Brachet, 1942; Caspersson, 1941). This conclusion was not accepted by biochemists until they could show, by work on cell homogenates with labelled amino acids, that the RNA-rich poly-ribosomes are indeed the agents of intracellular protein synthesis.

Cytochemical studies, initiated on oocytes and eggs, had thus provided us, already in 1940, with all the elements of Crick's 'fundamental dogma' of molecular biology (a dogma, which like so many others, is now under fire):

DNA $\longleftarrow$ DNA $\longrightarrow$ RNA $\longrightarrow$ Protein
　Replication　　　Transcription　　　Translation.

However one had to wait for the explosive growth of molecular biology before our poorly understood *hypotheses* became solid, well understood *facts*.

In the forties, very few people believed in these hypotheses, in particular in the participation of nucleic acids in protein synthesis: the accepted theory was that peptide bond formation, as well as peptide bond breakdown, results from protease activity.

Oocytes and eggs also had their share in the demonstration that, in contradiction to J. Loeb's (1899) old theory, the nucleus is not the centre of cellular oxidations: anucleate fragments of sea-urchin eggs obtained by centrifugation have a higher oxygen consumption than their nucleate counterparts (H. Shapiro, 1935); both the $O_2$ consumption and the $CO_2$ production of germinal vesicles isolated from amphibian oocytes are negligible when compared to the metabolism of whole oocytes; in addition, the respiration of enucleated oocytes remains at a normal and constant level for hours (Brachet, 1937).

We just mentioned nucleate and anucleate fragments of sea-urchin eggs. These have played an important role in elucidation of the molecular mechanisms of protein synthesis stimulation at fertilisation: it results from the translation of previously 'masked' mRNAs stored in the egg cytoplasm. This maternal mRNA store will be discussed by Dr E. Davidson in this book; for molecular embryologists, it is the equivalent of what the natural philosophers called preformation in the eighteenth century; synthesis of new mRNA species after fertilisation brings to the egg the fresh information required for epigenesis.

# II
# Why biochemists failed to crystallise the organiser's inducing factor

(For details, see J. Needham's *Biochemistry and Morphogenesis* and J. Brachet's *Chemical Embryology*.)

By 1930, all embryologists were convinced of the major importance for amphibian development of Hans Spemann's *organiser* concept. He had recently summarised his evidence for the view that the dorsal lip of the blastopore (chordomesoblast) is the *inductor* for nervous system formation in a Croonian lecture (1927). A little later (1938), he published his famous book *Embryonic Development and Induction*; in the meantime, he had received a Nobel Prize (1935), the only one ever to be awarded to an embryologist.

The now-classical experiments of Spemann and his school are known to all embryologists; among them, two techniques proved particularly useful for chemical embryologists. One of them, O. Mangold's (1928) *Einsteckung* technique, is particularly easy to perform: all one has to do is to introduce the material to be tested into the blastocoel cavity through a slit made at the animal pole of a young gastrula. As pointed out by J. Needham (1942) 'this method has been of inestimable help as a routine test'. The other fruitful approach is J. Holtfreter's (1931) explantation technique: isolated pieces of young gastrula ectoderm differentiate into epidermis; they do not form neural structures unless they are placed in contact with an inducing agent.

Two trends of thought were in opposition, around 1930, about the *nature of the organiser*: Is it a region of high metabolic activity? Or is induction due to a chemical substance specifically localised in the organiser?

In 1928, Child had tried to include the organiser concept in his general theory of *physiological gradients*: he conceived of a primary gradient decreasing from animal to vegetal pole, which would be replaced at gastrulation by a new physiological gradient whose most active region would be the organiser. The dorsal lip of the blastopore would, by virtue of a higher respiratory rate, constitute a '*dominant*' region capable of controlling the development of other regions.

One may wonder today why Child placed the emphasis on differences in respiratory rates: the reason is that there was no alternative at the time. These were the days when cellular respiration was in the forefront for almost all biochemists: O. Warburg and H. Wieland were fighting a gigantic battle about its mechanisms; F. G. Hopkins and particularly D. Keilin gave the correct solution to the problem around 1930–35. Everybody was interested in cellular oxidations, almost nobody in proteins or nucleic acids. This is why Professor Dalcq insisted that I should study the respiration of frog eggs and embryos.

The experimental evidence for Child's hypothesis rested on experiments by his student Bellamy (1919, 1922) on '*differential susceptibility*': when frog gastrulae are treated with a variety of chemicals, including KCN, cytolysis occurs first at the animal pole and in the organiser region. Although this evidence was too indirect to carry conviction, Child's theory appealed to a number of leading embryologists: for A. Brachet (1931) the organiser is 'some kind of activator, wherein there is an intense metabolism, the energy from which may be transmitted to neighboring regions'. In 1934, J. S. Huxley and G. R. De Beer attempted to explain morphogenesis in terms of a metabolic gradient. A little later, A. Dalcq and J. Pasteels tried to explain amphibian morphogenesis in terms of the distribution of egg constituents rather than in terms of a metabolic gradient: a *vitelline* (yolk) *gradient*, most active at the vegetal pole, would interact with a *cortical field* with its focal point situated dorsally; the result of this interaction would be the production of a morphogenetic substance, the *organisin*. Differentiation would result from the gradient distribution of this substance and from *thresholds* in its concentration. This 'field-gradient-threshold' theory was based on a number of classical experiments on frog eggs (inversion of fertilised eggs, centrifugation of fertilised eggs and blastulae, etc.).

But in 1932 a sensational paper by H. Bautzmann, J. Holtfreter, O. Mangold & H. Spemann came out: it showed that a killed organiser retains its inducing activity. Induction must therefore be due to a specific *chemical substance* located in the organiser. The next surprise was that ventral ectoderm, which is devoid of inducing activity, becomes an excellent inducer when it has been killed: denaturation of the proteins presumably liberates the inducing agent from an inactive complex (Holtfreter, 1933). A little later, Holtfreter (1934*b*) discovered that many different tissues, dead or alive, are able to induce neural tubes after grafting into the blastocoel. Thus the active inducing substance is widely distributed in animal tissues of both vertebrates and invertebrates; even yeast extracts gave good results in our hands. Since the inductions obtained after grafting killed tissues seldom have the perfection of those obtained after grafting a living organiser, J. Needham, C. Waddington and D. M. Needham proposed to name the active substances present in killed tissues *evocators* : these heterogeneous inductors stimulate formation of complex and abnormal organs in contrast to the axial organs induced by the living organiser.

There is no reason to go further into the numerous embryological problems raised by primary (neural) induction since they are handled in this book by Saxén & Toivenen; they were already treated, in a masterly way, by J. Needham in *Biochemistry and Morphogenesis* (1942). I shall limit myself here to a summary of the experiments done between 1930 and 1940 on the metabolism of the organiser and on the chemical nature of the evocator.

Between 1934 and 1939, I tried to establish whether, at the early gastrula stage, the dorsal lip of the blastopore has a higher *respiratory metabolism* than the ventral marginal zone; a number of techniques and animal species were used for this work which has been described and discussed in detail in *Chemical Embryology* (Brachet, 1950). The general conclusion for this and similar work done in other laboratories is that, in general, the organiser has a slightly higher oxygen consumption than the corresponding ventral regions; but, in short-term experiments, the difference between the two is negligible in gastrulae from some species. On the other hand, much larger differences in favour of the organiser were found when the

$CO_2$ production was measured, suggesting that the organiser has a higher respiratory quotient ($CO_2/O_2$) than the ventral marginal zone. This was proven by the elegant experiments of Boell, Koch & Needham (1939) who succeeded in measuring simultaneously both $O_2$ consumption and $CO_2$ production of the same explants. These experiments, as well as studies on glycogenolysis and glycolysis conducted mainly in J. Needham's and L. G. Barth's laboratories, led to the conclusion that the organiser region is characterised by a high carbohydrate metabolism. Work in several laboratories showed that, in early gastrulae, there is also a sharply decreasing animal–vegetal gradient in respiratory rate. The most recent work on the subject is now 30 years old (Sze, 1953): it confirms that, in early gastrulae, there is a sharp respiratory animal–vegetal gradient (due to the accumulation of yolk at the vegetal pole) and a much weaker dorso–ventral one. This agrees with what Child had proposed in 1928, except that the respiration of the organiser is much lower than that of the ectoderm.

Already in 1940, serious doubts about the importance of carbohydrate metabolism for induction had been expressed: an artificial increase in this metabolism is not sufficient to stimulate neural induction in ectoderm and a lowering of the rate of oxidations does not prevent normal induction. Glycogenolysis also occurs in the non-inducing ventral lip of the blastopore: enhanced glycogen breakdown seems to be correlated with morphogenetic movements rather than with neurogenic induction. On the whole, the general conclusions to be drawn from all this work were disappointing: animal–vegetal and dorso–ventral gradients in respiratory metabolism (and in ribosome distribution, as I showed in 1942) undoubtedly exist in gastrulae, but there is no evidence that they are directly involved in neural induction.

More exciting, but still more frustrating, is the long story of the many attempts made in order to identify the *chemical nature* of the inducing agent. Biochemists were full of hope at the start: even before it was known that a killed organiser is still active, J. Needham (1930) wrote: 'it is probably not fantastic to picture, no doubt in the remote future, the amphibian organizer in a crystalline state'. These were the days when an increasing number of hormones and vitamins, active at very low concentrations, were crystallised; Nobel Prizes were given for such achievements. It was not realised at the time that a crystalline hormone is nothing but a valuable tool and that the real problem is to find out how it is acting on target tissues. Chemical embryologists should have been warned against excessive confidence by the parthenogenesis story: many chemically unrelated substances may induce the same biological response.

Since a number of hormones are sterol derivatives, J. Needham, C. H. Waddington & D. M. Needham (1934) looked for inducing activity in sterol fractions from amphibian gastrulae and other tissues: neural inductions were obtained with them, but their percentage was small. Waddington & D. M. Needham (1935) further showed that synthetic polycyclic hydrocarbons are also evocators. However, in other laboratories (in particular in that of F. G. Fischer), neural inductions were obtained with non-sterol chemicals: fatty acids, nucleoproteins, nucleic acids, adenylic acid, ATP, etc. were among the active substances (Fischer *et al.* 1935). This led the German workers to propose that induction, at least by dead tissues, results from a stimulation of the ectoderm by an acid of any sort (*Saüre-Reiz*). A lively controversy went on for several years about the dangers of sterol contamination by acids and *vice-versa*. It ended when it was found that *basic* substances totally unrelated to lipids were good evocators: that was the case for several vital dyes. All this work led Waddington, Needham & Brachet (1936) to the conclusion that numerous, chemically unrelated substances will be evocators because they liberate the true inducing substance from an inactive complex by an 'unmasking' action. This left little hope of isolating the true inducing agent, which was supposed to be present in its active form only in the organiser.

A little later, L. G. Barth (1939) showed that lipid-free killed organisers are still excellent inducers: this suggested proteins or nucleic acids as the inducing agents. But he emphasised a worrying, unpleasant fact: any substance which induces local cytolysis can act as an evocator. The final blow came when Holtfreter (1944) showed that a short shift in the pH (produced by adding $CO_2$ or ammonia) of the salt solution in which explants of ectoderm are cultivated induce their 'spontaneous neuralisation'; no less distressing was the fact that simple salts such as LiCl or even NaCl, or implantation of inert substances (Teflon, crushed glass) could lead to nervous system formation. All these agents were supposed to produce a sublethal, eventually reversible, cytolysis and to set free an inducing agent present in masked form in the ectoblast. We were facing the same difficulties as the old embryologists who had discovered parthenogenesis.

Since even $CO_2$ or ammonia can produce neuralisation of ecto-derm explants, one had to accept the possibility that the organiser might act by simply producing the end products of its metabolism (that this is not true was shown later by transfilter induction experiments). All those working in the field realised that to prove that an active substance is *the* natural inducer and not a factor acting in an indirect way was an impossible task. Hopes had been too high at the start; disappointment was too deep, as evidenced by the later work of Toivonen, Tiedemann and others (see Saxén's & Toivonen's chapter).

Then came the war which stopped research in embryology (and in many other fields). Needham and Waddington served their country and became interested in problems other than the identification of the elusive inducing factor. In occupied Belgium, I did the same thing as the Abbé Sieyès during the French Revolution: 'I sur-vived'. However our interest in embryology also survived: both Needham's *Biochemistry and Morphogenesis* and my *Embryologie chimique* were written during the war.

Soon after the war, the molecular biology revolution took place; it attracted, more than embryology, many bright young scientists. Around 1950, Jacques Monod once said to me: 'Why are you sticking to your eggs and embryos? We shall never understand the molecular mechanisms of development. Join us and work with bacteria if you wish to know how RNA controls protein synthesis'. He was right: the solution to that problem – and to larger ones such as heredity – came from work done on bacteria, phages, and *in vitro* systems of protein synthesis, not from studies on eggs and embryos. But I was still interested in morphogenesis and nucleocytoplasmic interactions: bacteria did not appeal to me.

Today the situation has completely changed: more and more molecular biologists are now fascinated by morphogenesis and the *Xenopus* oocyte, thanks to J. B. Gurdon, has become an ideal test tube for those who are interested in gene expression. As we shall hear from Dr Saxén, few scientists have been attracted to the organiser problem, perhaps because chemical embryologists had failed to identify the inducing substance. However, since vertebrate mor-phogenesis would be impossible without inductive processes, the problem remains of paramount importance. I am confident that a new generation of molecular embryologists, bringing in new ideas and techniques, will ultimately solve the organiser problem.

# References

Barth, L. G. (1939). The chemical nature of the amphibian organizer. III. Stimulation of the presumptive epidermis of *Ambystoma* by means of cell extracts and chemical substances. *Physiol. Zool.* **12**, 22–30.

Bautzmann, H., Holtfreter, J., Spemann H. & Mangold, O. (1932). Versuche zur Analyse der Induktionsmittel in der Embryonalentwicklung. *Naturwissenschaften* **20**, 971–4.

Bellamy, A. W. (1919). Differential susceptibility as a basis for modification and control of early development in the frog. *Biol. Bull.* **37**, 312–61.

Bellamy, A. W. (1922). Differential susceptibility as a basis for modification and control of development in the frog. II. Types of modification seen in later developmental stages. *Am. J. Anat.* **30**, 473–502.

Boell, E. J., Koch, H. & Needham, J. (1939). Morphogenesis and metabolism: studies with the Cartesian diver ultramicromanometer. IV. Respiratory quotient of the regions of the amphibian gastrula. *Proc. Roy. Soc. B* **127**, 374–87.

Brachet, J. (1929). Recherches sur le comportement de l'acide thymonucléique au cors de l'oogenèse chez diverses espèces animales. *Arch. Biol.* **39**, 677–97.

Brachet, J. (1931). *L'oeuf et les facteurs de l'ontogénèse*. Paris: Doin.

Brachet, J. (1933). *Arch. Biol.* **44**, 519–76.

Brachet, J. (1937). Some oxidative properties of isolated amphibian germinal vesicles. *Science* **86**, 225.

Brachet, J. (1940). La localisation de l'acide thymonucléique pendant l'oogénèse et la maturation chez les Amphibiens. *Arch. Biol.* **51**, 151–65.

Brachet, J. (1942). La localisation des acides pentosenucléiques dans les tissus animaux et les oeufs d'Amphibiens en voie de développement. *Arch. Biol.* **53**, 207–57.

Brachet, J. (1950). *Chemical Embryology*, Transl. L. G. Barth. N.Y.: Interscience Publ.

Brachet, J. (1957). *Biochemical Cytology*. N.Y.: Academic Press.

Brachet, J. (1967). Effects of hydroxyurea on development and regeneration. *Nature* **214**, 1132–33.

Brachet, J. (1968). Some effects of deoxyribonucleosides on sea urchin egg development. *Currents in Modern Biology* **1**, 314–19.

Callan, H. G. (1967). The organization of genetic units in chromosomes. *J. Cell Sci.* **2**, 1–7.

Caspersson, T. (1941). Studien über den Eiweissumsatz der Zelle. *Naturwissenschaften* **29**, 33–43.

Child, C. M. (1928). The physiological gradients. *Protoplasma* **5**, 447–76.

Dische, Z. (1930). Über einige neue charakteristische Farbreaktionen der Thymonukleinsäure und eine Mikromethode zur Bestimmung derselben in tierischen Orgranen mit Hilfe dieser Reaktionen. *Mikrochemie* **8**, 4–32.

Feulgen, R. & Rossenbeck, H. (1924). Mikroskopisch-chemischer Nachweis einer Nucleinsäure vom Typus der Thymonucleinsäure und die darauf beruhende elektive Färbung von Zellkernen in mikroskopischen Präparaten. *Z. Physiol. Chem.* **135**, 203–48.

Fischer, F. G., Wehmeier, E., Lehmann, H., Jühling, L. & Hultzsch, K. (1935).

Zur Kenntnis der Induktionsmittel in der Embryonal-Entwicklung. *Berichte Chem. Gesell.* **68**, 1196–9.

Gall, J. G. & Callan, H. G. (1962). $H^3$ Uridine incorporation in lampbrush chromosomes. *Proc. Nat. Acad. Sci. USA* **48**, 562–70.

Godlewski, E. (1908). Plasma und Kernsubstanz in der normalen und der durch äussere Faktoren veränderten Entwicklung der Echiniden. *Arch. Entw. Mech. Org.* **26**, 278–328.

Holtfreter, J. (1931). Über die Aufzucht isolierter Teile des Amphibienkeimes. II. Zuchtung von Keimen und Keimteilen in Salzlösung. *Wilhelm Roux Arch. EntwMech. Org.* **124**, 404–66.

Holtfreter, J. (1933). Eigenschaften und Verbreitung induzierender Stoffe. *Naturwissenschaften* **21**, 766–70.

Holtfreter, J. (1934a). Der Einfluss thermischer mechanischer und chemischer Eingriffe auf die Induzierfähigkeit von Triton-keimteilen. *Wilhelm Roux Arch. Entw Mech. Org.* **132**, 225–306.

Holtfreter, J. (1934b). Über die Verbreitung induzierender Substanzen und ihre Leistungen im Triton-keim. *Wilhelm Roux Arch. Entw Mech. Org.* **132**, 307–83.

Holtfreter, J. (1944). Neural differentiation of ectoderm through exposure to saline solution. *J. Exp. Zool.* **95**, 307–44.

Huxley, J. S. & De Beer, G. R. (1934). *Elements of Experimental Embryology*. Cambridge: Cambridge University Press.

Levene, P. A. & Mori, T. (1929). The carbohydrate group of ovomucoid. *J. Biol. Chem.* **84**, 49–61.

Levene, P. A. (1931). *Nucleic Acids*. N.Y.: Chem. Catalog. Co.

Loeb, J. (1899). Warum ist die Regeneration kernloser Protoplasmertücke unmöglich oder erschwert? *Wilhelm Roux Arch. Entw Mech. Org.* **8**, 689–93.

Loeb, J. (1910). Über den autokatalytischen Charakter der Kernsynthese bei der Entwicklung. *Biol. Zentralblatt.* **30**, 347–9.

Mangold, O. (1928). *Methodik der wissenschaftlichen Biologie II Physiologie.*

Masing, E. (1910). Über das Verhalten der Nucleinsäure bei der Furchung des Seeigeleis. *Z. Physiol. Chem.* **67**, 161–73.

Needham, J. (1930). *Chemical Embryology.* 3 vols. Cambridge: Cambridge University Press.

Needham, J. (1942). *Biochemistry and Morphogenesis.* Cambridge: Cambridge University Press.

Needham, J. & Needham, D. M. (1930). On phosphorus metabolism in embryonic life. I. Invertebrate eggs. *J. Exp. Biol.* **7**, 317–48.

Needham, J., Waddington, C. H. & Needham, D. M. (1934). Physico-chemical experiments on the amphibian organizer. *Proc. Roy. Soc. B.* **114**, 393–422.

Shapiro, H. (1935). The respiration of fragments obtained by centrifuging the egg of the sea urchin, *Arbacia punctulata. J. Cell. Comp. Physiol.* **6**, 101–16.

Spemann, H. (1927). Organizers in animal development. *Proc. Roy. Soc. B* **102**, 177–$7.

Spemann, H. (1938). *Embryonic Development and Induction.* New Haven, Connecticut: Yale Univ. Press.

Sze, L. C. (1953). Respiration of the parts of the *Rana pipiens* gastrula. *Physiol. Zool.* **26**, 212–23.

Waddington, C. H. & Needham, D. M. (1935). Studies on the nature of the amphibian organization centre. II-Induction by synthetic polycyclic hydrocarbons. *Proc. Roy. Soc. B* **117**, 310–17.

Waddington, C. H., Needham, J. & Brachet, J. (1936). Studies on the nature of the amphibian organization centre. III-The activation of the evocator. *Proc. Roy. Soc. B* **120**, 173–98.

# Primary embryonic induction in retrospect

LAURI SAXÉN AND SULO TOIVONEN

Departments of Pathology and Zoology, University of Helsinki, Finland

## Introduction

Primary embryonic induction, the series of events guiding early embryogenesis, was detected sixty years ago and has since been analysed thoroughly by many scholars. However, after the initial, fundamental discoveries, development in the field has been slow and disappointing. This led us to ask recently: 'Why do the scientists investigating embryonic induction lag behind their brilliant colleagues in many other fields of biology, in which the 1960s and 1970s have witnessed many great victories and discoveries of fundamental importance?' (Saxén, Toivonen & Nakamura, 1978). We will seek answers to this question in the following and examine the past history of 'primary induction' to search for mistakes, misinterpretations, and erroneous premature judgements which might have led the research onto wrong tracks or dead-end roads. We may also ask why the great enthusiasm faded after the Second World War. Before the war the problem was intensively studied in practically every embryological centre in Europe, whereas today only a handful of persistent developmental biologists struggle with the unsolved fundamental problems of induction.

In the first part of this presentation, we will describe in some detail the three basic experiments by Hans Spemann and his school which laid the foundation for our ideas on embryonic induction. We will examine the validity of these experiments and the possible misinterpretations of their results in the light of our present knowledge. We will then discuss briefly the crusade of the 1930s to unravel the chemical basis of induction, and, finally, we will comment on the achievements, difficulties, and attitudes of the postwar period. For details of the experiments and their results mentioned only in passing, the reader is referred to the following monographs: Spemann (1936) covering the early days of research in this field, Saxén & Toivonen (1962) summarising mainly the achievements until 1960, and Nakumura & Toivonen (1978) dealing with more recent developments.

# Three fundamental experiments and their retrospective evaluation

## Discovery of the organiser

A morphogenetically significant interaction, induction, between two tissues was first demonstrated by Spemann in 1901 in the eye, between the optic cup and the presumptive lens ectoderm. It took, however, twenty years before this single observation could be broadened to a unifying concept of the early organisation of an embryo. In 1921, Hilde Pröscholdt performed the following experiment in Spemann's laboratory (Fig. 1): while making interchange experiments of various parts of gastrulae between the pale *Triturus cristatus* and the pigmented *Tr. taeniatus*, she grafted a piece of the dorsal blastoporal lip of *cristatus* to the ventrolateral ectoderm of a *taeniatus* gastrula. As a result, a supernumerary neural plate developed at the site of the graft, but only a narrow, central part of this was of the donor origin while most of the cells were heavily pigmented. This striking result led Spemann & H. Mangold (Pröscholdt) (1924) to the following conclusion: 'The piece of the upper blastoporal lip . . . exerts an organizing effect on its surroundings in such a way that when transplanted into an indifferent location of another embryo, it leads to the formation of a secondary embryo anlage.'

This basic experiment and the main conclusion have, however, been criticised recently by Jacobson (1982) who stresses two defects in the experimental design. Accordingly, the authors had not shown that the cells of the ventral ectoderm converted to the neural plate did not already have a neurogenic bias, and, further, the secondary neural structures were not shown to contain nerve cells or to 'belong to specific regions of the CNS' (Jacobson, 1982). Strictly taken, the first point might be justified as far as this particular experiment is concerned, but we believe that various subsequent studies have confirmed the conclusions drawn from the above results. Especially

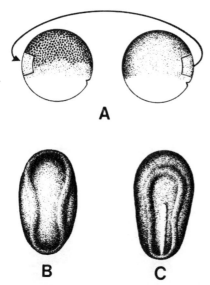

Fig. 1. The transplantation experiment by Spemann & Mangold (1924): a piece of the dorsal blastoporal lip from a pale *Tr. cristatus* gastrula was transplanted to the ventral ectoderm of a pigmented *Tr. taeniatus* gastrula (A). A normal medullary plate developed in the host (B) but, in addition, a supernumerary neural plate was formed at the site of the graft (C). The latter was only partially pale and consisted mostly of pigmented cells of the host.

experiments performed with in vitro techniques and an isolated ventral ectoderm as the test object have repeatedly shown how this competent, multipotent target cell population can be converted exclusively into mesodermal structures with no indication of preneuralised cells or neural crest cells migrating into the ventral location. As to Jacobson's other point, many subsequent studies along this line have described the secondary CNS (central nervous system) structures in detail (e.g. Nieuwkoop *et al.* 1952). We therefore believe that the basic conclusion by Spemann & Mangold (1924) is still valid, and, hence, could be applied in subsequent approaches.

## Discovery of the regional specificity of the inductor

Early transplantation experiments had suggested that the invaginating mesoderm, the archenteron roof, not only led to the neural determination of the ectoderm but also caused the regional specialisation of the induced neural plate. The decisive experiment was performed by O. Mangold (1933). He dissected the archenteron roof of a young neurula into four consecutive segments and *implanted*

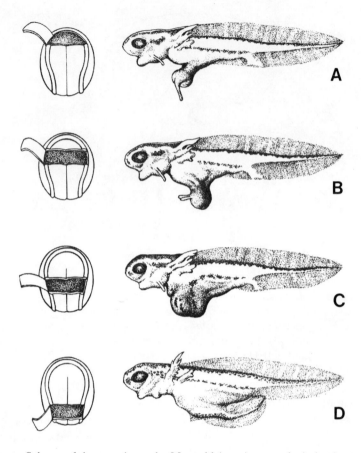

Fig. 2. Scheme of the experiment by Mangold (1933) to test the inductive capacity
of various territories of the archenteron roof at an early neurula stage. The roof was
dissected into four transverse segments, and the fragments were implanted into the
blastocoel of an early gastrula. The most cranial segment induced head structures
with balancers (A); the second, head with eyes and forebrain (B); the third,
hindbrain structures with ear vesicles (C); and the most caudal territory, tail structures with spinal cord (D).

these into the blastocoel of young gastrulas – a new method to
examine the inductive action of any tissue ('Einsteckmethode').
The results illustrated in Fig. 2 indicate clearly how the different
territories of the inductor induced different secondary structures in a
cranio-caudal fashion – also including different regions of the CNS.

These results have been confirmed in subsequent, similar
experiments (Ter Horst, 1948; Sala, 1955), but the basic study
could be criticised by using some of Jacobson's (1982) arguments:

the target cells could be in a 'predetermined' state, cells from the living inductor might have contributed to the secondary structures, or cells with neurogenic potential might have migrated into the region. However, these alternative explanations could be excluded by various observations of the past years:

– An experimentally neuralised ectoderm can be transformed into various regions of the CNS by certain second-step interactions, which thus excludes the possibility of a pre-existing regional bias (Saxén, Toivonen & Vainio, 1964; Toivonen & Saxén, 1968).

– The results can be mimicked by the use of various devitalised tissues or their mixtures that certainly do not contribute cells to the secondary structures (Saxén & Toivonen, 1961).

– Similar results have been obtained *in vitro* by using isolated ectodermal explants to which cells cannot migrate (Toivonen & Saxén, 1955).

Hence we conclude that it is still safe to believe that the normal inductor tissue, the archenteron roof, is responsible for the regional segregation of the neural plate.

Recently, two important reports have been published fully confirming our above conclusions. As they illustrate recent history in the field of neural induction well, we shall quote them briefly.

In December 1983 Gimlich & Cooke published their results obtained in *Xenopus* embryos by labelling individual blastomeres at the 32-cell stage with fluorescein-conjugated dextran. The progenitor cells were followed up both in normal embryos and in those where an inducing dorsal lip had been grafted according to the original technique of Spemann. The results showed conclusively that the secondary, induced neural structures were made up of ectodermal cells lying in appropriate spatial relationship to the graft. These cells would normally give rise to the belly epidermis but would not contribute to the host's central nervous tissue. Hence, the authors conclude that their results fully support Spemann's original explanation. Three months later Jacobson's own re-interpretation and new observations were published (1984). By using the original horseradish peroxidase label, he repeated the experiments of Gimlich & Cooke (1983) and obtained exactly the same results which 'may be regarded as a vindication of Spemann's theory of the organizer'. This conclusion seems to put an end to this short but confusing episode in the modern history of embryonic induction.

## Discovery of the chemical nature of induction

In 1932, Spemann's group published a short, four-page report entitled 'Versuche zur Analyse der Induktionsmittel in der Embryonalenentwicklung', which, in many respects, determined the main lines of research on induction for years to come. The following important observations were reported (Bautzmann, Holtfreter, Spemann & Mangold, 1932):

– Heat-killed normal inductor retains its inductive capacity to a certain degree. Killing by drying, freezing or by short-term alcohol treatment similarly retained some of the inductive action.

– The active material could be deposited in agar: when a piece of the neural plate was kept in contact with agar, then removed and the agar implanted into a gastrula, weak inductions were observed.

These observations led the team to two conclusions, both affecting our thinking profoundly: (1) induction is chemical in nature, and (2) it is carried by diffusible compounds. Retrospectively, these somewhat elementary experiments can be criticised (e.g. the completeness of the removal of the inductor from the agar), but again subsequent studies have lent support to the original conclusions: (1) induction of an isolated, competent ectoderm can be brought about by purified, apparently non-toxic protein preparations in physiological conditions (reviews: Tiedemann, 1976, 1981), and (2) separation of the normal inductor from its ectodermal counterpart by filters preventing cell contacts does not inhibit neural induction (Saxén, 1961; Toivonen, 1979).

## The confusing thirties

In the nineteen thirties, the three major discoveries reviewed above resulted in a vigorous search for the active compounds, the molecules apparently carrying the inductive messages. The basic methods were simple: either the compound to be tested was introduced into the blastocoel (implantation), or an isolated piece of ectoderm was exposed to it *in vitro* (explantation). At first, progress was stimulating, and scientists in various laboratories could report successful experiments with various purified compounds. There is no need to review this extensive literature, but the reader is referred to the monographs of Brachet (1950) and Needham (1950).

In brief, three schools could soon be distinguished in this competitive effort to find the 'inductor': the Cambridge school with Needham and Waddington collaborating with Bracket, the German School with Lehmann, Wehmeier and Fischer, and Barth with his students. The Cambridge group reported results suggesting that the active compound might be a *sterol*, while the German school obtained inductions with various acidic preparations, including nucleic acid, oleic acid, and linoleic acid, and ended with the '*acid stimulus theory*'. Barth, on the other hand, reached the conclusion that the normal inductor in the blastoporal lip might possess features of a *protein* (see Brachet, this volume).

This multitude of active compounds, together with the many controversial views and observations of the investigators, naturally had to result in great confusion and a gradual fading of interest in the entire complex problem. The mood of these brilliant scientists of this era is reflected well in Joseph Needham's concluding comments in his overview from 1939:

In conclusion, it may be said that although the progress made in the last ten years in these fields has been very great, we can nevertheless see now that owing to the special difficulties of the subject, especially perhaps the presence of evocator in ventral ectoderm, it may be more like fifty years before we can expect to have certain knowledge concerning the chemical nature of the naturally-occurring substances involved in embryonic induction. Like so many other biological problems, this has turned out to be more complex than the first explorers thought.

Today we can fully agree with the conclusion of Needham. The fifty years which Needham in 1939 allotted to developmental biologists are coming to the end, and still our knowledge of the molecular events underlying induction is meagre, and chemical characterisation of the inducing factors has proceeded only very slowly.

## The target

Much of the confusion created by the above conflicting results can be explained, when a closer analysis of the target cells, the competent ectoderm, is made. There is still abundant evidence for the *multipotency* of these cells during early gastrulation, and by experimental manipulation they can be converted into practically any cell type of the organism. The basic question is then how much of this differentiative bias has been created *prior* to the classic study situation, the early gastrula. In other words, to what extent are these cells already *predetermined*, maybe requiring only some additional, permissive stimuli to stabilise subsequent overt differentiation?

The question was first extensively analysed by Holtfreter (1944, 1945, summarised in 1955). Briefly, this author could show – what already could be expected from previous, occasional findings (see Needham, 1939) – that the ectoderm can express its differentiation capacities in the absence of any specific inductors, after various unphysiological treatments and insults. The following list gives examples of such 'triggers': high and low pH, treatment with alcohol, ammonia or toxic compounds, and direct mechanical injuries. All these led in his experiments to 'autoneuralization' of the isolated ectoderm, and the author suggested that a neuralising agent, the neurogen, is released from the damaged cells and acts upon the still-viable cells with the observed consequences (Holtfreter, 1948).

These early findings of Holtfreter (1948) have been confirmed repeatedly, and autoneuralisation has been obtained even after minor changes in the ionic environment of the cultured cells (Barth & Barth, 1959). The apparent conclusion from such results is that while primary induction *in vivo* must be a specific, both temporally and spatially strictly controlled, decisive event, the prospective neuroectoderm is already predetermined to a certain extent. In fact, very recently the attention of several independent research teams has been focused on these target cells prior to the classic stage of induction. Membrane receptors especially, which may mediate or modify neural induction, have been explored with different techniques (Takata, Yamamoto & Ozawa, 1981; Duprat, Gualandris & Rouge, 1982; Grunz, 1983).

## Postwar depression: mental and physical

The somewhat shocking news of Holtfreter (1944, 1945) reached the scientific community at an unfortunate moment. Research conditions all over the world were poor. Many of the great scientists of the 1930s were gone, or they were searching for new avenues for their interrupted research. The new generation was cut off from the traditions of the 1930s, and many of them could not even read the classic German literature. In this situation it was understandable that Holtfreter's (1944, 1945) reports did not stimulate young scientists, and when the new era of biology rose in the 1950s, embryonic induction was not among the popular and rapidly advancing fields. We will return briefly to this question at the end of this presentation.

## Slow recovery

With flourishing biochemistry and the dawn of molecular biology in sight, a few groups of embryologists decided to attack the problem of the chemical basis of induction again. With the lesson from the 1930s and the awareness of the risk of autodifferentiation by these frequently toxic agents, most of these teams abandoned the 'fishing expedition approach' of testing various purified chemicals directly on the ectoderm. Instead, they started to search for active compounds in tissues known to exert an inductive action on the competent ectodermal cells. Since the tissue mass of the normal inductor was too minute for such laborious fractionation experiments, heterogeneous inductors were made use of (since the late 1930s it was evident that practically every embryonic and adult tissue exerted an inductive action when tested in young gastrula as shown by Chuang, 1938, and Toivonen, 1938). In Finland, Toivonen and Kuusi performed extensive fractionation experiments on guinea-pig tissues and produced two active fractions with neuralising and mesodermalising actions, respectively (summarised by Toivonen, 1950). Barth and his collaborators focused on ribonucleic acids (summarised by Barth & Barth, 1954), as did the reborn German school in Cologne (summarised by Vahs, 1957) and the active team in Nagoya (summarised by Yamada, 1961). Somewhat later the first fractionation results were published by the Heiligenberg group, a direct descendant of Otto Mangold and subsequently directed by Heinz Tiedemann (summarised by Tiedemann, 1978, 1981). Tiedemann's team, now in Berlin, has marshalled abundant data on the vegetalising protein and its chemical characteristics.

Our present knowledge of the inductive signals can be very briefly summarised as follows: There are apparently at least two active molecules or groups of molecules, one causing neuralisation of the ectoderm and another leading to mesodermal and endodermal differentiation (vegetalisation) (Toivonen, 1950; Tiedemann, 1976, 1981). The former, apparently smaller molecule is transmitted by diffusion (Saxén, 1961) while the other might require actual cell-to-cell contacts (Toivonen, 1979). Moreover, the neuralising factor seems to act upon receptors on the cell membrane, while the vegetalising factor is active only if entering the intracellular compartment

(Tiedemann & Born, 1978). Practically all experiments have, how-ever, been performed on fractions of heterogeneous origin, and it is not clear whether these fractions are identical with the factors operating during normal embryogenesis. The actual molecular mechanism of the induction process has, by and large, remained unknown. Altogether, progress since 1945 has been slow, and despite the many active efforts, our knowledge on the whole induc-tive event is most incomplete. However, even a retrospective analysis cannot pinpoint any single reason for this exceptionally tardy development of a field, which, undoubtedly, should be con-sidered most central in biology today. Let us, however, close this discussion with some thoughts of factors which possibly affected research in the field of induction.

## Attitudes

As already mentioned, the scientifically sound but unsuccessful chemical approach of the 1930s and the subsequent demonstration of autodifferentiation of the ectoderm made the field unattractive to many young scientists who searched for ways towards rapid scientific victories in the reshaping postwar research world. Hence, *the total manpower* left to study this problem *remained small* as compared to many other fields in biology and medicine. Moreover, the emphasis given to molecular biology and to the great discoveries made in this field further recruited young scientists who had no direct contacts with the pioneers of the field of induction and who, therefore, were lacking a scientific tradition in this discipline.

## Biology – biochemistry

In retrospect, the postwar period of research on induction shares one common feature with the 1930s: at that time we also were looking for molecular short cuts to the extremely complex process of induction apparently forgetting the wise prediction of Needham (1939) quoted above. In the nineteen thirties, scientists were seeking for *the* active compound, the hypothetical organiser or evo-cator, until direct biological experiments employing heterogeneous inductors demonstrated that there must be several such factors (Chuang, 1938; Toivonen, 1938). When the fractionation studies became activated in the 1940s and 1950s, we were still uncertain as to what kind of *biologically active* compounds should be searched for, and research went on more or less in the manner of a 'fishing expedition'. One example shall illustrate this point.

Since it was (and still is) customary – mostly for practical reasons – to distinguish between an archencephalic, a deuterencephalic, and a spinocaudal *response*, specific transmitters for these three were actively sought. Again, pure biological experimentation with heterogeneous inductors and their mixtures soon indicated that the deuterencephalic response was due to a *combined action* of two active principles operating consecutively, rather than to a specific 'deuter-encephalic' inductor which had been sought in vain by biochemists (Toivonen & Saxén, 1955; Saxén & Toivonen, 1961; Saxén *et al.* 1964). This could be confirmed in chemical separation and combination experiments (Tiedemann & Tiedemann, 1959). Thus, it appears that a biochemical/molecular approach sometimes bypasses our biological and morphogenetic knowledge and so tends to solve problems not yet clearly defined by 'old-fashioned' embryologists.

## Methods

While biochemical technology has advanced tremendously in the past years, our test methods to study an inductive effect have hardly changed at all in the last 60 years. By definition, such an action is monitored as *differentiation* of the target cells after a considerable lag period. Analysis of an immediate response of the target cells has so far proved difficult and often inconclusive so far as initiation of morphogenesis is concerned. This is probably due to the fact that we are dealing with an extremely complex *chain of events* rather than a single process. Probably our thinking has been obscured by the simplified concept that induction could be such a single, one-step event that can easily be exposed to experimental tests in artificial conditions. Today we are becoming aware of this misconception and have already learned that the inductors act upon predetermined cells, and that the subsequent determination is a multistep process.

Another point of concern is the frequently emphasised fact that most chemical approaches have employed heterogeneous tissues and their fractions – simply because it is very difficult to collect large enough quantities of amphibian embryonic material. Although such foreign inductors lead to a response closely mimicking normal development this does not necessarily imply that the mediators are the same – perhaps they are not even related to the 'normal' inductors acting *in vivo*. An extreme, though probably exaggerated, attitude would be that we are repeating the basic mistake made by our colleagues in the 1930s by exposing our predetermined, biased cells to unspecific, permissive stimuli in the belief that we are studying normal embryogenesis!

## Conclusions

The history of the research on primary embryonic induction seems to have taught us some lessons which should be taken into consideration in future research:
– The problem itself is still timely and holds a most fundamental concept of embryogenesis.
– The progress in the field has been slow owing to unfortunate circumstantial conditions. Approaches at the purely biological and molecular levels have not been synchronised. Therefore, premature efforts and conclusions have been made at both levels.
– The above features are not uncommon in biological research, but become accentuated when a process of extreme complexity, like embryonic induction, is explored.
– The slow progress and the conflicting results have led to a basic mistake, namely in disbelief and scepticism about the whole existence of inductive phenomena. This, in turn, has become the major retarding factor in the field.

## References

Barth, L. G. & Barth, L. J. (1954). *The Energetics of Development*. New York: Columbia University Press.
Barth, L. G. & Barth, L. J. (1959). Differentiation of cells of the *Rana pipiens* gastrula in unconditioned medium. *Journal of Embryology and Experimental Morphology* **7**, 210–22.
Bautzmann, H., Holtfreter, J., Spemann, H. & Mangold, O. (1932). Versuche zur Analyse der Induktionsmittel in der Embryonalentwicklung. *Naturwissenschaften* **20**, 971–4.
Brachet, J. (1950). *Chemical Embryology*. New York and London: Interscience Publishers.
Chuang, H.-H. (1938). Spezifische Induktionsleistungen von Leber und Niere im Explantationsversuch. *Biologische Zentralblatt* **58**, 472–80.
Duprat, A. M., Gualandris, L. & Rouge, P. (1982). Neural induction and the structure of the target cell surface. *Journal of Embryology and Experimental Morphology* **70**, 171–87.

Gimlich, R. L. & Cooke, J. (1983). Cell lineage and the induction of second nervous systems in amphibian development. *Nature* **306**, 471–3.

Grunz, H. (1983). Differentiation of ectoderm of *Xenopus laevis* after treatment with neuralizing and vegetalizing factors. In *The Role of Cell Interactions in Early Neurogenesis*, ed. A. M. Duprat, A. C. Kato & M. Weber, pp. 21–38. International Workshop, Cargese, 1983.

Holtfreter, J. (1944). Neural differentiation of ectoderm through exposure to saline solution. *Journal of Experimental Zoology* **95**, 307–40.

Holtfreter, J. (1945). Neuralization and epidermization of gastrula ectoderm. *Journal of Experimental Zoology* **98**, 161–209.

Holtfreter, J. (1948). Concepts on the mechanism of embryonic induction and its relation to parthenogenesis and malignancy. *Symposia of the Society for Experimental Biology* **2**, 17–48.

Holtfreter, J. (1955). Studies on the diffusibility, toxicity and pathogenic properties of 'inductive' agents derived from dead tissues. *Experimental Cell Research* (Suppl.) **3**, 188–209.

Horst, J. Ter (1948). Differenzierungs- und Induktionsleistungen verschiedener Abschnitte der Medullarplatte und des Urdarmdasches von Triton im Kombinat. *Archiv für Entwicklungsmechanik der Organismen* **143**, 275–303.

Jacobson, M. (1982). Origins of the nervous system in amphibians. In *Current Topics in Neurobiology*, ed. N. C. Spitzer, pp. 44–99. New York: Plenum Press.

Jacobson, M. (1984). Cell lineage analysis of neural induction: origins of cells forming the induced nervous system. *Developmental Biology* **102**, 122–9.

Mangold, O. (1933). Über die Induktionsfähigkeit der verschiedenen Bezirke der Neurula von Urodelen. *Naturwissenschaften* **21**, 761–6.

Nakamura, O. & Toivonen, S., ed. (1978). *Organizer – A Milestone of a Half-century from Spemann*, 379 pp. Amsterdam, Oxford, New York: Elsevier/North-Holland Biomedical Press.

Needham, J. (1939). Biochemical aspects of organizer phenomena. *Growth* (Suppl.) 45–52.

Needham, J. (1950). *Biochemistry and Morphogenesis*. Cambridge: University Press.

Nieuwkoop, P. D. and others. (1952). Activation and organization of the central nervous system in amphibians. I. Induction and activation. II. Differentiation and organization. III. Synthesis of a new working hypothesis. *Journal of Experimental Zoology* **120**, 1–108.

Sala, M. (1955). Distribution of activating and transforming influences in the archenteron roof during the induction of nervous system in amphibians. *Proceedings of the Academy of Sciences of Amsterdam*, series C, **58**, 635–47.

Saxén, L. (1961). Transfilter neural induction of amphibian ectoderm. *Developmental Biology* **3**, 140–52.

Saxén, L. & Toivonen, S. (1961). The two-gradient hypothesis in primary induction. The combined effect of two types of inductors mixed in different ratios. *Journal of Embryology and Experimental Morphology* **9**, 514–33.

Saxén, L. & Toivonen, S. (1962). *Primary Embryonic Induction*. London: Logos Press, Academic Press.

Saxén, L., Toivonen, S. & Nakamura, O. (1978). Concluding remarks – primary embryonic induction: an unsolved problem. In *Organizer – A Milestone of a Half-century from Spemann*, ed. O. Nakamura & S. Toivonen, 315–20. Amsterdam, Oxford, New York: Elsevier/North-Holland Biomedical Press.

Saxén, L., Toivonen, S. & Vainio, T. (1964). Initial stimulus and subsequent interactions in embryonic induction. *Journal of Embryology and Experimental Morphology* **12**, 333–8.

Spemann, H. (1901). Entwicklungsphysiologische Studien am Triton-Ei. I. *Archiv für Entwicklungsmechanik der Organismen* **5**, 224–64.

Spemann, H. (1936). *Experimentelle Beiträge zu einer Theorie der Entwicklung.* Berlin: Springer.

Spemann, H. & Mangold, H. (1924). Über Induktion von Embryonalanlagen durch Implantation artfrämder Organisatoren. *Archiv für Entwicklungsmechanik der Organismen* **100**, 599–638.

Takata, K., Yamamoto, K. Y. & Ozawa, R. (1981). Use of lectins as probes for analyzing embryonic induction. *Wilhelm Roux Archives of Developmental Biology* **190**, 92–6.

Tiedemann, H. (1976). Pattern formation in early developmental stages of amphibian embryos. *Journal of Embryology and Experimental Morphology* **35**, 437–44.

Tiedemann, H. (1978). Chemical approach to the inducing agents. In *Organizer – A Milestone of Half-century from Spemann*, ed. O. Nakamura & S. Toivonen, pp. 91–117. Amsterdam, Oxford, New York: Elsevier/North-Holland Biomedical Press.

Tiedemann, H. (1981). Pattern formation and induction in amphibian embryos. *Fortschritte der Zoologie* **26**, 121–31.

Tiedemann, H. & Born, J. (1978). Biological activity of vegetalizing and neuralizing producing factors after binding to BAC-cellulose and CNBr-Sepharose. *Wilhelm Roux Archives of Developmental Biology* **184**, 285–99.

Tiedemann, H. & Tiedemann, H. (1959). Versuche zur Gewinnung eines mesodermalen Induktionsstoffes aur Hühnerembryonen. *Hoppe-Seylers Zeitschrift für Physiologische Chemie* **314**, 156–76.

Toivonen, S. (1938). Spezifische Induktionsleistungen von abnormen Induktoren in Implantatversuch. *Annales Societatis Zoologica-Botanica Fennica Vanamo* **6**, 1–12.

Toivonen, S. (1950). Stoffliche Induktoren. *Revue Suisse de Zoologie* **57**, (Suppl.), 41–56.

Toivonen, S. (1979). Transmission problem in primary induction. *Differentiation* **15**, 177–81.

Toivonen, S. & Saxén, L. (1955). The simultaneous inducing action of liver and bone-marrow of the guinea-pig in implantation and explantation experiments with embryos of *Triturus*. *Experimental Cell Research* (Suppl.) **3**, 346–57.

Toivonen, S. & Saxén, L. (1968). Morphogenetic interaction of presumptive neural and mesodermal cells mixed in different ratios. *Science* **159**, 539–40.

Vahs, W. (1957). Experimentelle Untersuchungen am *Triturus*-Keim über die stoffliche Mittel abnormen Induktoren. *Archiv für Entwicklungsmechanik der Organismen* **149**, 339–64.

Yamada, T. (1961). Chemical approach to the problem of the organizer. In *Advances in Morphogenesis* **1**, 1–53. New York and London: Academic Press.

# Structural and dynamical explanations in the world of neglected dimensions

ROBERT OLBY

Department of Philosophy, The University of Leeds, Leeds, LS2 9JT, UK

The chief point for the theory of heredity is, however, that protoplasm always offers us certain historical characters besides physical and chemical properties. It is to these that it owes its peculiarity. . . .
. . . The historical characters demand a molecular structure of such complicated nature that the chemistry of the present time fails us entirely in our attempts at an explanation.

> H. de Vries, *Intracellular Pangenesis*, 1889 (1910), p. 43

That 'function presupposes structure' has been declared an accepted axiom of biology. Who it was that so formulated the aphorism I do not know; but as regards the structure of the cell it harks back to Brücke, with whose demand for a mechanism, or organisation, within the cell, histologists have ever since been attempting to comply. But unless we mean to include thereby invisible, and merely chemical or molecular, structure, we come at once on dangerous ground . . .
. . . Of late years especially, an immense importance has been attached to these various linear or fibrillar arrangements, as they occur (*after staining*) in the cell-substance of intestinal epithelium, of spermatocytes, of ganglion cells, and most abundantly and most frequently of all in gland cells. Various functions, which seem somewhat arbitrarily chosen, have been assigned, and many hard names given to them; for these structures now include your mitochondria and your chondriokonts (both of these being varieties of chondriosomes), your Altmann's granules, your microsomes, pseudochromosomes, epidermal fibrils and basal filaments, your archeoplasm and ergastoplasm, and probably your idiozomes, plasmosomes, and many other histological minutiae . . .
. . . The outstanding fact, as it seems to me, is that physiological science has been heavily burdened in this matter, with a jargon of names and a thick cloud of hypotheses; while, from the physical point of view we are tempted to see but little mystery in the whole phenomenon, and to ascribe it, in all probability and in general terms, to the gathering or 'clumping' together,

275

under surface tension, of various constituents of the heterogeneous cell-content, . . .

D'Arcy Thompson, *Growth & Form*, 1917, pp. 160, 284–6.

# I Introduction

## Structural and dynamical explanations in the world of neglected dimensions

It has been customary in studying the history of the life sciences to organise the material in terms of conflicting approaches, the one reductionist, the other anti-reductionist. In this paper an attempt is made to escape from these categories and to offer an alternative in the form of (1) the dynamical explanations of the biophysical tradition, drawing upon James Clerk Maxwell and developed by W. B. Hardy, d'Arcy Thompson, Joseph Needham, and Sir Rudolf Peters, (2) the structural explanations of the tradition of classical cytology and histochemistry developed by Garnier, Caspersson, Brachet, Bensley and Claude, (3) the molecular explanations of the X-ray crystallographic tradition of Astbury, Bernal, Pauling and Hodgkin.

The history of biology, it is claimed, cannot be understood in terms of the triumphs of one of these traditions and the defeat of the others, but rather as a synthesis of elements from all three. Nor can the program of support for experimental biology by the Rockefeller Foundation be viewed merely as a means for the instrument-based domination of biologists by physicists. Rather it is viewed in the context of revealing the structural organisation of the cell in those regions between the molecular and microscopic dimensions. Its chief success is considered to have been its role in effecting the fusion of cytological, histochemical and biophysical/biochemical programs. Thompson's *Growth and Form* and Needham's *Order and Life*, which have often been considered anachronistic contributions, are interpreted within the context of the biophysical tradition and in opposition to the classical cytological tradition.

It is hoped that this broad analysis at the level of cell biology will be of value in approaches to the history of embryology.

## On the place of dimensions in physiological explanations

It is well known that organisms are confined to a range of dimensions which is narrower than the range of dimensions of the physical objects which we encounter in our life on the earth. The discussion of a lower limit to the size of organisms by Clerk Maxwell, raised questions about structural explanations of the forms of organisms, the development of those forms, and their inheritance, and about the relation between molecular structure and morphological structure. The two-and-a-bit orders of magnitude which lay between microscopic structures, visible under the light microscope, and what were considered the largest molecules (see Table 1) were populated with a variety of structures depending upon the models available from physical and colloid chemistry and subsequently macromolecular chemistry. In this paper I explore the ways in which attempts were made to fill this range of dimensions, and I examine the role of instrumentation in achieving this goal. It is therefore appropriate to begin with some comments on the current debate over the motivation and justification for the introduction of instrumentation into biology.

Warren Weaver of the Rockefeller Foundation's program for natural sciences claimed that his role in supporting physicists and chemists in the 1930s and 1940s who would work in biology and use instruments requiring their skills, was crucial to the development of molecular biology. Further he pointed out that it was he who coined the term 'molecular biology' to describe this program in 1938.[1] Robert Kohler has emphasised the directing role which Weaver played in his *managerial* capacity as a director of a grant agency.[2] Pnina Abir Am has opposed Weaver's account vigorously and the versions offered by those 'who credited Weaver with creating the new discipline of molecular biology by his sheer managerial ingenuity'. The Foundation in the 1930s, she suggested, was not anticipating the molecular biology of the 1960s. Instead Weaver was aiming at 'technology transfer' from physics and chemistry to biology. In this policy 'physical technologies became instruments of power leading to the domination of biology by physicists who lacked any real interest in biological problems'. Effectually it amounted to finding biological applications for a battery of tools, and it served the imperial claims of physics and chemistry to colonise biology. Significantly, she noted, 'none of Weaver's long-term grantees pioneered a redefinition of biology along molecular lines. . . .'[3]

Elsewhere Fuerst, Bartels, and I have examined these claims critically.[4] It suffices here to point out that the program funded in the 1930s did not need to 'anticipate' the molecular biology of the 1960s in order to lay the foundations upon which those later successes

Table 1. *Dispersed systems* (adapted from Ostwald, K. W. W. 1917, p. 20)

DISPERSED SYSTEMS

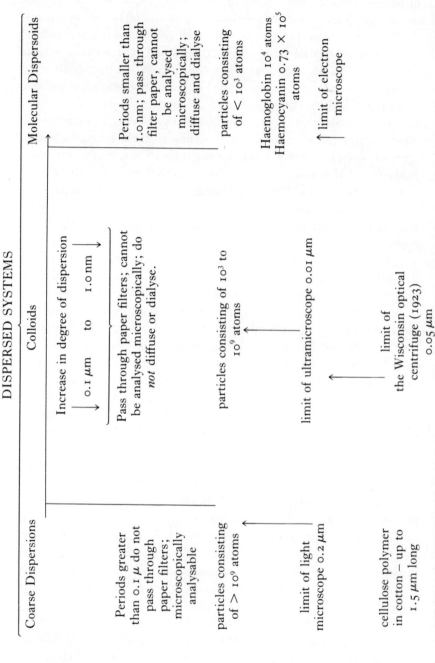

| Coarse Dispersions | Colloids | Molecular Dispersoids |
|---|---|---|
| | Increase in degree of dispersion $0.1\ \mu m \to 1.0\ nm$ | |
| Periods greater than $0.1\ \mu$ do not pass through paper filters; microscopically analysable | Pass through paper filters; cannot be analysed microscopically; do *not* diffuse or dialyse. | Periods smaller than $1.0\ nm$; pass through filter paper, cannot be analysed microscopically; diffuse and dialyse |
| particles consisting of $> 10^9$ atoms | particles consisting of $10^3$ to $10^9$ atoms | particles consisting of $< 10^3$ atoms |
| limit of light microscope $0.2\ \mu m$ | limit of ultramicroscope $0.01\ \mu m$ | Haemoglobin $10^4$ atoms Haemocyanin $0.73 \times 10^5$ atoms limit of electron microscope |
| cellulose polymer in cotton – up to $1.5\ \mu m$ long | limit of the Wisconsin optical centrifuge (1923) $0.05\ \mu m$ | |

could be built. Nor are the contributions of the 1930s and 1940s to our knowledge of the proteins invalidated as inputs to the molecular biology of the 1960s because they were focused on proteins rather than on nucleic acids, for there was a transfer of techniques and concepts from the one class of compounds to the other. Furthermore I have never claimed that DNA research was 'the principal focus of molecular biology'[5], but rather that this focus was provided by protein–nucleic acid relationships.[6] These relationships underwent a striking transformation between the 1930s and late 1950s, but within an overall program of structural analysis based on the instrumentation which Weaver, among others, promoted.

It is suggested that the role and context of the introduction of instruments has not yet been adequately explored. True, they offer special claims for grant applications. To an extent, the instrument defines the problem, as for instance one of structure rather than process, in the case of X-ray diffraction, the ultracentrifuge, the ultramicroscope, and electrophoresis, but one of process in the case of radioactive tracers. Yet many instruments were invented to solve specific, extant problems. In the case of X-ray diffraction analysis the experimental arrangement was first set up to investigate the nature of X-rays, but it was quickly appreciated that it also offered a means of studying molecular structures. The ultracentrifuge and the electron microscope, on the other hand, were invented to examine the physical parameters of *amicroscropic* structures such as were believed to exist in cells, and were used for that purpose. In research programs to develop and utilise such instruments, I wish to argue, there was a group of underlying assumptions. These involved some form of materialism[7] for our purposes, meaning thereby that functions and structures of organisms are to be attributed to the properties (including interactions) of their structured components *going right down to the atomic level*. Whether one believes that the properties which emerge at successive higher levels of organisation *are irreducible or reducible has been less important, to the type of research program carried out, than the views held as to the persistence or transience in time, from generation to generation, of the structured elements believed to exist*. Frederick Churchill long ago described Weismann's transition from the Haeckelian view of unstructured protoplasm to the continuity of structured germinal material, a change which had considerable implications for his subsequent research.[8] This view of the cell as analysable by investigation of its fine structure may be just 'common sense' *today*, as Professor Davidson told us, but there have been considerable and extended debates and rival research programs on the subject.

In order to establish this point I shall refer to d'Arcy Thompson's *Growth and Form*. Although this work has been discussed by a number of very able authors[9] Thompson's opposition to structured elements in the cell, assumed to persist, and function by virtue of their particular constitution, has not, to my knowledge, been sufficiently recognised. At the same time his reliance on changing structured elements based on a dynamic conception of colloid systems deserves noting. An analysis of his work brings to our attention the existence of a controversy over the *kinds of structures present in the cell and the extent of their persistence*.

When Theodor Schwann described the cell in 1839[10] he concentrated its functions largely on the cell membrane and considered its contents in very simple terms. Then came Brücke in 1861 with his famous paper on 'The Cell as an Elementary Organism', in which he claimed:

To the living cell, apart from the molecular structure of the organic compounds which it contains, there must be attributed yet another structure complicated in a different manner, and it is this to which we give the name Organisation.[11]

Then in 1899 William Bate Hardy, working in Cambridge on the production in colloidal gelatin of structures, like those seen in cytoplasm by the cytologists, claimed that no structures like the networks, fibrils, foams, and granules depicted by cytologists following Brücke's call had been established.[12] Furthermore Hardy was able to produce such structures in colloids using the cytologists' fixatives. Therefore they were artefacts of fixation.

The debate over the significance of the cytologists' structures is a long one. Miescher extracted his Nuclein (nucleoprotein) from cell nuclei because he wanted a *chemical criterion* for establishing the identity of nuclei, hitherto identified on morphological criteria.[13] This did not prevent him confusing yolk granules with nuclei.[14] In our century the debate over the chemical significance of the Feulgen stain raged into the 1950s.[15] Hence the question of the status of microscopic (visible under the light microscope), *ultramicroscopic* (visible under the ultramicroscope), and *amicroscopic* structures (not visible even under the ultramicroscope), has been a pervasive one in biology. Curiously the ultraviolet microscope which Köhler used in 1904 to demonstrate chromosomes was not exploited until three decades later by Caspersson.[16] The ultramicroscope was used chiefly to demonstrate colloidal particles, but Seifriz used it to show longitudinal striations in protoplasm. It is common knowledge that the ultracentrifuge, like the electron microscope, has had a tremendous impact on biology, but it was built to investigate amicroscopic structures in colloidal solutions.[17]

Now behind the programs to investigate the hidden region of the amicroscopic there was a conviction that there was an unbroken series of structural entities upon which the properties of living and non-living systems depended. This belief we may call *structural continuity* throughout the scale of dimensions. It is seen in such examples as Wolfgang Ostwald's *Introduction to Theoretical and Applied Colloid Science* (Table 1), in Frey Wyssling's *Ultrastructural Cytology* (Table 2), and in the second edition of d'Arcy Thompson's *Growth and Form* (Table 3). Also Bernal expressed such a view at the first meeting of the Cambridge Theoretical Biology Club. As Donna Haraway tells us in *Crystals, Fabrics, and Fields* (1976) 'Bernal outlined a scale of form reaching from quantum-mechanical systems to the metazoa, and he called for a "neo-cytology of proteins".'[18]

This thesis of structural continuity provided the context for the application of such instruments to biology. In Weaver's mind this view may have been strongly structural rather than dynamic, thus predisposing research into structural questions and supporting structural schools at Cambridge and Caltech, for in 1937 he wrote: 'Gradually there is coming into being a new branch of science – molecular biology – which is beginning to uncover many secrets concerning the ultimate units of the living cell.' 'Modern tools', he wrote, were revealing new facts about the structure and behaviour of the 'minute intercellular substances.'[19] That his views were at first tentative and amateurish Weaver readily acknowledged, but it was clear that the physical sciences could offer 'a whole battery of analytical and experimental procedures capable of probing into nature with a fineness and with a quantitative precision that would tremendously supplement the previous tools of biology'. Even at that time, he recalled, 'one could identify some of the procedures and the instruments that were ready to be applied more intensively to basic biological problems.'[20]

That this was not a starkly structural view may be gathered from his support of Schönheimer's use of radioactive tracers which gave

Table 2. *The domains of cytology* (from Frey Wyssling, A. & Mühlethaler, K., *Ultrastructural Plant Cytology*, with an Introduction to Molecular Biology. Amsterdam etc., 1965, p. 3, Table II)

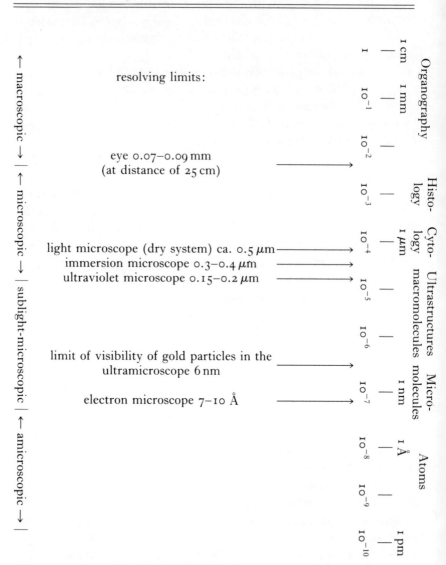

Table 3. *Linear dimensions of organisms and other objects* (from Thompson, *Growth & Form*, 2nd edn, 1942, reprinted 1959, vol. **i**, p. 66)*

| | | |
|---|---|---|
| (10 000 km) | $10^7$ | A quadrant of the earth's circumference |
| (1000 km) | $10^6$ | Orkney to Land's End |
| | $10^5$ | |
| | $10^4$ | Mount Everest |
| (km) | $10^3$ | |
| | $10^2$ | Giant trees: *Sequoia* |
| | | Large whale |
| | $10^1$ | Basking shark |
| | | Elephant; ostrich; man |
| (metre) | $10^0$ | Dog; rat; eagle |
| | $10^{-1}$ | Small birds and mammals; large insects |
| (cm) | $10^{-2}$ | Small insects; minute fish |
| (mm) | $10^{-3}$ | |
| | $10^{-4}$ | Minute insects |
| | $10^{-5}$ | Protozoa; pollen grains |
| (micron) | $10^{-6}$ | Large bacteria; human blood corpuscles |
| | $10^{-7}$ | Minute bacteria |

Cells

| | | |
|---|---|---|
| | | Limit of microscopic vision |
| | $10^{-8}$ | Viruses, or filter-passers |
| | | Giant albuminoids, casein, etc. |
| (nm) | $10^{-9}$ | Starch molecule |
| Angström unit | $10^{-10}$ | Water molecule |

Colloid particles

* (In *Growth and Form* the column of tens is erroneously headed cm instead of m. Here the original mμ has been replaced by nm for $10^{-9}$ of a metre.)

strong evidence in favour of the dynamic conception of metabolism. Nor was Weaver unsympathetic to the program which Needham wished to develop at Cambridge based upon the view of the cell in structural, dynamic, and holistic terms.[21] We know that Weaver was not committed to such a strong reductionist line as was later presented to him by Crick.[22] Furthermore, the use of instruments like the ultracentrifuge and electron microscope did not dissolve morphological entities like microsomes, mitochondria, and plastids; rather such instruments were used to bring into relationship morphological and chemical entities and the cytological and biochemical points of view. The sharp isolation and insularity of the biochemists and cytologists towards each other, so well described by Professor Brachet in this volume, were thus overcome. Therefore it deserves stressing that there has been no necessary and direct connection between the use of these instruments and a commitment to the 'reduction' of biology in the strong positivist sense of that term. However, it has been closely associated with belief in the existence of structural continuity between the molecular and the microscopic levels.

## II Maxwell's paradox

Let us turn to the paradox raised by Clerk Maxwell in 1875 concerning the degree of minuteness of submicroscopic structures. In the latter half of the nineteenth century Kekulé's conception of giant 'net-like and sponge-like' molecules based on linear and branching chains was exploited by a number of writers to account for morphological structures of both minute and large size. Among these were Hermann's inogen molecule[23], Pflüger's giant protein molecules of protoplasm[24], Pfitzner's molecular chromomeres of the chromosomes[25], Verworn's biogen molecules[26] and the many molecular models proposed by Montgomery[27] and Hörmann[28].

These models brought the dimensions of molecules right across the gap traditionally recognised between organic molecules and microscopic structures, but they did not survive alongside the rapid development of colloid chemistry in the first three decades of this century. Nor were they assimilated by the new school of biochemistry whose doyen, F. G. Hopkins, based his research program on the role of 'comparatively simple molecules' in the cell, and who rejected the alleged giant molecules of protoplasm, treating them as one of the 'grand heresies of biochemical thought'.[29] It is well known that strong opposition to the concept of polymeric molecules arose in organic chemistry as well as in biochemistry which lasted until the late 1920s.[30]

Also popular in the latter half of the nineteenth century were speculations about submicroscopic entities in the cell held responsible for development and heredity. Such were the physiological units of Spencer, the micelles of Naegeli, the gemmules of Darwin, the plastidules of Elsberg and Haeckel, the biophores of Weismann, and the bionts of Oskar Hertwig. All, save Haeckel, considered the largest chemical molecules to be much smaller than the giant molecules of Pflüger et al. The dimensions of their morphological units were conceived as lying between those of molecules and those of subcellular structures just visible under the light microscope. As such they constituted a *hierarchy of structures above the level of organic molecules, whose role was to account for those physiological and morphological properties of the cell, for which mere protein molecules with an assumed average diameter of 50 nm were found inadequate.*

Weismann reckoned his biophores contained 1000 protein molecules each, but Darwin gave no estimate of the size of the gemmules. He stated that 'their number and minuteness must be something inconceivable'. This reflection might seem to constitute a formidable objection, he admitted, until we ponder the 64 million eggs laid by a single *Ascaris* or the million seeds produced by a single orchidaceous plant.

In these several cases, the spermatozoa and pollen-grains must exist in considerably larger numbers. Now, when we have to deal with numbers such as these, which the human intellect cannot grasp, there is no good reason for rejecting our present hypothesis on account of the assumed existence of cell-gemmules a few thousand times more numerous.[31]

From this passage we infer that Darwin assumed some nine orders of magnitude difference between cells and gemmules. It was this assumption which prompted Maxwell to declare:

Thus molecular science sets us face to face with physiological theories. It forbids the physiologist from imagining that structural details of infinitely small dimensions can furnish an explanation of the infinite variety which exists in the properties and functions of the most minute organisms.[32]

Two different organisms each develop from a microscopic germ. Are all their differences referable to 'differences in the structure of the respective germs?' And he added:

Even if we admit this as possible, we shall be called upon by the advocates of Pangenesis to admit still greater marvels, for the microscopic germ, according to this theory is no mere individual, but a representative body, containing members collected from every rank of the long drawn ramifications of the ancestral tree, the number of these members being amply sufficient not only to furnish the hereditary characteristics of every organ of the body and every habit of the animal from birth to death, but also to afford a stock of latent gemmules to be passed on in an inactive state from germ to germ, till at last the ancestral peculiarity which it represents is revived in some remote descendant.

Some of the exponents of this theory of heredity have attempted to elude the difficulty of placing a whole world of wonders within a body so small and so devoid of visible structure as a germ, by using the phrase structureless germs.* Now one material system can differ from another only in the configuration and motion which it has at a given instant. To explain differences of function and development of a germ without assuming differences of structure, is, therefore, to admit that the properties of a germ are not those of a purely material system.[33]

Hugo de Vries appeared not to have been aware of Maxwell's critique when he revised Darwin's theory in his famous book *Intracellulare Pangenesis*, 1889. There he assumed the existence of as many kinds of pangenes as there are independent hereditary characters, presumably a much smaller number than Darwin required for each cell or group of cells. Each pangene was composed of 'innumerable chemical molecules'[34] and hence much larger than they, yet invisible under the microscope. That was also the opinion of Weismann, Naegeli, and Hertwig, for their particles.

In 1903 the subject of the limit to the minuteness of organisms was discussed by Leo Errera (1858–1905) with specific reference to Maxwell. This Belgian physiologist came to the conclusion that the smallest bacterium then known to science *Micromonas progrediens* (0.15 μm) contained between ten and thirty thousand molecules of protein. Therefore he concluded that organisms *much* smaller than the smallest known bacteria were unlikely to exist. The filterable viruses could be only a little smaller than *Micrococcus*, and organisms several hundred times smaller 'would already be an impossibility'.[35]

# III D'Arcy Thompson's view of structural and dynamic theories

In his renowned book *Growth and Form*, 1917, D'Arcy Thompson referred four times to Maxwell's discussion of the invisible structure of the germ, and he accepted Maxwell's rejection of Darwin's numerous and minute gemmules. However, he believed that Maxwell had implied first, that 'molecules more complicated by far than the chemist's hypotheses demand' were needed to explain the phenomena of life, and second, that nothing of the nature of *physical* causality could ever explain all of life.[36] He denied the first and it was

---

* This is a reference to Galton's paper. 'On blood-relationships', *Proc. Roy. Soc.* **20** (1872), 394–401.

the central theme of his book to weaken the second by appealing to physical forces as the agents of organic form.

Also central to his theme was the denial of preformation in its modern version, as manifest in structural conceptions of the cell contents. Referring to the desire to find fine structure in the cell he wrote:

In an earlier age, men sought for the visible embryo even for the *homunculus*, within the reproductive cells; and to this day we scrutinise these cells for visible structure, unable to free ourselves from that old doctrine of 'pre-formation'.[37]

Thompson reminded his readers that in 1861 C. Brücke had called for the discovery of supramolecular organisation in the cell. But Maxwell had later shown us that unless we mean 'merely chemical or molecular structure we come at once on dangerous ground'. Did not recent research show that the simplest microscopic structure 'is that of a more or less viscous colloid, or mixture of colloids, and nothing more. . . . If we cannot assume differences in structure, we must assume differences in *motion*, that is to say, in *energy*.'[38] In this way Thompson ruled *structural theories out of his strategy, just as had W. B. Hardy, and ruled dynamic theories in*. Attempts to locate vital functions within certain regions of the cell were products of a structuralist view. 'From the moment that we enter on a dynamical conception of the cell', he declared, 'we perceive that the old debates were in vain as to what visible portions of the cell were active or passive, living or non-living. For the manifestations of force can only be due to the interaction of the various parts, to the transference of energy from one to another.'[39] '. . . how a thing grows and what it grows into, is a dynamic and not a merely material problem; so far as the material substance is concerned, it is so only by reason of the chemical, electrical or other forces which are associated with it.'[40] There is no escaping the evidences of the physicists' influence on Thompson. More than once he referred to Clerk Maxwell's paper: 'On Reciprocal Figures, Frames, and Diagrams of Force', and quoted with enthusiasm Maxwell's reference to the term *skeleton* when describing the frame constituted of points 'mutually acting on each other with forces whose directions are those of the lines joining each pair of points'.[41] As Thompson confessed: 'we [naturalists] . . . never look at a standing quadruped without contemplating a bridge', and conversely, the engineer who wrote *Treatise on Bridge-Construction* described his subject under the

heading 'The Comparative Anatomy of Bridges'.[42] At the microscopic level of the cell the story was the same; the shapes of cells,
their aggregation during cleavage of the embryo, were not 'specific
characters of this or that tissue or organism', but products of forces
and the partitioning of space.[43]

Admittedly Thompson did include a long discussion of the forms
of inorganic deposits such as otoliths, spicules, radiolarian skeletons
and the conostats of diatoms. What Thompson sought to establish
was that such structures are the product of a *process of crystallisation*
in which the precipitates in coming out of solution pass through a
liquid phase. Their temporary sojourn in this phase 'leaves its
impress in the rounded contours which surface tension brought
about while the little aggregate was still labile or fluid....'[44] It
cannot be too strongly emphasised that *we should cease to view
Thompson as an idiosyncratic figure. We should rid ourselves of any
notion that his obsession over form implies any sympathy for an
Aristotelian ontology.* Instead we should locate him very much
within his period, *as an exponent of dynamical explanations in the
tradition set by Clerk Maxwell in physics and by Wilhelm His in
embryology.* Molecules, as fields of forces, fell within that explanatory program. In short they were to be understood in terms of the
*physical forces* operating in colloidal solutions. Even the polarity
evident in the polyhedral forms of radiolarian skeletons might be
understood by reference to Lehmann's *Gestaltungskraft* which
aggregated molecules into a definite configuration in the fluid state –
Lehmann's liquid-crystals.[45] The forms of living cells, like those of
their dead skeletal products, were also to be attributed to the action
of mechanical forces:

Thus we can at once define and explain, from the point of view of surface-
tension, the spherical form of the cell in a simple unicellular organism.
When equal cylindrical cells are set together, we can understand how mechanical pressure, uniformly applied, converts the circular outlines into a
pattern of regular hexagons, as in the hexagonal facets of the insect's eye or
the hexagonal pigment-cells of our own retina. On similar lines PAPPUS,
MACLAURIN, and many other mathematicians have contributed to elucidate the hexagonal outline and the rhombic dodecahedral base of cells of
the honeycomb; while a more extended application of the theory of surface-
tension, on the lines laid down by PLATEAU, gives us a complete understanding of the froth-like conglomeration of cells which we observe in
vegetable parenchyma and in many other cellular tissues.[46]

Thompson's underlying strategy is clear. It was physicalist and holist; holist not in any mystical sense like Bergson but in terms of holistic physical analogies such as the field of forces, the surface tension of a rain drop, the ripples on the sea shore. In contrast to physics, chemistry was not yet a science, but it would become one when it can 'deal with molecular mechanics, by the methods and in the strict language of mathematics, as the astronomy of Newton and Laplace dealt with the stars in their courses'.[47] Morphology had to do likewise. This position, he expressed clearly in his Presidential Address to the Zoological Section of the British Association, in 1911[48], but it can be traced to his anti-selectionist views, which surfaced in *Some difficulties in Darwinism* of 1894.[49] We know that he was an early advocate of causal analysis in embryology and adopted a position very close to the physicalist stance of Wilhelm His.[50] It would be misleading to describe him as a member of the school of *Entwicklungsmechanik*, but he did echo their call for the investigation of 'formative forces or energies'. Now Bergson had argued that physicochemical laws were miserly and brutal, where they reigned there was only incoherence and chaos. Thompson rejected this view, and pointed to waves, ripples and clouds as 'so many riddles of form, so many problems of morphology . . . nor is it otherwise with the material forms of living things.'[51] And further, if organic form was to be understood in terms of forces, if the segmentation of the egg was but an exquisite froth, 'would it wholly revolutionise our biological ideas?' he asked in 1911. 'It would greatly modify some of them, and some of the most cherished ideas of embryologists . . . biologists of today are in no self-satisfied and exultant mood. The reasons and the reasoning that contented a past generation call for re-inquiry.'[52]

Thompson's targets for destruction were the two most popular types of explanation of form: structural theories like those of Weismann, which postulated an invisible organisation of matter as the hereditary basis of the characteristics of an organism, and hereditary determination by the mere matter of the nucleus or chromosomes. As Thompson remarked:

. . . in dealing with the facts of embryology or the phenomena of inheritance, the common language of the books seems to deal too much with the *material* elements concerned, as the causes of development, of variation or of hereditary transmission. Matter as such produces nothing, changes nothing, does nothing . . . the spermatozoa, the nucleus, the chromosomes or the germ-plasm can never *act* as matter alone, but only as seats of energy and as centres of force.[53]

His other target was the Darwinian approach to adaptation and the biogenetic law which Ridley has already so ably described in this volume. With regard to adaptation it was Thompson's aim to show wherever possible that form has been the product of physical forces, thus making a selectionist account redundant. His classic example was the oval-pointed shape of birds' eggs which he attributed to the forces squeezing the egg down the oviduct.[54]

Thompson's alternative to structuralists, recapitulationists and selectionists was his dynamic conception of form as 'a diagram of forces' and as a function of growth. Where very small organisms were involved the form might be determined purely by *molecular* forces, as in the raindrop and the shape of a free-swimming protozoan; in larger organisms form was due to growth and hence to larger-scale or molar forces. Therefore, he concluded: 'It is the ratio between the rates of growth in various directions by which we must account for the external forms of all, save certain very minute, organisms.'[55] In his 1914 address to the Royal Society of Edinburgh on *Morphology and Mathematics*, Thompson urged the value of describing form mathematically. This was in all but a few cases exceedingly difficult. Fortunately it was often possible to compare related forms by 'recognising in one form a definite permutation or *deformation* of another', using the Cartesian method of co-ordinates.

This once demonstrated it will be a comparatively easy task ... to postulate the direction and magnitude of the force capable of effecting the required transformation. Again, if such a simple alteration of the system of forces can be proved adequate to meet the case, we may find ourselves able to dispense with many widely current and more complicated hypotheses of biological causation.[56]

It would seem that he regarded this effect of forces as a means employed by Nature herself in transforming one species into another.[57]

As Steven J. Gould has noted recently[58], and Ridley has reminded us in this volume, there are parallels between d'Arcy Thompson and William Bateson. Both men were products of Cambridge biology under Balfour; both came to oppose the biogenetic law in the 1890s. They scorned Weismannism. Thus Thompson wrote:

But if we speak, as Weismann and others speak, of an hereditary *substance*, . . . we can only justify our mode of speech by the assumption that that particular portion of matter is the essential vehicle of a particular charge or distribution of energy, . . .[59]

They were both attracted to processes yielding symmetry, Bateson being happy to use different rates of cell division of different recombinants rather than the Sturtevant/Morgan scheme of linkage and crossing-over, to account for departures from random (Mendelian) assortment.[60] Both men were impressed by the analogy between repeated structures in organisms and the ripples in sand on the seashore, and they stressed the fundamental nature of the division of the cell, but not the longitudinal division of the chromosome.[61] Yet they chose different strategies by which to tackle the problems of evolution, Thompson studying growth and Bateson studying variation. As for heredity, Bateson devoted his career from 1900 onwards to it whereas d'Arcy Thompson explained that he deliberately avoided any discussion of it, his aim being to emphasise 'the direct action of causes other than heredity'.[62]

# IV The world of neglected dimensions

Between the largest molecules accepted by classical organic chemists at the turn of the century and the smallest objects visible under the light microscope lay what Wolfgang Ostwald called 'The World of the Neglected Dimension'. This was the world of the *colloidal state*, which was bounded on one side by coarse suspensions in which the particles were greater than $1/10$th $\mu$m in size and visible in the microscope, and on the other by molecular solutions, the molecules being less than 1 nm in diameter. Ostwald wrote that

Between matter in mass and matter in molecular form there exists a realm in which a whole world of remarkable phenomena occur, governed neither by the laws controlling the behaviour of matter in mass, nor yet those which govern materials possessed of molecular dimensions. We did not know that this middle country existed. . . . We have only recently come to learn that every structure assumes special properties and a special behaviour when the particles are so small that they can no longer be recognised macroscopically while they are too large to be called molecules. Only now has the true significance of this region of the colloid dimensions

– THE WORLD OF NEGLECTED DIMENSIONS – become manifest to us.[63]

It was this world of the size range around 1/100th of a micron (or ~100 Å) that would reveal the structural level *above* the largest molecules, a level ruled *not* by traditional chemistry (the laws of simple and multiple proportions, the law of mass action, Gibbs phase rule), but by the laws of colloid chemistry. The peculiarities of protoplasm were not simply due to special chemical compounds – the proteins – but to a *special state* – the colloidal state. In this state the compounds in question failed to respond to the techniques of the organic chemist. Their solutions were atypical. Their viscosity was remarkably high; it did not vary uniformly in relation to concentration, and they readily formed gels. Their composition varied with the technique used in their preparation, and they were susceptible to ageing. In colloidal solutions the substance distributed in the liquid, the disperse phase, consisted of a range of differently sized particles, because the molecules were aggregated to varying degrees. Any figure for the degree of dispersion was therefore an average value and the system was *poly-disperse*, in contrast to molecular solutions in which (barring aggregation and dissociation) there was only one size and the system was *mono-disperse*. Powerful support for the theory of the molecular aggregate came from crystallography where it was generally conceded that the chemist's molecule was either the same size as the crystalline molecule (or unit cell) or it was smaller. With the advent of X-ray diffraction analysis this opinion was reinforced. Thus W. H. Bragg claimed that X-ray analysis shows that the unit cell nearly always contains the substance of more than one molecule; generally of 2, 3 or 4.[64] When fibre diagrams were discovered in 1920 their interpretation gave small unit cells. Surely, then, the long fibres of cotton, silk, and rubber must be made up of molecules the same size as the unit cell, but held together by secondary aggregating forces?

The supporters of the colloid aggregate based their claims for the significance of their science not only on its economic importance in industry and agriculture – metallurgy, paper-making, photography, leather-work, food, dyeing, histology, steel, textiles, cement, glass, ceramics, smelting and the manufacture of paints – but also on its central importance to the study of life. Zsigmondy declared that colloids are 'extremely widely distributed on the terrestrial globe.' All living creatures, animals as well as plants, are for the greater part constructed of colloids. Without colloids no living thing is possible; cells consist of colloids, their content and their membranes; blood serum, plant sap are essentially colloidal solutions; gelatin, obtained from leather and bones, is a typical colloid.[65]

Bechold, pioneer of the technique of ultrafiltration to estimate the size of colloid particles including viruses, declared that if there were men on Mars, they might have unconventional constituents, but one thing we could be sure of was that they would be in the colloidal state.[66] Höber claimed that the colloid was for physiology what the electron was for physics – a centre of crystallisation around which the many scattered and disordered facts would become organised.[67] Four years later C. M. Child supported Höber's rejection of Pflüger's hypothetical labile molecules which left the scientist in a *cul de sac* 'from which the only possible way out is retreat – the more we know concerning colloids the less possible it becomes to conceive anything similar to what we regard as life apart from them. Whatever else it may be, it seems certain that the organism is a colloid system.'[68]

The connections between colloid chemistry and dynamical theories in physiology and embryology are clear. The forces operating in colloids were appealed to by d'Arcy Thompson repeatedly in *Growth and Form*. L. Rhumbler gave a prominent place to colloidal properties in his discussion of protoplasm in 1914.[69] Martin Fischer and Wolfgang Ostwald saw the close identity between the methods used in the artificial parthenogenesis of the egg and the methods of coagulating protein colloids as evidence that fertilisation was a process of coagulation. So were the asters of the dividing sea-urchin egg.[70] The experiments of P. Harting on what he called '*Synthetic Morphology*' (1872) were interpreted in terms of colloid chemistry by Thompson and other authors.[71]

Colloid chemistry represented an extension of chemistry and physics to cover the special case of molecular aggregates where classical organic and physical chemistry were considered inappropriate. It was attractive to those holding a dynamical conception of the cytoplasm, because it concerned general properties of a state of matter, rather than the special properties of a particular compound. In such processes as gel formation and swelling, it offered models for the water relations of the cell and its growth which were dynamic in nature. The reversibility of coagulation in many colloidal solutions, and its polyphase nature, again offered models for change of the cell in development. As Hopkins remarked: '*Its life* [the cell's] *is the expression of a particular dynamic equilibrium which obtains in a polyphasic system.*[72]

# V The macromolecule and the monodisperse system

The colloid-aggregate theory came under attack in the late 1920s chiefly because of the work of Hermann Staudinger on synthetic and natural polymers, and because of Theodor Svedberg's development of the ultracentrifuge. Whereas Staudinger worked in the discipline of organic chemistry, and was considered as no friend of colloid chemistry, Svedberg was a leading member in that subject having worked in that field from the start of his career.

Staudinger arrived at the conception of macromolecular substances when he investigated the production of synthetic polymers. He found that a small six-membered ring could be cleaved and the resulting short chains linked up to produce progressively longer and longer chains. Apart from the insolubility of the product there seemed to be no limit to the process. Yet at each stage the addition of another link was a conventional chemical reaction. Moreover, the products could be acetylated, methylated, hydrogenated etc., in a straightforward chemical reaction and the polymer analog produced had the same particle size as the unreacted polymer. Clearly these polymer particles were not aggregates of smaller molecules but macromolecules. They were behaving like any ordinary chemical substance.

Central to his program was his use of synthetic polymers as models for macromolecular natural products: polystyrene for rubber, polyoxymethylene for cellulose, polyethylenoxide for starch, and polyacrylic acid for protein.[73] Since the synthetic polymers were produced in a whole range of chain lengths he thought that the so-called molecular weight of a macromolecule was an average and that there was no unique value for a given macromolecular compound.

In contrast with Staudinger, Svedberg arrived at the concept of macromolecular species through colloid chemistry itself. He invented the ultracentrifuge in order to discover the distribution of the differently sized particles in a colloidal solution, where those particles were below the range of the ultramicroscope – they were amicrons. Such particles were not heavy enough to sediment under gravity, but could be affected by a strong centrifugal force. If he could study such finely dispersed particles he hoped he could learn about the *formation* of colloidal solutions. Hence it was no accident that the instrument he invented showed the degree of uniformity in particle size.

It is well known that he tried the effect of ultracentrifugation upon the proteins, beginning in 1924. This was because he considered them to be biologically important polydisperse colloids. At the suggestion of the haematologist, Robin Fåhraeus, who was on a research visit to Uppsala from the Karolinska Institute in Stockholm, he tried haemoglobin. To their amazement it turned out to be a chemical species with uniform particle weight of nearly 67 000 daltons.[74] There followed the extensive study of the haemoglobins of other mammals, birds and fish[75], and of the respiratory proteins of the invertebrates, starting with haemocyanin (1928) with its remarkably sharp sedimentation boundary[76] (see Fig. 1). From such studies Svedberg in 1934 drew up a chemical taxonomy of 100 respiratory pigments paralleling the known systematic relationships of the species concerned.[77] Thus the haemoglobins of mammals, birds and fish had approximately the same sedimentation rates, but haemoglobins from the different species were stable at different pHs, they dissociated at different rates on dilution, and their electrophoretic mobilities differed. These subtle differences had to be due to differences in amino acid composition, to the presence of one or another active group.[78]

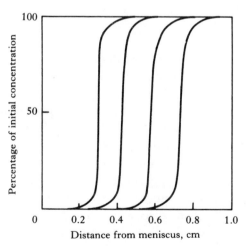

Fig. 1. Haemocyanin (from Svedberg & Pederson, *The Ultracentrifuge*, 1940, plate facing p. 33).

On the other hand there was a notable clustering of all globular proteins around a few sedimentation rates, which were 1, 2, 3 or 6 times 35 000 [35 200] or they were 12, 24, 48, 96, 192 times as in the invertebrate respiratory pigments, the haemocyanins and the erythrocruorins (see Table 4). By the time Svedberg and Pederson

Table 4. *The molecular weights of respiratory proteins* (from Svedberg and Eriksson-Quensel in *Tabulae Biologicae*, 11 (1936), 351–4)

| Name | Origin | Mol. wt obs. | | |
|------|--------|--------------|--------------|--------------|
| | | Sed. rate | Sed. equilibr. | Theoretical value as multiple of $36\,400 = M$ |
| Erythrocruorin | *Lampeter* | 18 300 | 19 000 | $M \times 0.5 = 18\,200$ |
| | *Chironomus* | — | 31 400 | $M \times 1 = 36\,400$ |
| Haemoglobin | horse | 69 000 | 68 000 | $M \times 2 = 72\,800$ |
| | human | 63 000 | | |
| Erythrocruorin | *Palinurus* | 446 000 | 460 000 | $M \times 12 = 400\,000$ |
| Haemocyanin | *Nephrops* | 766 000 | — | $12\,M \times 2 = 800\,000$ |
| | *Octopus* | 2 785 000 | — | $12\,M \times 7 = 2\,800\,000$ |
| Erythrocruorin | *Lumbricus* | 3 190 000 | 2 946 000 | $12\,M \times 8 = 3\,200\,000$ |
| Haemocyanin | *Helix* | 6 630 000 | 6 706 000 | $12\,M \times 16 = 6\,400\,000$ |

published their great work *The Ultracentrifuge* in 1940 myoglobin was known to have a molecular weight of $\sim$17 500 (and cytochrome c 15 600) so the '*periodic table*' of the proteins was expressed as multiples of 17 600 instead of 35 000. Svedberg interpreted this evidence of limited sizes of proteins as a product of the procedures for synthesis of these giant molecules in which nature proceeds along 'a very limited number of main lines'.[79] Therefore they are built on a general plan and the variety appears in the specialisation of details. It is remarkable that he had come to this conclusion at least as early as 1927, possibly as soon as he discovered the monodisperse character of haemoglobin in 1926. Svedberg and his school thus arrived at a conception of proteins as giant molecules with a pretty regular morphology based upon a construction from subunits of a standard length. As is shown in Table 4 the size of the respiratory pigments which he studied stretched beyond the so-called molecular disperse particle size and into Ostwald's 'World of Neglected Dimensions'. Svedberg also confirmed Staudinger's assertions as to the great lengths of the cellulose chain molecules which completely bridged the gap between molecular and microscopic dimensions.

Thus we have here an example of an instrument, the theoretical basis of which was that colloidal particles are kinetic units just like ordinary molecules, and will fall in accordance with Stokes' Law under the influence of gravity as worked out by Warren Weaver.[80] They will also diffuse, as a result of Brownian motion, like any other kinetic unit. It was a tool of physical chemistry (the first model was an adaptation of a cream separator) which quite unexpectedly demonstrated the molecular character of protein colloids. Once a tool like this is invented one can look for problems it might solve. It was soon applied to the study of the cell and has played a decisive part along with the electron microscope in repopulating the cell with organelles, with structural entities over which speculation had raged in the history of cytology long before.

# VI Structure according to Astbury

At the same time that Svedberg was developing his conception of the nature of proteins Astbury was developing his conception of the molecular morphology of the proteins. The technique here was X-ray diffraction analysis, which he applied to wool, following the work of the German School on cotton and silk. From these studies he developed the fundamental concept of the extended chain as the structural basis to the kmf group of proteins, and the folding of the chain as the molecular basis to the contraction of the protein fibre. These studies led him to a conception of molecular biology in which he was happy to embrace the evolution and morphology of macromolecules centred on proteins as the 'first things'.[81]

'This structural plan', he said, 'traces back to no narrow special locus but, through tissues of endo-, meso-, and ectodermal original alike, to the embryonic epithelial cell. It represents a fundamental power of the cell of synthesizing fibrous proteins of a standard molecular form, yet whose details can be adapted to a variety of ends as the processes of differentiation demand'.[82]

This passage comes from Astbury's contribution to the d'Arcy Thompson *Festschrift*. He saw himself doing something 'to bring physics and biology closer together in their common aim, the study of the structure of matter'. He was, he said, throwing 'a tentative bridgehead from molecules to organized structures. . . .'[83] In view of Thompson's physicalist stance towards biology it is not surprising that he was delighted with Astbury's contribution. He told Astbury: 'I have lapped it up, and am still inwardly digesting it. It is so plain and simple, there is really nothing to discuss. It is plain sailing from shore to shore'. That muscle contraction might be based on the folding of long-chain molecules, he wrote, 'seems to me nothing short of magnificent, and a "glimpse of the obvious" on a very large scale.'[84] At first sight Thompson's enthusiasm here might be thought to undermine the view developed in this paper of Thompson as an opponent of the structuralists. On closer examination, however, it is clear that he saw muscle contraction as a dynamical process, as folding at the molecular level, and since molecules were physical, not classical cytological and embryological entities, Astbury's beautiful survey of molecular form delighted him.

The world of neglected dimensions was now populated, not only with monomolecular layers of membranes, but with giant molecules possessing their own morphology, their own dynamics and, to an increasing extent, occupying the territory once reserved for the dynamic colloidal state. What was new about this transformation was that the dynamic relations of molecular structures were seen to be finely tuned and dependent upon their architectures, which were in turn related to composition and constitution (sequence).[85] Astbury pictured the proteins as bulky ships, structured on a few basic designs, carrying their specific groups, much as a ship carries its guns.[86] His work with Preston on the construction of cell walls and with Harrison on the deposition of molecules in the embryo shows that the study of structure can lead to the study of process, or the former can be paralleled by the latter, as in J. D. Bernal's study of both the structure of paracrystalline tobacco mosaic virus and of the process of gel formation by liquid crystalline virus preparations.

# VII Monomolecular surfaces and the cytoskeleton

The study of the colloidal state led to the recognition of the importance of surface phenomena, and to the role of molecules in forming monomolecular layers at the interfaces between the two liquids. To many students of the cell the structural organisation in the amicroscopic region was to be pictured in terms of such molecular layers. Thus in 1930 Sir W. B. Hardy recalled: 'When Emil Fischer displayed the real configuration of proteins, biologists expected great things. They were disappointed to an unexpected degree, just because the idea of orientation at interfaces was wanting. I am still optimist enough to look to that idea for the great advance in biology.'[87] Similarly, Rudolph Peters pictured a complete structural organisation of the cell,

not only in respect of its grosser parts such as the nucleus, but also in regard to the actual chemical molecules of which it consists. Owing to the microheterogeneous nature of the system, surface effects take precedence over ordinary statistical, mass action relationships, and become in the ultimate limit responsible for the integration of the whole and therefore the direction of activities. It is believed that the directing effect of the internal surfaces is displayed predominantly by an organised network of protein molecules, forming a three-dimensional mosaic extending throughout the cell. The enzymes would form part of this structure, their activity being largely controlled by the mosaic. . . . The cell must be considered as a reflex entity structurally organised so far as even its chemistry is concerned, with chains of chemical substances acting as it were as reflex arcs. The surfaces presented by the more highly organised molecules and particularly the proteins, I regard as a kind of central nervous system. I have here purposely avoided reference to the organised structure of cells which is visible to the microscope, because attention to this tends to obscure the real issue relating to the parts below the limits of microscopic vision.[88]

Like Peters, Joseph Needham saw the need for levels of organisation between the molecular and microscopic so that morphology would be extended into biochemistry, thus bridging 'the gulf between the so-called sciences of matter and the so-called sciences of form'.

Biochemistry and morphology should, then, blend into each other instead of existing, as they tend to do, on each side of an enigmatic barrier. The whole life work of Hardy, continued in certain aspects by Peters, showed that the chemical structure of molecules, the colloidal conditions in the cell, and the morphological patterns so arising, are inextricably connected.[89]

Needham's own work was concerned with fusing biochemistry with *Entwicklungsmechanik*, but not with cytology and histochemistry. He accepted Woodger's notion of hierarchical order, and his conception of the continuity of hierarchies between molecules and cell organelles relied heavily upon the surface phenomena studied by W. B. Hardy. Although gradients, fields, and topographical landscapes figured in Needham's Terry Lectures, he was cautious as to their promise, his enthusiasm being reserved for surface phenomena. Such attention was well justified, for there had to be some 'geography' to the innumerable enzyme-catalysed reactions in the cell, some organisation to make what Peters called 'co-ordinative biochemistry' possible. Needham expressed the problem thus:

How can living matter preserve its space pattern? How can it be the seat of so many peculiar and special chemical processes, how maintain within itself sinks and sources of energy? Its multitude of suspended particles can be swept aside by the centrifuge, leaving a hyaline optical vacuum. The complexity of this appears to be the complexity neither of a machine nor of a crystal, but of a nebula. Gathered into the nebula are units relatively simple but capable by their combinations of forming a vast number of dynamical structures into which they fall as the distribution of energy varies.[90]

What is so strange about his account of cell organisation is the complete absence of any reference to the organelles described by the cytologists and histochemists. Where are the mitochondria, known from 1865, named by Benda in 1897[91], and isolated *en masse* by Robert Russell Bensley in 1934[92]? What of the specialised zone of the cytoplasm which Garnier called the ergastoplasm in 1897[93], or the microsomes of von Hanstein[94], and the much-debated Golgi apparatus? Hardy had long ago shown that such morphological entities might well be artefacts. Thompson had gleefully dismissed them[95], but others were more receptive to the claims of classical cytology. Especially in cancer research, the isolation of cell organelles made possible the investigation of their chemical constitution and biological activity. The work of Albert Claude at the Rockefeller Institute, beginning in 1937[96], is justly renowned, as is that of his countryman, Jean Brachet, who localised RNA in the cytoplasm where it was involved in 'the formation of the *ergastoplasm*, that is to say, basophylic formations of lamella structure present in very active secretory cells (Garnier and Bouin, Prenant). . . .'[97] Looking back on the chequered history of the ergastoplasm, Francoise Haguenau saw that it provided 'an excellent example of the necessity for the conjoined exploitation of varied disciplines in modern research. . . .' What had been a 'morphological problem of classical microscopy is

being transformed into a biochemical (microsomes) and histochemical (detection of RNA) problem'.[98]

With the aid of new instruments – the ultracentrifuge and electron
microscope – and the new technique of freeze-drying, two organelles
were recognised, the microsomes (later known as ribosomes) of
Claude[99] and the endoplasmic reticulum of Porter[100], the latter
showing many of the features formerly attributed by the 'Nancy
School' to ergastoplasm. It is noteworthy that the classical cytologists' morphological conception of cell organelles as self-
replicating structures was applied by Claude to the microsomes and
he remarked:

If our hypothesis regarding the mode of origin of the microsomes is correct,
then these small particles would share with the cell itself, and within the
cell, with the chromosomes, the centrioles, the plastids and possibly the
tonoplasts, the most general law of living matter, that of genetic
continuity.[101]

Although Needham's vision of the organism in *Order and Life* was
holistic and hierarchical, the levels of order recognised were biophysical not morphological in the sense in which a trained biologist
would have understood such levels. Hence we doubt that Abir Am
was justified in singling out the program of research which Needham
presented to the Rockefeller as 'the only project embodying a conception of molecular biology in the 1930s' and to claim that it
minimised the 'threat of reductionism'.[102] On the contrary the anti-
histological, pro-biophysical stance of *Order and Life* ruled out the
very morphological entities which have since come to assume such
central importance. Organelles such as the mitochondria and
ribosomes are built of just the monomolecular layers that Needham
and Peters invoked. But as morphological entities, characterised on
the basis of staining reactions of 'fixed' material, they had long been
treated with suspicion. Hence it was not in Cambridge, or in Britain,
that an attempt was made to synthesise this histological tradition
with cell biochemistry, but in Belgium, the Rockefeller school in the
USA, and at Villejuif in France.

# VIII Conclusion

In conclusion there was a world of neglected dimensions between the
ultramicroscopic and the molecular, and a belief in the existence of
structures therein which should be accessible to study if suitable

instruments could be designed. We can identify three approaches to the nature of these structures: (1) The biophysical tradition popularised by Wolfgang Ostwald, developed by the Cambridge colloid scientist, W. B. Hardy, and supported and admired by d'Arcy Thompson and Needham. It turned to physics for causal agents, to engineering for models, and to colloid science for simulations. The co-ordination of biochemical activity in the cell was attributed to surface phenomena. There was a marked scepticism towards the morphological structures of the cytologist. (2) The histochemical and cytological tradition represented in the work of Charles Garnier, Tobjorn Caspersson, and Jean Brachet, which stressed the localisation of biochemical activity to specific cell organelles such as the nucleolus and ergastoplasm. Co-ordinated biochemistry was possible because of the distinctive functions of these organelles and their specific location in the geography of the cell. This tradition drew upon histochemistry, cytology and biochemistry but not upon colloidal biophysics. (3) The X-ray crystallographic tradition pioneered in biology by Astbury and Bernal and centring upon the study of proteins and carbohydrates; it rejected the old colloid-aggregate theory, but supported the importance of surface phenomena. It was not in opposition to the existence of persistent cell organelles but viewed such microstructures as products of macromolecular structure.

To those who incline to view the evolution of our modern conception of the cell as a process of emergence from the uncertain and vague conceptions of the past into the definitive light of the present this paper offers an antidote. It views the history in terms of the first two conflicting programs, each complemented and revised to differing extents by the successes of the third. Within any one of these programs the *type* of explanation of cell function to be achieved was clear, and techniques and instruments were chosen or invented with that aim in mind. Not that this form of analysis can be simply assimilated to a conflict of programs and a victory of one over the other with the help of a third. Colloid science had positive results. Surface phenomena were emphasised, our conceptions of molecular size and the alleged uniqueness of the Avogadro molecule were revised. The histochemical tradition, in its turn, assimilated the positive fruits of colloid science. Yet it has to be said that much of pre-Svedberg colloid science has been rejected and the many phenomena explained by that science are now re-interpreted in terms of macromolecular chemistry. Equally the persistent morphological structures of histochemistry and cytology are now defined in macromolecular terms and pictured as arrangements of monomolecular layers. But the persistence of these structure as morphological entities is not in question.

As for the claims made for and against Weaver's role in establishing the emerging field of molecular biology, we can see that the 'instrumentation of biology' which he supported was to prove crucial in the merging of the conflicting biophysical and cytological approaches to the cell. This is not to say that Weaver explicitly set out to achieve such a merger nor that his conception of molecular biology was the same as ours; clearly it was not, for his view assumed the primacy of the proteins, the nucleic acids being subordinated to them. It serves no purpose to picture Weaver as *anticipating* the modern conception of the relations between proteins and nucleic acids, which he did not prefigure. But neither was the Rockefeller program of 'molecular biology' a mere device for the subordination of biologists to physicists. Behind it lay a belief in the structural continuity between molecular and microscopic dimensions, and the conviction that the functions of the cell are to be investigated in terms of the structures revealed by those instruments. This approach was necessary, though not sufficient, for the molecular biology which emerged later. To what extent Weaver merely absorbed this approach from the scientists he met and the literature he read is a question this paper has not explored, for the chief aim has been not to defend Weaver, but to undermine the historiographical claims of analyses in terms of reductionists and anti-reductionists. A better approach, it is suggested, is offered by the distinction between structuralists and dynamicists, the former recognising the persistent structures of classical cytology and histochemistry, the latter rejecting them.[103] The structural and dynamical approaches are offered here as categories appropriate to the historiography of biology in the late nineteenth and first half of the twentieth century. It should not be forgotten, however, that both categories were opposed to the vitalist approach of Hans Driesch.[104]

# Endnotes

1. Weaver, W., Molecular biology: the origin of the term, *Science* **170** (1970), 581–2.
*Scene of Change*, New York, 1970.

2. Kohler, R., A new policy for the patronage of scientific research: the reorganization of the Rockefeller Foundation, 1921–1930, *Minerva* **14** (1976), 279–306.
   Warren Weaver and the Rockefeller Foundation Program in molecular biology ... a case study in the management of science. In Reingold, N. (ed.), *The Sciences in the American Context: New Perspectives*, Washington, 1980, pp. 249–93.

3. Abir Am, P., The discourse of physical power and biological knowledge in the 1930s: a reappraisal of the Rockefeller Foundation's 'Policy' in molecular biology, *Social Studies of Science* **12** (1982), 341–82.

4. Fuerst, J. A., The definition of molecular biology and the definition of policy: the role of the Rockefeller Foundation's policy for molecular biology, *Social Studies of Science*, **14** (1984), 225–237;
   Bartels, D., The Rockefeller Foundation's funding policy for molecular biology: success or failure? *Ibid*, 238–243;
   Olby, R. C., The sheriff and the cowboys: or Weaver's support of Astbury and Pauling, *Ibid*, pp. 244–247.

5. Abir Am, op. cit. (ref. 3) p. 345.

6. Olby, *The Path to the Double Helix*, London, 1974, pp. xii, 434.

7. This usage is deliberately broad, and includes dialectical materialism. It should not be equated with reductionism in the strict positivist sense, as in: Schaffner, K. Chemical systems and chemical evolution: the philosophy of molecular biology, *Am. Scien.* **57** (1969), 410–20.

8. Churchill, F. B., August Weismann and a break from tradition, *J. Hist. Biol.* **1** (1968), 91–112.

9. Gould, S. J., D'Arcy Thompson and the science of form, *Boston Studies in the Philosophy of Science* **27** (1976), 66–97.
   Medawar, P. B., Postscript: growth and form in Thompson, R. D'Arcy. *D'Arcy Wentworth Thompson. The Scholar Naturalist 1860–1948*. London, 1958, pp. 219–33.
   Bonner, J. T., Editor's Introduction to Abridgment of *Growth and Form*, Cambridge, 1966, pp. vii–xiv.

10. Schwann, T., *Mikroskopischen Untersuchungen über die Uebereinstimmung in der Struktur und dem Wachstum der Thiere und Pflanzen*, Berlin, 1839.

11. Brücke, C., Die Elementarorganismen, *Wiener Sitzungsberichte* **44** (1861), 386.

12. Hardy, W. B., On the structure of cell-protoplasm, *J. Physiol.* **24** (1899), 159.

13. Miescher, F. J., Ueber die chemische Zusammensetzung der Eiterzellen, *Med. Chem. Untersuchungen* **1** (1866–71), 441–60.
   He commented: 'But if, as seems likely, it should turn out that there is a deeper chemical contrast between nucleus and protoplasm which, despite a variety of exceptions, applies to all organisms, surely this would be a great stride forward towards the distant goal of understanding the significance of histological structural elements. The concept of the nucleus could then be freed from superficialities.... We shall be able to distinguish nuclei from questionable structures by those properties which are closely related to their physiological functions'.

14. Miescher, F. J., Die Kerngebilde im Dotter des Hühnereies, *Med. Chem. Untersuchungen* **1** (1866–71), 502–9.
15. Pearse, A. G. E., *Histochemistry, Theoretical and Applied*, 2nd edn, Boston, 1968, p. 193 ff.
    Clark, G. & Kasten, P. H., *History of Staining*, 3rd edn, Baltimore and London, 1983, pp. 246–51.
16. Olby, op. cit. (ref. 6), pp. 104–5.
17. Svedberg, T. & Rinde H., The Ultracentrifuge, a new instrument for the determination of size and distribution of size of particles in amicroscopic colloids, *J. Am. Chem. Soc.* **46** (1924), 2677–93.
18. Haraway, D. J., *Crystals, Fabrics & Fields: Metaphors of Organicism in Twentieth Century Developmental Biology*, Yale, 1976, p. 133.
19. Weaver, W., The natural sciences, *Rep. Rockefeller Fdn* (1938), 203–25.
20. Weaver, W., *Scene of Change*, pp. 60–1.
21. The failure of Weaver to support Needham's proposal seems to have had more to do with the attitude of Cambridge University and Needham's reputation as an experimentalist and 'leftist' than with the nature of his programme.
22. Weaver, op. cit. (ref. 19), p. 199.
23. Hermann, L., Ueber eine Bedingung des Zustandekommens von Vergiftungen', *Archiv. Anat. Physiol.* (1867), 64–74.
24. Pflüger, E., Ueber die physiologische Verbrennung in den lebendigen Organismen, *Pflüger's Arch. ges. Physiol.* **10** (1875), 251–367.
25. Pfitzner, W., Ueber den feineren Bau der bei der Zelltheilung auffreten fadenförmigen Differenzierungen des Zellkerns, *Morph. Jahrb.* **7** (1882), 289–311.
26. Verworn, M., *General Physiology. An Outline of the Science of Life*, London, 1899.
27. Montgomery, E., Ueber das Protoplasma einiger 'Elementarorganismen', *Jena Z. Naturw.* **17** (1885), 677–712.
28. Hörmann, G., *Die Kontinuität der Atomverkettung; ein Strukturprinzip der lebendigen Substanz*, Jena, 1899.
29. Needham, J., *Order and Life*, New Haven, 1936, p. 140. The phrase 'grand heresies of biochemical thought' is used by Needham. I cannot trace it in: Hopkins, F. G., *The Dynamic Side of Biochemistry*, (1913) reprinted in: Needham & Baldwin (eds.), *Hopkins and Biochemistry 1861–1947*, Cambridge, 1949, pp. 136–59. The most relevant passage is on p. 144.
30. Edsall, J. T., Proteins as macromolecules: an essay on the development of the macromolecule concept and some of its vicissitudes, *Archs. Biochem. Biophys.* (Suppl.) **3** (1962), 12–20; Olby, op. cit. (ref. 6), chap. i.
31. Darwin, C. R., *The Variation of Animals & Plants under Domestication*, 2 vols, London, 1868, vol. **ii**, p. 379. My inference about orders of magnitude which follows this quotation is, I admit, only a speculation.
32. Maxwell, C., 'Atom', *Encyclopedia Britannica*, 9th edn, Edinburgh, 1875, vol. **iii**, p. 42.
33. Ibid.
34. Vries, H. de, *Intracellular Pangenesis*, Jena, 1889, transl. C. S. Gage, Chicago, 1910, p. 70.

35. Errera, L., Sur la limite de petitesse des organismes, in his *Receuil d'oeuvres (Physiologie générale)*, Bruxelles, 1910, p. 332.
36. Thompson, d'Arcy W., *On Growth and Form*, 1st edn, Cambridge, 1917, p. 9. [Referred in future as G & F].
37. G & F, p. 159.
38. G & F, p. 160.
39. G & F, p. 198.
40. G & F, pp. 194–5.
41. G & F, pp. 691–2.
42. Fidler, T. C. *A Practical Treatise on Bridge-Construction*, London, 1st edn, 1887. 3rd edn, 1909. Cited in G & F, p. 691.
43. G & F, pp. 205, 385.
44. G & F, p. 426.
45. G & F, p. 485.
46. Thompson, D. W., Morphology and Mathematics, *Trans. R. Soc. Edinb.* **50** (1915), 859.
47. G & F, p. 1.
48. Thompson, D. W., Magnalia naturae: or the greater problems of biology, *Rep. B. Ass.* **80** (1911), 395–404.
49. Thompson, D. W., Some difficulties of Darwinism, *Nature* **50** (1894).
50. His, W., On the principles of animal morphology, *Proc. R. Soc. Edinb.* **15** (1887–8), 287–98. Thompson quoted from p. 294 in G & F, p. 56.
51. G & F, p. 7.
52. Thompson, op. cit. (ref. 48), pp. 402, 403.
53. G & F, pp. 14–15.
54. Thompson, D. W., The shape of eggs and the causes which determine them, *Nature* **78** (1908), 111–13.
55. G & F, p. 54.
56. Thompson, op. cit. (ref. 46), pp. 859–60.
57. G & F, p. 724 ff.
58. Gould, S. J., Irrelevance, submission and partnership: the changing role of palaeontology in Darwin's three centennials . . ., in: *Evolution from Molecules to Man*, edited by D. S. Bendall, Cambridge, 1983, p. 355.
59. G & F, p. 158.
60. Bateson, W. & Punnett, R. C., On gametic series involving reduplication of certain terms, *J. Genet.* **1** (1911), 293–302. See also: Cock, A. G., William Bateson's rejection and eventual acceptance of chromosome theory, *Ann. Sci.* **40** (1983), 19–59, especially p. 46 ff.
61. See Cock, op. cit. (ref. 60) p. 34, and Coleman, W., Bateson and chromosomes: conservative thought in science, *Centaurus* **15** (1970), 228–314.
62. G & F, p. 715.
63. Ostwald, K. W. W., *Introduction to Theoretical and Applied Colloid Science. The World of Neglected Dimensions*, transl. M. H. Fischer, New York, 1917, pp. 218–19.
64. Bragg, W. H., The significance of crystal structure, *J. Chem. Soc.* **121** (1922), 27–67.

65. Zsigmondy, R. A., *Kolloidchemie: ein Lehrbuch*, 3rd edn, Leipzig, 1920, p. 4.
66. Bechold, H., *Die Kolloide in Biologie und Medizin*, Dresden, 1912, p. 194.
67. Höber, R., *Physikalische Chemie der Zelle und der Gewebe*, 3rd edn, Leipzig, 1911, pp. 310–11.
68. Child, C. M., *Senescence and Rejuvenescence*, Chicago, 1915, p. 26.
69. Rhumbler, L., *Das Protoplasma als physikalisches System*, Jena, 1914.
70. Fischer, M. & Ostwald, K. W. W., Zur physikalisch-chemischen Theorie der Befruchtung, *Pflüger's Arch. ges. Physiol.* **106** (1905), 229.
71. G & F, p. 426.
72. Hopkins, op. cit. (ref. 29) p. 152.
73. Staudinger, H., *Die hochmolekularen organischen Verbindungen*, Berlin, 1932.
74. Svedberg, T. & Fåhraens, R., A new method for the determination of the molecular weight of the proteins. *J. Am. Chem. Soc.* **48** (1926), 430–8.
75. Svedberg, T. & Eriksson-Quensel, I.-B., Molekülkonstanten der Eiweisskörper, *Tabul. Biol.* **11** (1936), 351–5.
76. Svedberg, T. & Chirnoaga, E., The molecular weight of hemocyanin, *J. Am. Chem. Soc.* **50** (1928), 1399–411.
77. Svedberg, T. & Hedenins, A., The sedimentation constants of the respiratory proteins, *Biol. Bull.* **66** (1934), 191–223.
78. Svedberg, T., 'Opening Address', a discussion on the protein molecule, *Proc. R. Soc.* **170A** (1939), 40–56.
79. Svedberg, T. & Pederson, K. O., *The Ultracentrifuge*, Oxford, 1940, part iv.
80. Weaver, W., The duration of the transient state in the settling of small particles, *Phys. Rev.* **27** (1926), 499–503.
81. Astbury, W. T., The forms of biological molecules, in: Clark, W. E., Le Gros & Medawar, P. B. (eds.), *Essays on Growth and Form presented to d'Arcy Wentworth Thompson*, Oxford, 1945, p. 340.
82. Ibid., p. 326.
83. Ibid., p. 309.
84. Thompson to Astbury, 20 October 1945.
85. For information on this point see: Olby, The origins of modern biochemistry: a retrospect on proteins (Review), *History and Philosophy of the Life Sciences* **4** (1982), 159–68.
86. Astbury, op. cit. (ref. 81) p. 334.
87. Hardy, W. B., 'Conclusion'. Colloid Science applied to Biology. A General Discussion, *Trans. Faraday Soc.* **26** (1930), 865.
88. Peters, R., Co-ordinative biochemistry of the cells and tissues, *J. St. Med.* **37** (1929), 278. See also: Teich, M., from 'Enchyme' to 'Cytoskeleton': the development of ideas on the Chemical organization of living matter. In: Teich, M. & Young, R. (eds.), *Changing Perspectives in the History of Science Essays in Honour of Joseph Needham*, London, 1973, chap. 20.
89. Needham, J. op. cit. (ref. 29), pp. 139–40.
90. Needham, J. op. cit. (ref. 29), p. 136.
91. Benda, *Verhandl. Physiol. Ges.* Berlin (1897), 14.
92. Bensley, R. R. & Hoerr, N. L., Studies on cell structure by the freezing-

drying method, *The Anatomical Record* **60** (1934) 449–55. Lehninger wrote that: 'It was Bensley's pioneering work that pressaged the confluence of cytological work on mitochondria with biochemical work on respiration'. In Lehninger, A. L., *The Mitochondrion. Molecular Basis of Structure and Function*, New York & Amsterdam, 1964, p. 4.

93. Garner, C., Les filaments basaux des cellules glandulaires, *Biblio Anat.* **5** (1897), 278–89.

94. Hanstein, J. von, *Das protoplasma als Träger der pflanzlichen und tierischen Lebensverrichtungen*, Heidelberg, 1880.

95. G & F, p. 286.

96. Claude, A., Properties of the causative agent of a Chicken Tumour XIII. Sedimentation of the Tumor Agent, and Separation from the Associated Inhibitor, *J. exp. Med.* **66** (1937), 59.

97. Brachet, J., *Embryologie Chimique*, Paris & Liège, 1944, p. 222.

98. Haguenau, F., The ergastoplasm: its history, ultrastructure and biochemistry, *Int. Rev. Cyt.* **7** (1958), 425–6.

99. Claude, A., The constitution of protoplasm, *Science* **97** (1943), 451–6.

100. Porter, K. R., Claude, A. & Fulham, E., A study of tissue culture cells by electron microscopy. Methods & preliminary observations, *J. exp. Med.* **81** (1945), 161.

101. Claude, op. cit. (ref. 99), p. 456.

102. Abir Am, op. cit. (ref. 3), pp. 353 and 361.

103. Farley, J., *Gametes and Spores*, chap. 6 for account of Loeb's opposition to classical cytology.

104. Ravin, A. W., The gene as catalyst; the gene as organism, *Studies in History of Biology*, **1** (1977), 1–45.

# Form and strategy in biology: reflections on the career of C. H. Waddington

EDWARD YOXEN

Department of Science and Technology Policy, University of Manchester,
Manchester, UK

One of the classic concerns of historians of science is the reconstruction of the sequence of intellectual moves that preceded and facilitated major conceptual change. They seek to recover how the imaginative activity of a particular individual was ordered by past scientific experience and stimulated by particular contextual cues. Often the analysis is thus doubly individualised, being restricted to one person and one instance of his or her research. In this paper I want to consider the enduring themes in one man's work, but not to dwell in any detail on a particular achievement. The person under discussion is the British embryologist and geneticist, Conrad Waddington. For me the form of his career, with its evident phases, but with an underlying continuity of interests, is at least as instructive as particular experiments or publications. In particular Waddington's attempts to bring cybernetics and biology together seem to me well worth historical analysis.

Undoubtedly this derives from my perspective as an outsider. Working embryologists may find the study of specific pieces of work more interesting. I find it easier and more satisfying to consider the unfolding of Waddington's career in historical context and to consider his movement between research problems, guided by a commitment to an enduring set of concerns. In an earlier draft of this paper I tried to analyse Waddington's career as a series of investments of what the French sociologist, Pierre Bourdieu, has called 'symbolic capital', whereby scientists risk accumulated status and resources in the scientific market, playing both against competitors and a refractory nature.[1] Whilst this perspective on the practice of research has proved very illuminating, as in the work of Pinch, Kohler, Green and Latour and Woolgar,[2] it requires a wealth of detail concerning 'investment' calculations that is only available through interview or the most extended archival research. It has seemed more sensible to be less ambitious and to chart the phases of Waddington's career, without attempting a complete explanation of each move that he made. In fact I would suggest that sociological models of this kind can be misleading, if applied without detailed knowledge of the scientific context, in that what seems now as if it would have been a large and risky move between disciplines may in fact have been an entirely conventional shift of perspective.

I shall try to demonstrate that Waddington was a man of contrasts. Firstly he had an enduring theoretical concern with one of the fundamental issues in biology since the mid-nineteenth century, the mechanism of evolution. Yet not only did his work call into question some aspects of the received theory of evolution by natural selection, he came to state his ideas in a highly abstract form, far from the idiom of experimental and field biologists. In fact the mathematical symbols and equations turn out to be far less important than the formal ideas they represent. Secondly he constantly sought to reason abstractly and to develop new languages based on a commitment to a holistic epistemology in which to formulate enduring problems. Yet his involvement in experimental work placed real limits on the utility of such speculation and attempts at conceptual clarification. Waddington's career can be seen as a valuable example of how much, or for some people, how little, can be achieved by theoretical studies in biology. Thirdly though he acquired considerable power and distinction, his career was not a saga of tenacious institution building. He helped to create a large research institute in Edinburgh after 1945, but he moved so rapidly from project to project that he neglected its consolidation. The legacy of his writing and research is considerable but it is very diffuse. Finally his intellectual development represents something of a historiographical puzzle, which I do not claim to have solved. Waddington's contact with the life sciences began in the 1920s and his intellectual preference for holism, his tendency to criticise existing concepts as imprecise date from that period. But he was clearly also influenced by the rise of cybernetics and information theory in the 1940s and by the experience of wartime operations research. One can either place the emphasis on the 1920s and 1930s, and see subsequent conceptual changes as mere terminological innovation, or place the emphasis on the 1940s and 1950s, and argue that the shift from 'form' to strategy, evident in Waddington's language, marks a decisive epistemic change.

## The search for authentic ways of knowing nature

Conrad Waddington was born in 1905, the son of Quaker tea planters in India.[3] Most of his early years were spent in Britain away from his parents, and the evidence of a governess, attendance at prep school and then at Clifton College suggests a comfortable but not extravagantly wealthy family background. Certainly a research career, which, as Werskey has reminded us, required some private income in inter-war Cambridge, was a feasible option.[4]

His experience of family life as a child was not one of a stable, nuclear structure, but of movement between relatives' households, with consequent changes in fields of authority and influence. But in such precarious settings his youthful interest in science was indulged and rewarded. In later life he had the confidence, or the emotional need, to break down the barrier between work and home in creating a communal household for his research group.

In 1923 he went up to Sidney Sussex College, Cambridge, taking the Natural Sciences Tripos, and graduating with a first class degree in geology in 1926. Originally he had intended to be an oil geologist and presumably would have worked abroad as his parents had done. Instead he began research in palaeontology, a very classical and academic retreat from the scientific service of an expanding international industry. Moreover, the funding for his postgraduate studies came partly from a philosophy studentship, which further underlines the academicism of this period of work, and suggests, as do other events in his career, a certain confidence to make unusual and risky academic choices.

Waddington's social circle as a research student included a number of young scientific researchers, embarking on what were to prove distinguished careers. They included Gregory Bateson, son of the geneticist William Bateson, Evelyn Hutchinson, Robin Hill, Joseph Needham, Edward Bullard and Barton Worthington. The tea-club of the Department of Zoology provided a convenient meeting place for some of these people. Waddington's association with the younger Bateson was close, and they remained in contact for many years.[5] In the mid-1930s when Bateson, by then a social anthropologist, was writing his classic work, *Naven*, Waddington read the manuscript. There are highly suggestive parallels between Gregory Bateson's discussion of 'schismogenesis' – the control of divisions within social groups – and Waddington's discussion of opposing forces or tendencies in morphogenesis.

It is also tempting to consider the elder Bateson, the most enthusiastic proponent of the Mendelian study of heredity in Britain in the first two decades of this century, as an influence on Waddington, but I know of no evidence to sustain it. William Bateson died in 1926, having failed to establish a significant research school. So even though Waddington wanted to study genetics in the late 1920s, and shared William Bateson's sympathies for holism in biology, the interaction between them can only have been very indirect and transient.[6] What is interesting and deserves further study is the strength of interest in holistic thinking, epitomised by the work of Alfred North Whitehead, amongst biologists of that generation. Waddington certainly claimed that much of his work, including that in genetics, was based on metaphysical assumptions taken directly from Whitehead.[7] I illustrate the evident indebtedness below.

His years as a research student in the late 1920s were, as is often the case, a time of shifting intellectual goals. His friendship with Gregory Bateson deepened his interest in genetics: his interest in fossil systematics waned and his thesis on ammonites was never completed. The younger Bateson's path was similarly disordered at this point and one has the impression of two highly intelligent, playful, rather sophisticated young men casting about for projects that would satisfy their aesthetic and philosophical sensibilities in anthropology and biology respectively. Such research would capture not the mere details of function but the inner formal relatedness of nature *and* culture. It would reveal the immanent processes that underly form, stability and discontinuous change in both domains. At the same time it had to offer them a job. Identifying such a project was a search for authenticity. But Waddington discovered that to work as a geneticist in Britain in the late 1920s, and to entertain these noble ideas about nature, was not feasible.[8]

## Establishment in a research niche

Even if Waddington's work with fossils did not lead to a Ph.D., it was not a disaster. He published a paper on it and by 1929 he was able to relocate himself at the Strangeways Research Laboratory. His hope was that he could apply the tissue culture technique in use there to the study of development in warm-blooded animals. This was a daring move to make. In the event he was successful. Dame Honor Fell, who became Director of the Laboratory in 1929, after the death of its founder, writes of this move:

Waddington was an acquaintance of our physicist Miss Sidney Cox. Miss Cox told me of this bright young palaeontologist (who also had a scholarship in philosophy) who had been reading Spemann's papers and wondered if the organ culture method developed here could be used for the experimental study of avian and mammalian embryos. I was much impressed with the idea . . . So he came and stayed in a quite unplanned way without any references merely because I was interested in his project.[9]

His research at the Strangeways Laboratory, beginning in 1929, was initially supported by a senior research grant of the Department of Scientific and Industrial Research (DSIR). In 1930 he was able to report promising results in a note in *Nature*[10] and gave a paper at the International Congress of Experimental Cytology in Amsterdam. At that meeting the German geneticist, Richard Goldschmidt, invited him to Berlin and in the spring of 1931 Waddington applied to the Medical Research Council (MRC) for a Rockefeller Medical Fellowship to meet the expenses of this trip.[11] Apparently he worked not with Goldschmidt, but with Mangold in Berlin.[12] He was to go only for six months, since the Demonstratorship in Experimental Zoology, offered him by James Gray, was to begin in October of that year. Whilst this could have been a full-time post, Waddington chose, on the advice of Sir Walter Fletcher, the Secretary of the MRC, to hold it half-time and to supplement his salary with a personal research grant from the MRC for work at the Strangeways. The advantage of this arrangement was held to be that Waddington would act as liaison between the small group at the Strangeways and James Gray's department.[13] The very form of his appointment was carefully fitted into the strategy developed by Fletcher and Gray for experimental zoology in Cambridge. Whilst it represented a vote of confidence in his research potential, it also made his financial and institutional position somewhat precarious. Whilst R. F. Tisdale, a Rockefeller Foundation official in Europe, who met him at this time, '. . . seemed to regard Mr Waddington as a rather inexperienced young man with unduly ambitious ideas'[14], clearly Fletcher, Gray, Goldschmidt and Honor Fell did not think so.

In 1931 he published a long and detailed paper on 650 embryo

experiments in the *Philosophical Transactions of the Royal Society*.
This paper provides evidence not only of real experimental skill, but
also of a preparedness to think incisively about theoretical questions
and to revise imprecise terminology. This emerges in his discussion
of differentiation, where as his transplantation experiments had
shown, the development potential of specific tissues changes with
time.

Since, as this work shows it appears to be necessary to specify the early
conditions under which a 'self-differentiation' takes place, the idea of self-
differentiation must be discarded as an exact concept for theoretical discussion. We retain then the three concepts, non-competent, competent to
differentiate to the tissues in question under certain conditions and
determined. This terminology has the advantage that it makes, even by
implication, no assumptions as to the specificity of any given set of conditions, such as the roof of the primitive gut, or an implanted ear-vesicle for
the limb-mesoderm at a later stage.
Very little is known about the origin of competence. It is naturally
limited by the restriction of potentialities involved in the last process of
determination which the tissue has undergone.[15]

This concern with the adequacy of concepts is a recurring theme
in Waddington's work. So also is the emphasis on potentiality. This
particular discussion is strikingly similar in its method and tone to
that of the embryologist Joseph Woodger, with whom Waddington
came into frequent contact in the early 1930s through the meetings of
the Theoretical Biology Club. Woodger taught biology to medical
students at Middlesex Hospital, and he worked as a descriptive
embryologist, but in the 1920s he developed an intense interest in
the philosophy of science. On a visit to Austria, he came into contact
with members of the Vienna Circle and through them intensified his
concern with the philosophy of language, with theory-neutral observation languages and the analysis of biological predicates.[16] His ideas
were set out in his book *Biological Principles*, published in 1929. He
clearly saw his role as one of educating biologists to think in philosophically more precise ways, in Britain at least. In 1930 he wrote:

It requires investigations which go below the specifically biological level
into problems concerning 'knowledge about knowledge itself' and it is the
difficulties attendant upon an intellectual upheaval of this kind, which are
largely responsible for the backwardness of the revolution in biological
thinking ... Among modern writers who have felt the need for such an
undertaking few are better equipped than A. N. Whitehead, and the reader

cannot do better than read the splendid first two chapters of that author's *Concept of Nature*.[17]

Here then was a program of intellectual reform through terminological purification, and here too the influence of Whitehead. Woodger went on:

To sum up: the crucial problems of embryology seem to be those concerning the three principal modes of elaboration [Woodger's term, substituting for differentiation, EY] of cellular, non-cellular and of cell parts. And it seems perfectly clear that it is the *primary* duty of the embryologist to discover how to interpret the development process in terms of the intrinsic properties of the various 'genetic types' of cells and the relational properties they exhibit in the several hierarchic organic relations which arise in the course of the elaboration of non-cellular parts and which are accessible to our observation.[18]

In this paper Woodger went on to describe abstractly a hierarchy of 'organizing relations' in biological organisms, which he sees as a means firstly to concentrate biologists' attention on relational properties and secondly to provide a theory-neutral framework for conceptual discussions. At this abstract level he endorses genetics, and indeed in his later work tried to axiomatise the discipline. This then is a very abstract, positivistic, methodological holism, that clearly had some influence on Waddington.

In 1930 Waddington read the manuscript of Joseph Needham's *Chemical Embryology*, and by 1933 they were in active collaboration in research on the biochemistry of induction, or as Needham renamed it, evocation. In the same year Waddington obtained a Fellowship at Christ's College.

Each of the five years up to 1937 saw at least four papers and sometimes more on induction, morphogenesis and the chemistry of the evocator, many of them written in collaboration with Joseph and Dorothy Needham, in the Biochemistry Department. In 1934 Waddington wrote to the Secretary of the Medical Research Council in the following confident terms:

Research is being, and will be carried out on various problems concerning the functioning of organisation centres in embryonic development. I have recently defined what seems to me to be the three most important problems in experimental embryology of the present day. They are, the nature of the agents by which an organisation centre produces an induction (the evocators); the nature of the materials on which the organisers work (the competencies of the reactioning tissue) and the nature of the agencies which build up a single individual from the various tissues provided by the inter-action of the organisers and competent tissues (the individuation fields).[19]

The MRC support ended in 1934, having reached a prescribed time limit. Instead Waddington received an 1851 Exhibition Senior Studentship and he and Needham began to receive support from the Rockefeller Foundation for technical assistance.

## Attempted Institutionalisation

By 1935 Needham and Waddington were confident enough of the intellectual potential of their approach to biological organisation to think of trying to institutionalise it, outside or alongside the struc-ture of departments in Cambridge. Their hope was to establish an Institute of Physico-Chemical Morphology in Cambridge with support from the Rockefeller Foundation matched by university funds. The roots of this idea lay in the discussions of the so-called Theoretical Biology Club. The rise and fall of this proposal for an institute have been described by Pnina Abir-Am.[20] She shows that the plan was for a research collective organised in a non-hierarchical way. The Director of the Natural Sciences Division of the Rockefel-ler Foundation, Warren Weaver, was interested in the general notion of physicochemical studies of development, which could have com-plemented that more reductive, general project in the life sciences that Weaver was funding.[21] There were good reasons at the time to suppose that Rockefeller support for an Institute of Physico-Chemical Morphology in Cambridge would be forthcoming. But Weaver was advised by a number of leading scientists, including Sir Henry Dale that the research program was too ambitious. He was also advised by others that it should be supported. In the end he with-drew his offer of support. Cambridge University was also not keen to back the Institute. Abir-Am represents this as a missed opportunity to support a genuine project in 'molecular biology', essentially because of Weaver's *idiosyncratic* conception of how the life sciences could be revolutionised solely by the transfer of research technology from physics and chemistry. The implication is that Weaver occasioned the emergence of 'molecular biology' – a term he created in 1938 for his own purposes – indirectly and unintentionally. Abir-Am implies that Weaver's apparent fixation on research tech-nology seriously distorted his judgement. Robert Kohler, John Fuerst and I have argued that Weaver's interests were not simply in technology *per se*, but extended to concepts, methods and standards of argument, which should be imported (from physics) to the life sciences.[22] Robert Olby has also suggested that an interest in research technology was a very sensible and understandable choice, for someone interested in the study of protein in the 1930s.[23] Moreover I believe that Abir-Am's own archival data actually support the conclusion she rejects, that Weaver's policy of selective funding of reductionist physicochemical studies was well thought out, relatively successful in its own terms and not particularly rigid or exclusive.

Thus there is evidence that he would have supported the establishment of an institute as part of his 'molecular biology' program, had the signals he received from members of the British scientific establishment been favourable.[24] It would also be ingenuous not to see an element of political censorship in this unfavourable peer review. For Weaver the over-riding consideration was that if he were to ignore the judgement of his distinguished referees, who included Sir Edward Mellanby, Secretary of the Medical Research Council, then he would be unable to approach them again for advice. Although Weaver was very tempted to back the Institute, in the event the Foundation did not do so.

Nonetheless, at a lower level of support Waddington and the others received Rockefeller money to travel to Europe and America. One striking instance was a trip to Copenhagen for a transdisciplinary meeting with Niels Bohr, Max Delbrück, Pierre Auger, Hermann Muller, Cyril Darlington, J. D. Bernal and others on the structure of the gene.[25] Even though the Institute, which would have been the culmination of their plans for a multifaceted attack on the problems of development was denied, members of the Theoretical Biology Club did receive some Rockefeller support and legitimation for the enterprise of model building and interdisciplinary theoretical cogitation. One might say, they were licensed to operate as an avant garde in science, even if their institution-building was blocked.

Although Waddington published no technical papers in genetics between 1931 and 1939, through this period he maintained an interest in research in genetics. He was in contact with geneticists in Edinburgh and elsewhere and offered thoughts on gene structure to the private meetings of the TBC. Donna Haraway describes these conclaves of the Theoretical Biology Club as useful for developing the terminology and metaphors of a new organicist paradigm.[26] I think one can also see them as a means through which the assurance to work as self-appointed theoreticians was sustained. Within this private forum new ideas could be tried out on friendly critics, before career capital was placed at risk by offering theoretical contributions to a wider, much more sceptical audience.

For Waddington, amongst others, the habit of theorising and of organising private convocations with this general purpose endured. What this discussion group of the 1930s gave him was a way of continuing an interest in genetics and a form of intellectual activity that allowed him to adapt to a new environment, when the niche represented by the study of the evocator had to be abandoned. By 1938 the research program of the search for the organiser was in deep trouble, as it became clear that a whole range of chemical substances could organise embryonic tissue. With that the whole notion of characterising *the* evocator chemically lost its value.

## Domains of stability in a landscape of possibility

In October 1936 Waddington wrote to the Paris office of the Rockefeller Foundation mentioning the possibility of a research visit to the United States.[27] In fact it was 1938 before he crossed the Atlantic. He went first to the laboratory at Cold Spring Harbor, then to Yale to meet Ross Harrison. From September 1938 until the end of the year he worked at Columbia with L. C. Dunn, making trips to see Curt Stern in Rochester and Streeter and Metz in Baltimore. By January 1939 he had reached Pasadena to work at the California Institute of Technology, where he stayed for three months, eventually leaving America on 6 April 1939. At that time Waddington wished to return to America in the summer of that year and discussed with Rockefeller officials the possibility of a permanent move to the United States.[28]

In 1938 Waddington completed his textbook on genetics, presumably before he left for America that autumn.[29] In the preface he acknowledges that he is crossing a disciplinary boundary, although clearly he had been reading the genetics literature carefully for some while. It is also interesting to see how he contrasts the insights to be obtained from genetics and embryology. In this passage he introduces his well known landscape metaphor, although without elaborating it.

If we want to consider the whole set of reactions concerned in a developmental process such as pigment formation, we therefore have to replace the single time–effect curve by a branching system of lines which symbolizes all the possible ways of development controlled by different genes. Moreover, we have to remember that each branch curve is affected not only by the gene whose branch it is but by the whole genotype. We can include this point if we symbolize the developmental reactions by branching valleys on a surface.[30]

He goes on:

The vertical dimension in our geological model represents probability, so that the valley bottom is really a representation of an equilibrium. This is rarely apparent in genetic experiments, but is much more clear in experimental embryology, where an experiment usually consists of carrying the point representing the state of a process into some unnatural place in the landscape defined by the genotype, and from this place the process may run down into the normal valley (i.e. equilibrium is restored, regulation occurs). The equilibrium aspect of the valley only appears in genetics if by chance the valley bottom becomes flat, i.e. there are several equally probable values for the process.[31]

And, adumbrating the methodology he was to use two years later, he goes on:

But if the genotype really provides alternative methods of reacting, or alternative valleys along which the developmental processes may go, it should be possible to reveal these potentialities without using the 'realizer' genes which normally bring them into action. This requirement can be fulfilled. Goldschmidt, in particular has shown that if normal flies are subject to severe environmental influences (sub-lethal temperatures etc.) a small percentage of modifications appear which exactly parallel the changes produced by genetic factors. The so-called phenocopies are evidence that the alternative valleys are given even in the normal genotype.[32]

Thus before he left for the United States Waddington was thinking seriously about the complementarity of embryology and genetics and of the kinds of genetical experiments that could throw light on the potentiality of development. In Pasadena he began a series of experiments on the genetic control of wing development in *Drosophila*, working with Sturtevant and Dobzhansky. A preliminary note of this work appeared in the *Proceedings of the National Academy of Sciences* in 1939; a much fuller description appeared in the *Journal of Genetics* the following year.[33] In meticulous detail he had studied the processes of wing formation in the pupal phase of *Drosophila*. First he identified and characterised four pre-pupal phases, and four pupal phases and then went on to consider the aberrations of morphogenesis in 24 different mutants. But the presumption was, and this must have been the direct influence of Whitehead, as Waddington himself suggested,[34] that gene action had to be considered holistically.

We shall find that the most important elements in the wing from the developmental point of view, are the epithelial sac, the veins and the margins. It might at first seem sensible to study successively the genes which affect these different elements, but in fact this would be a misplaced simplification. The three are not separate and independent entities, but are bound together in a continual interplay of causal relations. It has therefore seemed better to approach the matter extensively rather than intensively and to investigate a number of genes.[35]

He then goes on to a very detailed discussion of defects of wing formation, which are explained in structural and mechanical rather than physiological terms. The work of Richard Goldschmidt, who favoured physiological mechanisms, is repeatedly criticised, often for its sloppiness as well as explanatory inadequacy. Not surprisingly Goldschmidt was not pleased by the paper. Waddington was clearly unhappy with biochemical explanations and writes instead in the tradition of D'Arcy Thompson, whose work is cited, and James Gray. It is also a very empirical paper. Nowhere is the developmental landscape mentioned.

But whilst in Pasadena, Waddington also began a more theoretical monograph, *Organisers and Genes*, that was published in 1940. This book draws out the connections between embryonic induction, as a phenomenon under genetic control, and gene action in general. In it he presents a model of development as a progression between states of instability. Writing of the variable competence of tissues to form specific organs, he says:

In the first place, it is a state of instability, since it involves a readiness either to react to an organiser and follow a certain developmental path, or not to react and to develop in some other way. Moreover, this instability occurs in a system which, both before and after the competent period, is in process of developmental change. This change must be brought about by the interactions of chemical substances, and physical conditions, within the tissue. The whole progress of development may therefore be considered as resulting from an unstable configuration of substances which leads the embryonic tissue to change towards a more stable state, and the periods of competence are secondary instabilities when there are two or more alternate modes of progression towards stability . . .[36]

This leads in turn to a consideration of how such unstable chemical systems could be produced, and Waddington argues that they must arise from the action of the genes. He concludes:

Owing to our comparative ignorance of the effects of specific genes on competence, we must place more reliance on approaching the question from another angle, by investigating the effects of genes in general. We shall find that a discussion of gene action leads to the formulation of a scheme very similar to that which we suggested from the embryological angle; that is, development will appear as a result of a complex of reactions between substances which will form an unstable mixture, which may at certain times have two or more alternative modes of change open to it.[37]

This is theoretical biology to be sure. But what is fascinating is the move to consider development *abstractly* as structured movement between conditions of instability. An analogy encapsulates the conceptual shift that Waddington is negotiating. This is presented in a chapter on 'The temporal course of gene reactions' in which he concludes:

The system of developmental paths has been symbolised in two dimensions as a set of branching lines. Perhaps a fuller picture would be given by a system of valleys diverging down an inclined plane. The inclined plane symbolises the tendency for a developing piece of tissue to move towards a more adult state. The sides of the valleys symbolise the fact that developmental tracks are, in some sense, equilibrium states . . . This symbolic representation can be spoken of as 'The epigenetic landscape'.[38]

At the front of *Organisers and Genes* there is a reproduction of a picture by Waddington's friend, John Piper, commissioned specially to create a pictorial image of this landscape of analogy. It is a fascinating step to see in landscape a representation of movement between prestructured unstable states. But the shape of the resulting landscape only has significance as a representation of probability. The hills of Piper's painting belie their own appearance as solid forms. They are illusory and should properly be seen only as surfaces in probability or phase space. The observer is misled by the apparent substantiality of a magical or surreal landscape that embodies pathways of possibility under the influence of conflicting forces. The particular aesthetic of Piper's work is bent to Waddington's own intellectual purposes. In later years Waddington drew attention to the similarity between his epigenetic landscape and Sewall Wright's fitness surface, described in 1932. Whether this was the source of Waddington's analogy is not clear. The concern with the developmental stability posed interesting questions for evolutionary theory. The concept of 'canalization', which he introduced in 1942 was designed to bring the study of evolution and development together. Thus he wrote:

The main thesis is that developmental reactions as they occur in organisation submitted to natural selection, are, in general, canalized. That is to say, they are adjusted so as to bring about one definite end result regardless of minor variations in conditions during the course of the reaction. . . . At the same time, it is clear that this canalization is not a necessary characteristic of all organic development, since it breaks down in mutants, which may be extremely variable, and in pathological conditions, when abnormal types of tissue may be produced. It seems, then, that the canalization is a feature of the system which is built by natural selection and it is not difficult to see its advantages, since it ensures the production of the normal, that is, optimal type in the face of the unavoidable hazards of existence.[39]

The form of the landscape of developmental possibility, Waddington suggests, makes evolutionary sense. It conduces to the reproduction of optimality in a varying environment *and* can explain the appearance of novel types when development is switched by mutation or environmental stimulus into other similarly stabilised pathways. These thoughts link a model of development to the theory of evolution, and they indicate an enduring concern with evolution from his time as an apprentice palaeontologist in the late 1920s. The field of palaeontology at that time was a stronghold of anti-Darwinian ideas. In subsequent years he became known as an evolutionary theorist with his own distinctive terminology and deviant interpretation of neo-Darwinism.

His early work on 'canalization' was done in the early years of the war, whilst Waddington was still in Cambridge, working on wound healing at the Strangeways. Many of his friends and acquaintances from the 1930s like C. P. Snow and J. D. Bernal were deeply involved in the war effort and in 1942 Waddington joined the Operational Research Section of RAF Coastal Command. There he worked on the strategic and tactical planning of the airborne battle against German U-boat attacks on Atlantic convoys. Waddington's history of this work makes clear what kind of mathematical analysis this group performed but the nature of his own contribution is totally effaced.[40] On the basis of this evidence alone one is tempted to infer that the mathematical apparatus for evaluating optimal search strategies, maintenance schedules, bomb size and so on was developed by others, but that Waddington's predilection for this kind of modelling helped him to work with it enthusiastically. After the war he was an energetic propagandist of the value of OR but as far as one can see he made relatively little direct use of it in his research, except for his work in connection with the improvement of milk yields in British dairy herds, which was an isolated case. Indeed

contrary to my expectations there is no evidence of a major conceptual shift in Waddington's research occasioned by his OR experience.
In some of his post-war writings, such as *The Strategy of the Genes*
there are obvious borrowings from this area, but they simply extend
or put a gloss on thoughts that had been formed and uttered earlier.

# A new environment and the strategy and tactics of biology

In 1943 the Agricultural Research Council and the Agricultural
Improvement councils set up a survey group on Animal Breeding
and Genetics, which reported in 1944. Its chairman was Sir James
Gray, Waddington's former head of department. In November 1944
Waddington was appointed Deputy Director and Chief Geneticist
of a new National Animal Breeding and Genetics Research Organisation.[41] In a remarkably short space of time, from his first major
publications in genetics in the late 1930s, he had come to hold a
senior position in genetics in Britain.

In 1945 Waddington was offered the chair of genetics in Edinburgh University. The previous incumbent and head of the Institute
of Animal Genetics, F. A. E. Crew, had known Waddington before
the war.[42] Also the director of the Strangeways Institute, Honor
Fell, had been a student in this department in the 1920s. When the
new animal Breeding and Genetics Research Organisation was
located in Edinburgh Waddington was able to hold both appointments. As in Cambridge then he divided his allegiance between a
university science department and an independent research institute, with a more applied orientation. But unlike the earlier period
he now had administrative and managerial responsibilities of the
kind involved in running, getting new funds for and building up a
research organisation. In 1947 he was made a Fellow of the Royal
Society. By the late 1940s then he had status, resources and a field of
operation. He could build a research school and a program of
research.

I find it very difficult to summarise or chart Waddington's
scientific work in the 1950s, but I think a basic division can be
perceived between experimental studies, often done collaboratively
– the use of radioisotope tracers, his use of electron microscopy, the
breeding experiments with *Drosophila* on the nature of genetic
assimilation – and more theoretical writing, often devoted to con-

ceptual clarification and criticism, about evolutionary questions. As a matter of strategy, I infer, the speculative model-building of the latter category seems to have been held in check by or referred back to the empirical researches in the former. Although it verges on a truism to say this, one could see Waddington's work through the 1940s and 1950s as the rational exploitation of two sets of skills, one involving experimental manipulation, dissection, microscopy, and biochemical analysis built up through his work as an experimental embryologist and the other involving model building, deductive reasoning and a facility with mathematical and statistical concepts, developed through contributions to population genetics and evolutionary theory. Investment of effort and resources exclusively in the former – in embryology – would have restricted access to the interesting genetical questions which he took up professionally in the late 1930s. Investment exclusively in the latter would have drastically reduced the value of his accumulated practical experience with the study of development. In that case Waddington would have become, if I may put it like this, just another geneticist; that is, his own distinctive concern with development would have been lost. His post-war career then can be seen as an attempt to balance and combine these different kinds of knowledge and skill. After 1965 his efforts and career capital were channelled more and more into theoretical work, epitomised by the organisation of the series, *Towards a Theoretical Biology* and his involvement with the journal, *Ekistics* and the Alpbach Symposium. It is also clear that, whilst his own entrepreneurial activities were often successful, as an administrator of a large research organisation his performance was very mixed.

Throughout his career two characteristics stand out; firstly he was prepared to name issues as 'fundamental' or 'strategic'; and secondly he was interested in developing new languages or idioms for attacking such questions. His sources of inspiration in this enterprise bear the marks of their period: anthropology (Gregory Bateson) and art (John Piper) in the 1930s; operations research and cybernetics in the 1940s and 1950s; computing, simulation and topology in the 1960s.

In the 1950s he set out what he thought were the important strategic questions in biology and what could be relegated to the level of tactical detail. The lesser tactical details included molecular biology. These ideas appeared in *The Strategy of the Genes*, published in 1957. This book was created from a selection of papers and articles written from the late 1940s. It is a summary of Waddington's earlier theoretical ideas, revised and updated to accommodate some of the terminology of systems theory and extended to take account of the work of classical geneticists like Dobzhansky, Schmalhausen, Thoday and Mather in the 1940s and 1950s. But in the first chapter Waddington makes clear where he thinks the fundamental questions lie.

In the context of development, a whole host of investigators is at present actively attacking the fundamental problem of how a gene operates ... The experimental investigation of evolutionary processes has also notoriously made very great strides in recent years ... Both these great modern developments are mainly concerned with what I should venture to call tactical questions ... I wish to discuss the strategic question, how does development produce entities which have form, in the sense of integration or wholeness; how does evolution bring into being goal seeking or directiveness.[43]

As the book goes on to show such a classical concern with form can be approached in a very modern and abstract way.

The second chapter, *The cybernetics of development* is a discussion of what characteristics models of developmental processes must have, what formal properties they must embody, such as canalisation, specificity of buffering, and tolerance to developmental noise. An attempt is made to bring together a system of differential equations, describing the behaviour of substances being synthesised from common substrates, and the epigenetic landscape. The influence of the work of people like Shannon and Weaver, Ross Ashby, and Turing is evident, but, as Waddington admits, the discussion does not really go anywhere. The result is not a model but parts of a possible model. The remaining three chapters are essentially about population genetics. They too show the heritage of the war, as in the revision of Sommerhof's discussion of adaptation, which was drawn from work on the computer-controlled anti-aircraft gun. The conclusion brings together the themes of developmental stability, embodied in the idea of a 'chreod', selection as mediated through the interaction of the environment and phenotype and selective advantage of structured adaptability. As Waddington puts it:

Natural selection, whose direct operations impinge on the phenotypes which result from the interaction of genotype and environment, favours systems of genes which respond to the local situation by producing well adapted organisms. It then does something more than merely favour organisms which are, for some reason, fitter than their fellows; it builds into the epigenetic system tendencies to be easily modifiable in ways that are adaptive.[44]

Evolutionary questions then are high on the agenda and the mere details of how proteins get made is seen as relatively unimportant. In this context then it is instructive to compare Waddington and Francis Crick, since both passed through the Strangeways Institute, both worked on theoretical questions, both worked in wartime OR, but they were separated by a scientific generation in their entry into biology. Given Waddington's years of immersion in evolutionary and developmental studies since the late 1920s, compared with Crick's entry into structural biology from physics in the late 1940s it is not entirely surprising that the drastic, but immensely useful, schematisation of protein synthesis as an information flow that Crick proposed should have seemed such a secondary issue to Waddington. Perhaps in the mid-1950s this was not such a conservative judgement as it now seems.

In another sense too Waddington's writing in *The Strategy of the Genes* shows its roots in the British intellectual subculture of the 1920s, for the book closes with a direct attack on simplistic Darwinist formulations of natural selection, driven by separate, random mutations selected by the environment as a sieve picks out pebbles, as exaggeratedly atomistic. The reference to Whitehead is explicit: and the language of chreods, homeorhesis and strategic and tactical organisation in biology is referred back to its roots in the 1920s.

# Conclusion

Various common themes stand out from this representation of Waddington's career; they include a commitment on his part to development and holistic analysis and the revelation of process or potentiality behind the appearance of fixity, an evolutionary humanism and the habit of writing speculatively and didactically about very abstract models of organised systems moving between states of instability.

What is remarkable is the way in which Waddington was able to link theoretical and experimental study so productively; and to use abstract analysis to such good effect. I have tried to present his work in embryology in just that way. Where the representation of potentiality and equilibrium by reference to his landscape metaphor led him from embryology to a novel set of research questions in genetics the 1930s saw him casting around for a research niche, which he exploited very successfully. He and his collaborators produced results of great interest, that led them to formulate a general program of research, although their plans for institutionalisation were stopped. Waddington then moved confidently onwards to research in genetics and this work led rapidly to the idea of canalisation, in the early years of the war, before his connection with operations research. But from then on the focus seems to weaken, and his interests become more and more abstract and the language changes. Whilst the idea of stabilised developmental pathways remains a powerful one, his legacy lies in the model building of others. The language of strategy and tactics, which replaced that of form and stability, may have been topical and suggestive, but it failed to generate the kind of empirically accessible problems that fuelled the earlier phase of his career.

I am happy to acknowledge the advice and comments of numerous people in the production of this paper, in particular, Tim Horder, Jonathan Harwood, David Edge, Ruth Clayton, Andre Glucksman, Dame Honor Fell, Joseph Needham, Alan Cock, Brian Goodwin, D. R. Newth, E. B. Worthington, and Mary Nicholas of the MRC Archives.

# Endnotes

1. P. Bourdieu, The specificity of the scientific field and the social conditions of the progress of reason, *Social Science Information* **14** (1975), 19–47.
2. T. J. Pinch, Theoreticians and the production of experimental anomaly: the case of solar neutrinos, In K. D. Knorr *et al.* (eds.), *The Social Process of Scientific Investigation, Sociology of the Sciences, vol. 4* (Dordrecht: Reidel, 1980), 77–106; R. E. Kohler, *From Medical Chemistry to Biochemistry* (Cambridge: Cambridge University Press, 1982); B. Latour, S. Woolgar, *Laboratory Life* (Beverley Hills, Sage: 1979); J. Green, *The Social Construction of the XYY Syndrome* (Unpubl. Ph.D. thesis, University of Manchester, 1983).
3. For biographical information I have drawn heavily on A. Robertson, 'Conrad Hal Waddington', *Biographical Memoirs of Fellows of the Royal Society* **23** (1977), 575–622.
4. P. G. Werskey, *The Visible College* (London; Allen Lane, 1978).
5. D. Lipset, *Gregory Bateson: The Legacy of a Scientist* (Englewood Cliffs, N.J.: Prentice-Hall, 1980). Lipset does not in fact discuss the intellectual contact between Bateson and Waddington, although he mentions that it was significant.
6. W. Coleman, Bateson and Chromosomes: Conservative Thought in Science *Centaurus* **15** (1970), 228–314; A. G. Cock, William Bateson's Rejection and Eventual Acceptance of Chromosome Theory, *Annals of Science* **40** (1983), 19–59.

7. C. H. Waddington, An autobiographical note, In *The Evolution and an Evolutionist* (Edinburgh; Edinburgh University Press, 1975); C. H. Waddington, Fifty Years On, *Nature* **258** (November 6, 1975), 20–1.

8. On the institutionalisation of genetics in the UK see D. J. Kevles, Genetics in the United States and Great Britain, 1890–1930: a review with speculations, *ISIS* **71** (1980), 441–55.

9. Dame Honor Fell, personal communication, July 29, 1983.

10. C. H. Waddington, Developmental mechanics of chicken and duck embryos, *Nature* **125** (1930), 924.

11. See correspondence in File No. 1532/19 *Rockefeller Medical Fellowships successful applicants 1931–2 I* Medical Research Council, London.

12. C. H. Waddington *The Nature of Life* (London: Allen & Unwin, 1961), p. 58.

13. Letter from Honor Fell to Sir Walter Morley Fletcher, 6 December 1930 File No. 1532/19; on Gray's own research program and plans for the department see H. W. Lissman, 'James Gray', *Geographical Memoirs of Fellows of the Royal Society* **24** (1978), 55–70.

14. Ibid. (MRC files).

15. C. H. Waddington, Experiments on the development of chick and duck embryos, *Philosophical Transactions of the Royal Society, Series B* **221** (1932), 179–230, at page 233.

16. Woodger's career is described in his Festschrift, J. R. Gregg, F. T. C. Harris, (eds.), *Form and Strategy in Science* (Dordrecht, Reidel 1964); see also Donna J. Haraway, *Crystals, Fabrics and Fields: Metaphors of Twentieth Century Biology* (London: Yale University Press, 1976); N. Roll-Hansen, 'E. S. Russell and J. H. Woodger – the failure of two twentieth century opponents of mechanistic biology' *J. Hist. Biol.* **17** (1984), 399–428.

17. J. H. Woodger, The 'Concept of Organism' and the relation between embryology and genetics, *Quarterly Review of Biology* **5** (March 1930), 1–22, at page 7.

18. Ibid, p. 21.

19. C. H. Waddington to Secretary of the Medical Research Council, 10 February 1934; MRC File No. 84A Strangeways Research Laboratory Vol 6.

20. P. Abir-Am, The discourse of physical power and biological knowledge in the 1930s: a re-appraisal of the Rockefeller Foundation's 'Policy' in molecular biology, *Social Studies of Science* **12** (1982), 341–82.

21. R. E. Kohler, The management of science: the experience of Warren Weaver and the Rockefeller Foundation Programme in Molecular Biology, *Minerva* **14** (Autumn 1976), 279–306.

22. Kohler, op. cit.; V. Fuerst, The role of reductionism in the development of molecular biology: peripheral or central? *Social Studies of Science* **12** (1982), 241–78; E. J. Yoxen, Life as a productive force: capitalising the science and technology of molecular biology, In L. Levidow, R. M. Young (eds.), *Science, Technology and the Labour Process*, (London: CSE Books, 1981), 66–122.

23. R. C. Olby, On the place of dimensions in physiological explanations (this volume).

24. E. J. Yoxen, Scepticism about the centrality of technology transfer in the Rockefeller Foundation Programme in Molecular biology, *Social Studies of Science* **14** (1984), 248–52.

25. C. H. Waddington, Some European contributions to the prehistory of molecular biology *Nature* **221** (1969), 318–21.
26. D. J. Haraway, op. cit.
27. Waddington to Dr Miller, October 26 1936. Rockefeller Foundation Archives, Record Group 1.1; Series 401, Box 43 Folder 537.
28. Rockefeller Foundation Archives, Fellowship Records. Notes by H. M. Miller, 6 April, 1939.
29. C. H. Waddington, *An Introduction to Modern Genetics* (London: Allen & Unwin, 1939).
30. Ibid, p. 182.
31. Ibid, p. 188.
32. Ibid, p. 191.
33. C. H. Waddington, Preliminary notes on the development of the wings in normal mutant strains of *Drosophila, Proceedings of the National Academy of Sciences* **25** (1939), 299–307; The genetic control of wing development in *Drosophila, Journal of Genetics* **41** (1940), 75–139.
34. See note 7 above.
35. Waddington, op. cit., p. 94.
36. C. H. Waddington, *Organisers and Genes* (Cambridge: Cambridge University Press, 1940), 44–5.
37. Ibid, p. 55.
38. Ibid, p. 92–3.
39. C. H. Waddington, *The Strategy of the Genes* (London: Allen and Unwin, 1957), p. 188.
40. C. H. Waddington, *Operational Research in World War II; O.R. against the U-Boat* (London: Elek Books, 1973).
41. Sir William Henderson, The establishment of the council and its role, In G. W. Cooke, (ed.), *Agricultural Research, 1931–1981*. (London: ARC, 1981), 43.
42. The history of the Institute of Animal Genetics from 1920 to 1945 is described in M. Deacon, 'The Institute of Animal Genetics at Edinburgh – the first twenty five years' (mimeo, Science Studies Unit, University of Edinburgh). I am grateful to David Edge for bringing this to my attention.
43. Waddington, op. cit. (note 39 above), p. 12.
44. Ibid, p. 188.

# Regeneration

H. WALLACE

Department of Genetics, University of Birmingham,
Birmingham B15 2TT, UK

The term regeneration has had much the same meaning for about a century, a rather restricted form of what Morgan (1901) defined as restorative regeneration – the formation of a whole organ or organism from part of the original one. We tend to prefer the term regulation when such a process occurs in the egg or embryo, so I shall confine myself to postembryonic events in Metazoa. The exclusion of Protozoa and plants betrays my ignorance, of course, but also gives me a sporting chance of producing a coherent account. Morgan separated our meaning of regeneration from physiological regeneration, the spontaneous replacement of skin, hair, feathers, teeth, antlers, etc. He also distinguished between two main types of regeneration: morphallaxis and epimorphosis. Morphallaxis implies a direct restructuring of the partial organism to form a complete individual, as in hydra or in embryonic regulation. Epimorphosis involves a local cellular proliferation and growth of a blastema which then develops into the missing structures, as in limb regeneration, while the rest of the body remains unaltered. It is a convenient distinction but perhaps not a fundamental one, especially as both epimorphosis and morphallaxis occur during the regeneration of planarians and annelids. In practice, it is often merely a case of whether remodelling or growth becomes obvious first. I have avoided other definitions such as heteromorphosis which Morgan adopted from his contemporaries, because they are rarely used now or not in the same way. They remind us that regeneration is not a teleological process, for it occasionally produces bizarre results which should provide some insight into its developmental rules.

Despite such errors, regeneration falls into the general category of regulative phenomena and is a more dramatic example than wound healing. Perhaps it is more akin to asexual reproduction in that there is a clear developmental process which leads to the typical adult form, but often by a different pathway from that used in embryonic development. It seems to be generally true that forms which reproduce by fission or budding are also capable of extensive regeneration, but the converse is not true. Some have argued that these regulatory powers represent a fundamental property of living matter (e.g., Vorontsova & Liosner, 1960), only lost if it impeded a more advantageous complexity or specialisation. Others have suggested regeneration itself is a specific adaptation (e.g., Goss, 1969), resulting from natural selection where injury is common but not lethal. Morgan (1901) was circumspect or sceptical of such general issues and we are no wiser now. According to Reyer (1977), a newt's eye seems quite as complex as a human one and even less likely to be damaged in non-fatal circumstances. Yet the eyes of some newts show an enviable power of regeneration.

## Ancient history

The physiological regeneration of antlers and reptile skin must have been known in prehistoric times. Anyone who regularly ate lizards or crustaceans could hardly avoid noticing their propensity to shed tails or legs, and the undersized replacements on some specimens. Such knowledge surely persisted as folklore even if ancient philosophers considered them less striking than regeneration of the phoenix. The authenticity of fishermen's tales was confirmed for once by René Antoine Ferchauld, Seigneur de Réaumur in 1714 (see Guyénot, 1957). He recorded the regeneration of crab, lobster and crayfish claws and commented on an even earlier report of a regenerating lizard tail by Thevenot in 1686. Réaumur (1683–1757) was undoubtedly the instigator of experimental regeneration, not only by his own research but because he inspired and encouraged the often more ingenious tests of Trembley and Bonnet. He invented the world polyp (by analogy to the octopus) and announced Trembley's discovery of regeneration of hydra to the Académie Royale des Sciences in 1741. He confirmed this case and showed worms had the same property in one of his mémoires on 'insects' in 1742. He was even instrumental in persuading other correspondents to find that marine hydroids and starfish could regenerate.

When Abraham Trembley (1710–84) discovered *Hydra viridis* in

1740, he could not decide whether it was an animal or a plant. He soon found that fragments of a polyp could each form a complete individual, shortly after noticing that hydra moved and before realising it was carnivorous but reproduced by forming buds. The ensuing consternation among naturalists was compounded by his celebrated experiment in 1741 of turning hydra inside out. Trembley's monograph of 1744 was followed a year later by Charles Bonnet's (1720–93) report of regeneration in freshwater worms. Lazaro Spallanzani (1729–99) created an even wider sensation by announcing in his *Prospectus* of 1768 that he had witnessed the regeneration of tails and legs by tadpoles and newts, and that decapitated snails could grow new heads (Fig. 1). Pallas recorded planarian regeneration for the first time in 1774. The list of examples has been extended more slowly since then and attempts to delimit it continue to this day. Regeneration of the newt eye, for instance, was reported by Bonnet in about 1780 but the specific formation of a new lens from the dorsal iris was first noticed by Colucci in 1891 and shown to follow removal of the original lens by Wolff in 1895. Only one family (Salamandridae) seems to have this ability (see Yamada, 1977) but almost all urodeles can grow reasonable replicas of limbs or tails. Immature insects, spiders and myriapods were all known to be capable of limb regeneration by the middle of the last century.

Fig. 1. Decapitated snails from plate 2 of Spallanzani's *Tracts*.

There is not much point in asserting that our knowledge of regeneration antedated embryology, or that it anticipated Entwicklungsmechanik as an experimental science. The naturalists of this period did not make any distinction between different aspects of reproduction, while production and generation have almost exchanged meanings since then. Bonnet was a notable exponent of preformation theory. He saw nothing in regeneration to contradict the ovist viewpoint (vindicated by his demonstration of parthenogenesis in aphids) which was as much the antithesis of spontaneous generation as of epigenesis. Invisible germs stationed throughout the body could evidently reproduce missing parts just as easily as they could propagate individuals of future generations. Repeating Spallanzani's tests of limb regeneration in newts during 1776–8, Bonnet confirmed that only the removed portion is replaced and that it can be replaced repeatedly (Fig. 2). He concluded that different kinds of germs are arranged strategically to repair adjacent defects, and these germs must be self-reproducing to ensure that regenerated limbs are as well stocked with them as was the original structure.

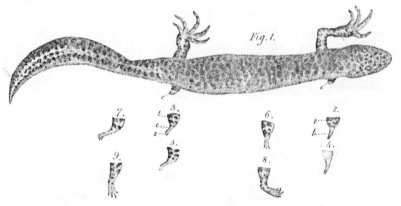

Fig. 2. Bonnet's repeated amputation of newt legs (*Tracts*, plate 6).

A few records still exist of the correspondence between Réaumur, Trembley & Bonnet (Baker, 1952). Bonnet contributed to Spallanzani's *Tracts*. Their astounding feats and genial disdain for the niceties of classification (Fig. 3), in addition to their mutual affection and extravagent compliments conjure up a version of The Three Musketeers with Spallanzani cast as D'Artagnan. Most of their discoveries had been made with scissors and a magnifying glass before 1780. Only intermittent reports of regeneration appeared during the following century, such as those on planaria listed by Brøndsted (1969). These are completely overshadowed first by the description of germ layers and then by the gradual development of the cell theory, as related elsewhere in this volume. Spallanzani had already described the appearance of frog embryos at cleavage and gastrulation, but could not appreciate their significance. In fact, he showed the impeccable logic of a Jesuit by arguing from the similar consistency of oviducal and fertilised eggs that tadpoles exist before fertilisation, and thus frogs are really viviparous. Spallanzani's artificial fertilisation of amphibian eggs and attempt at contraception (by fitting taffety wax breeches to an amorous male frog) took him to the brink of discovering the function of spermatozoa. His failure to do so could be construed as an act of faith or of scepticism; he faced a choice of belief and chose wrongly for excellent reasons.

## Modern history

By 1890 it was evident that both embryology and regeneration had to be investigated all over again in terms of cells and with better microscopes. The theories advanced by Weismann and Roux probably provided an extra motivation, or a convenient excuse, for such studies. Those concerning regeneration are best illustrated by Morgan's experience at the turn of the century. Morgan's (1901) *Regeneration* is a revelation of his temperament and preoccupations. He was not yet aware of Mendel's Laws and showed a sensible resistance to the more Panglossian aspects of the theory of natural selection. Consequently, he refused to regard regeneration as a specific adaptation. He accepted the cell theory without stressing it, for his own training and experience convinced him that a simple experimental intervention could be at least as informative as any painstaking description (see Allen, 1978). Over a period of about five years, Morgan had checked most of the classical examples of regeneration and added several experiments of his own. This preparation did not revolutionise the subject but lent considerable authority to his arguments, which are persuasive even when vaguely documented. His book shows minor blemishes, of course, such as the early mention of 'Abbé' Trembley – ostensibly a calculated insult to that moderate Calvinist, later compounded by Roundabush's (1933) reference to the 'Swiss priest Alexander Trembley'. I shall consider the general developmental issues which exercised Morgan in later sections, to see what progress we have achieved since then. Several issues specific to regeneration deserve some comment first, however, partly to complete this historical treatment and partly to show a concordance with advances in embryology.

The cell-lineage and other descriptive studies of embryos, which were in vogue during the early part of this century and have been resumed recently for *Caenorhabditis*, can be matched by increasingly detailed accounts of development in the major examples of regeneration. Partial descriptions now exist of biochemical, histochemical, immunological and ultrastructural changes. The experimental approach has become predominant in regeneration as well as in embryology but not to an excessive degree in either, for essentially descriptive studies are often called experiments nowadays. All this activity has produced a glut of information, often unrelated to any particular concept and including a proportion of suspect or contradictory interpretations. It is no longer possible to provide a comprehensive survey of these investigations. One symposium (Kiortsis & Trampusch, 1965) and the specialised monographs cited in this essay provide access to them, by bibliographies which greatly exceed the 300 or so articles which Morgan used to encompass regeneration of all animals and plants. I shudder to think what today's total would be.

The investment of time and effort in becoming familiar with one line of research, in addition to gaining enough expertise to contribute to it, ensures the investigator will be less aware of related topics and virtually ignorant of others. Empiricists are no better placed, for they are dependent upon our conclusions and so tend to perpetuate our mistakes. There is little significance in the common observation that we tend to concentrate either on embryology or on regeneration, for we tend to specialise much more than that yet form alliances across this supposed border. Part of the common ground established between adherents of amphibian limb regeneration and chick wing buddists surely lies in our mutual wonder over whether chopping planaria in half or sticking blastocysts together is the more esoteric occupation. Fortunately, there are a few general issues of developmental biology which unite our provincial pursuits, so that immersion in one may help us to appreciate another or even to switch topics at judicious intervals. It is getting harder to do so, I suspect. Spallanzani saw no obstacle to such diversions, for it was all natural history to him. Driesch and Morgan refused to be confined to either embryology or regeneration, which might be a matter of superior entelechy or genes. It is consoling to realise that several participants at this symposium have also ignored the slight barrier between our disciplines.

```
The author's neglect, or rather contempt of nomenclature,

occafions confiderable difficulties to thofe who ftudy his

works.  Indeed all the fyftems we have are fo brief and

indiftinct, that even the moft expert naturalift will fome-

times be at a lofs to difcover whether the animal there

named is truly the object of his fearch.
```

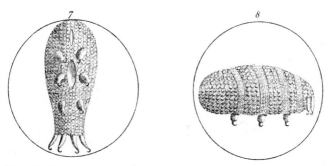

Fig. 3. Translator's complaint from Spallanzani's *Tracts* p. 437.

## Developmental concepts

Needham (1959) remarked that progress (in science generally) depends on a delicate balance of three things: speculative thought, accurate observation and controlled experiment. He observed that speculations tend to outlive their usefulness and considered that our obligations include not only choosing the right concepts but abandoning the wrong ones. That is a fine prescription, but I suspect more scientists have died clutching their misconceptions than have been persuaded to relinquish them. Needham did not exactly help matters by suggesting the balance would be improved mainly by theoretical advances, even mathematico-logical ones (a desperate remedy). The balance is a matter of opinion but surely we have quite enough theories to be going on with.

Apart from the grand unifying concepts of biology, such as cellular organisation and metabolism, evolution through natural selection and genetics, there are subsidiary developmental concepts which fall fairly neatly into two categories. Firstly, those which concern cellular differentiation and the allied topics of germ-layer, tissue and organ formation. Secondly, the concepts of spatial organisation concerned with growth and form. After a brief exposition of these concepts, I shall attempt to demonstrate how studies of regeneration have contributed to them.

## Cellular differentiation

This has long been the main preoccupation of embryologists and remains so with the modern emphasis on genetic regulation. Echoes of the old preformation–epigenesis controversy lingered in the distinction between mosaic and regulative development, but that converted a philosophical debate into an experimental one. Archetypal mosaic and regulative eggs are the rare or imaginary extremes of developmental modes, I suppose, and Driesch's (1908) analysis of prospective potency applies to both. The fertilised egg is totipotent by definition, to produce all the cell-types of the adult. Sooner or later, by ooplasmic segregation or inductions, the potency of each blastomere or region gradually diminishes until it is identical to the prospective value in fully determined cells. This analysis is no longer fashionable, yet many embryologists tacitly accept its main consequence: that determination leading to cellular differentiation is inevitably an irreversible process.

Studies of regeneration have persistently questioned that notion, usually without providing a definite answer but there are now at least five which refute it. The problem here is one of identifying the source of regenerating cells and tracing them, often through an unspecialised or undifferentiated stage of proliferation, to differentiated tissues in the regenerate. Successful identification may reveal any or all of three processes or minor concepts.

(1) Metaplasia (or Transdifferentiation) occurs if one cell-type is converted to another in the regenerate.

(2) Modulation occurs when cells in the regenerate revert to their previous kind of differentiation.

(3) Neoblasts (or Reserve Cells) are recognised as unspecialised cells of the body which are postulated to be undetermined. They could retain the potency to differentiate into any cell-type of a regenerate and might be capable of proliferating enough to form the bulk of the regenerate.

The classic example is Wolffian regeneration of the lens from the dorsal iris in newts. This shows a genuine dedifferentiation including depigmentation of the epithelial cells, which then proliferate before redifferentiating as lens fibres. Excluding the iris stromal cells as an alternative source has finally established this metaplasia beyond dispute (Yamada, 1977). Trembley's ingenious experiment of turning hydra inside out did not demonstrate metaplasia, partly by an amazing act of escapology (the polyps reverted while transfixed by a bristle, see Baker, 1952) or more prosaically because of cellular migration across the mesoglea (Roundabush, 1933), and partly because hydra is predominantly composed of interstitial cells (70–80% according to Gierer *et al.* 1972). The latter have been considered by some to be neoblasts and by others to be determined precursors of nematoblasts and nerve cells. Cultured fragments which lack interstitial cells, however, have provided two convincing cases of metaplasia. Isolated gastric endothelium of *Hydra viridis*, which consists entirely of digestive and gland cells, can regenerate the complete polyp (Normandin, 1960; Haynes & Burnett, 1963). Either epithelial striated muscle or endothelium from the anthomedusan *Podocoryne carnea* can be isolated in culture and then transforms into flagellated cells or nematocysts. When combined, they regenerate into a manubrium which contains at least seven cell-types including gametes (Schmid, Wydler & Alder, 1982).

Planaria contain an unspecialised mesenchyme between their major organs, including small neoblasts which may amount to 15–25% of the total cell population according to Baguña (1976). Ever since their original description by Randolph in 1897, neoblasts have usually been considered as at least the main source of the regenerate (Brøndsted, 1969). Dissent from that opinion seems to have increased during the last decade and recently Gremigni & Micelli (1980) have taken advantage of a natural cell marker to demonstrate metaplasia in a triploid race of *Dugesia lugubris*. Spermatogonia occupying scattered testes on each side of the body (but not the pharynx) become diploid and thus distinguishable from somatic cells. Regenerated pharynx muscle is mainly triploid but about 5% of its cells are diploid, implying a minor (perhaps proportionate) contribution from these germ-line cells.

Amphibian limbs also contain poorly differentiated cells in the connective tissue. These cells have rarely been considered as neoblasts, because dedifferentiation of other tissues is quite obvious, but limb regeneration might occur by a modulation of cells derived from these tissues with some extra neoblast activity. Definitive evidence of metaplasia was eventually obtained by grafting cartilage into irradiated limb stumps whose own tissues could not contribute to the mesenchymal blastema. Namenwirth (1974) traced triploid graft chondrocytes into all the main mesodermal tissues of the regenerate, except for muscle which was virtually absent. Wallace, Maden & Wallace (1974) used unmarked cartilage grafts in the same way with very similar results, but claimed some strips of muscle were present in a few regenerates and these must also have originated from chondrocytes. I have recently repeated this experiment using triploid cartilage grafts and obtained precisely the same result as Namenwirth. Most of the regenerates contained no muscle (in contrast to those derived from skin or nerve grafts), and the occasional strips of muscle at the base of a few regenerates seldom possess identifiable triploid nuclei (Fig. 4). In fact, I now wonder whether my earlier claim might not be based on mistaking tendon for muscle. Despite that failure, there is no doubt that axolotl chondrocytes are pluripotent and that some cells in several other tissues can transform into cartilage.

Now that the existence of metaplasia has been established in diverse animals and is well known in plants (Vasil & Vasil, 1972), we can reconsider the alternatives mentioned earlier. Modulation may often be an accurate description of cell changes during regeneration, relating the prospective value of cells but it can only infer their persistent determination under a limited set of conditions. Metaplasia shows that these cells have a greater potency to form other cell types as well, and hence subsumes modulation. Similarly, now that well differentiated cells are known to be pluripotent the concept of neoblasts has become redundant. Since both differentiation and determination are reversible processes, I fear the whole concept of prospective potency is tarnished. Driesch (1908, p. 84) anticipated this objection and promptly invented two kinds of regulation and

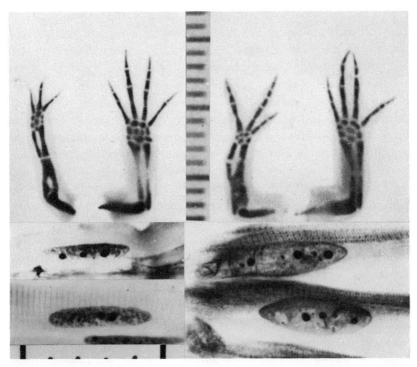

Fig. 4. Results of grafting triploid cartilage into an irradiated forearm stump. Top: 2 cases of experimental left arm regenerates with serially repeated radius and ulna and control right arm regenerates (mm scale). Bottom left: muscle nuclei of experimental regenerates rarely contain 3 nucleoli (10 μm scale). Bottom right: muscle nuclei from a triploid control regenerate.

potency: primary ones for embryonic development and secondary ones for regeneration and asexual reproduction. I suggest that such tampering in itself brings the concept into disrepute.

## Spatial organisation

The preformed axes found in most eggs and early embryos, with rather stereotyped morphogenetic movements, go a long way to explaining their basic pattern of development. Yet the determination of axial gradients has been studied as much in regeneration as in embryology. Physiological gradients of metabolic activity are matched by ones of regenerative rate, most obviously in coelenterates and planarians (Child, 1941). These gradients have also been interpreted in terms of competition, inhibition or dominance, but with surprisingly little argument over whether one or two gradients are required to establish a differential pattern along the main axis. A single gradient can explain how colonial coelenterates form a hydranth at one or both ends of an isolated stem. Tardent (1963) considered that a single gradient was also enough to decide that hydra fragments grow a head at the distal end and a foot at the proximal end. Most subsequent investigators have accepted his opinion. Similarly, head or tail regeneration in planarians is ascribed to a single initial gradient followed by a cascade of inductions and inhibitions which are seen as responsible for the spacing of internal organs (Brøndsted, 1969). This gradient might be simply expressed as intrinsic polarity, except that the polarity of the fragment is so commonly reversed in the regenerate. The polarity of hydra is stable during regeneration and most grafting experiments, but it can be reversed locally (Burnett, 1962; Müller, 1982). A small apical fragment of *Tubularia* can form a reversed regenerate like those on the proximal ends of long stems. Small terminal fragments and small intermediate slices of planarians are known to regenerate into mirror-image double heads or tails, as do earthworm tail fragments (Morgan, 1901). Similar mirror-image regenerates form regularly on the distal pieces of amphibian limbs and tails. In these cases, the polarity of the fragment seems unimportant, but its wound surface dictates the initial gradient level of the regenerate. The gradient should decline from there, I suppose, but then can hardly account for both Janus-headed and Janus-tailed planaria. A variety of inhibitory treatments which reverse polarity have been compiled by Brøndsted (1969) and Rose (1970), but I am not aware of a satisfactory explanation of this phenomenon either.

Axial polarity is obviously an abstraction even in coelenterates. The concept of the morphogenetic field merely extends gradient theory to two dimensions, with the implication that both the whole field and its gradients are self-regulatory – in contrast to a mosaic territory. Besides the obvious application to embryonic tissue layers such as the limb field of amphibia, flatworms have always been considered as practically two-dimensional. Their rate of head regeneration nicely illustrates the field concept, declining laterally as well as posteriorly from the brain region (Brøndsted, 1969). Although it is less easily envisaged in three dimensions, Weiss had already applied the field concept to amphibian limb regeneration in 1925. Partly out of spite, Guyénot advanced a relatively mosaic concept of regeneration territories soon afterwards and continued to explore it for the next twenty years. Mapping the extent of the limb-forming territory by deflecting brachial nerves to the flank demanded such persistence because of the reduced size and frequency of the induced limbs, as expected at the weak fringes of a field. Some of the axially reversed limbs encountered in these studies, however, can be interpreted as evidence for polarising centres like those demonstrated subsequently in amphibian and chick embryos (Wallace, 1981).

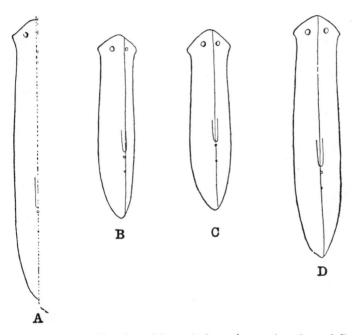

Fig. 5. A drawing modified from Morgan's fig. $13\frac{1}{2}$ – to show he and *Dugesia lugubris* were both capable of lateral thinking.

There has been ample criticism of the field concept, notably of its vagueness and superficiality in postulating gradients without enquiring into their nature (e.g. Child, 1941). Weiss' contention that a grafted blastema is induced to conform to its surrounding field turned out to be false. On the contrary, the blastema has an established organ-specificity and axes and is sometimes capable of influencing its surroundings. Testimony to this effect comes not only from amphibian limbs but also from the later morphallactic responses of regenerating planarians and annelid worms. The only virtue I have noticed in the later concept of positional information lies in its reassertion of the importance of individual gradients. Perhaps it has a wider scope, by including highly determinate cases such as the progress zone theory of chick wing bud development, in addition to the regulative examples to which the field concept was restricted. Even within regulative systems, unless I am mistaken, there is an important difference. Positional information is regarded as forming a gradient specified by points which are usually boundaries, whereas the corresponding reference point of a field is its centre of activity. Where boundaries can be discerned or inferred during development or regeneration, such as the segments or compartments of insects, we usually have little evidence that they serve as reference points or that they are established relative to prior boundaries rather than centres of activity. Can it be mere prejudice which provided us with so many examples of active centres – organisers, differentiation centres, eye and limb fields – which lack definite boundaries? Consequently, I remain rather sceptical about both concepts. Positional information could add much needed precision to the field concept, but only if it can be reconciled to a much wider spectrum of observations than at present. Wolpert's contribution to this volume will no doubt fortify the reader against my prejudice, and perhaps cause me to change my mind.

## Instant history

Positional information has spawned a deluge of models. The self-imposed restraint exercised by Morgan seems to have broken down, resulting in an indecent display of speculation and attendant gullibility. The essence of a model is surely to focus attention on certain aspects of a problem at the expense of neglecting others. Whether that leads to a sensible solution or not depends on the quality and accuracy of the model, but that is not all. Any virtues still have to be balanced against the tunnel vision imposed by the model. Even the best will limit or distort our perspective, so that mediocre models probably cause more mischief than blatantly inadequate ones (such as computer simulations). We should at least recognise this malign influence and so become more wary of the charms of any particular model. Furthermore, now that models have become a major currency in developmental biology, we must be alert to the possibility that they may be subject to Gresham's Law.

# References

Allen, G. E. (1978). *Thomas Hunt Morgan.* 447 pp. Princeton: University Press.

Baguña, J. (1976). Mitosis in the intact and regenerating planarian *Dugesia mediterranea* n. sp. *Journal of Experimental Zoology*, **195**, 53–64, 65–79.

Baker, J. R. (1952). *Abraham Trembley of Geneva.* 259 pp. London: Arnold.

Brøndsted, H. V. (1969). *Planarian Regeneration.* 276 pp. Oxford: Pergamon.

Burnett, A. L. (1962). The maintenance of form in hydra. In *Regeneration*, ed. D. Rudnick, pp. 27–52. New York: Ronald Press.

Child, C. M. (1941). *Patterns and Problems of Development.* 811 pp. Chicago: University Press.

Driesch, H. (1908). *The Science and Philosophy of the Organism*, vol. 1, 329 pp. London: Black.

Gierer, A. & 7 others (1972). Regeneration of hydra from reaggregated cells. *Nature New Biology*, **239**, 98–101.

Goss, R. J. (1969). *Principles of Regeneration.* 287 pp. New York: Academic Press.

Gremigni, V. & Micelli, C. (1980). Cytophotometric evidence for cell 'transdifferentiation' in planarian regeneration. *Wilhelm Roux' Archives* **188**, 107–13.

Guyénot, É. (1957). *Les Sciences de la Vie aux XVIIᵉ et XVIIIᵉ Siècles.* 482 pp. Paris: Michel.

Haynes, J. F. & Burnett, A. L. (1963). Dedifferentiation and redifferentiation in cells of *Hydra viridis. Science* **142**, 1481–3.

Kiortsis, V. & Trampusch, H. A. L., eds. (1965). *Regeneration in Animals and Related Problems.* 568 pp. Amsterdam: North-Holland.

Morgan, T. H. (1901). *Regeneration.* 316 pp. New York: MacMillan.

Müller, W. A. (1982). Intercalation and pattern regulation in hydroids. *Differentiation* **22**, 141–50.

Namenwirth, M. (1974). The inheritance of cell differentiation during limb regeneration in the axolotl. *Developmental Biology* **41**, 42–56.

Needham, J. (1959). *A History of Embryology*, 2nd end. 304 pp. Cambridge: University Press.

Normandin, D. K. (1960). Regeneration of hydra from the endoderm. *Science* **132**, 678.

Reyer, R. W. (1977). The amphibian eye: development and regeneration. In *Handbook of Sensory Physiology*, ed. F. Crescitelli, **VII/5.** pp. 309–90. Berlin: Springer.

Rose, S. M. (1970). *Regeneration.* 264 pp. New York: Meredith.

Roundabush, R. L. (1933). Phenomenon of regeneration in everted hydra. *Biological Bulletin* **64**, 253–61.

Schmid, V., Wydler, M. & Alder, H. (1982). Transdifferentiation and regeneration in vitro. *Developmental Biology* **92**, 476–88.

Spallanzani, L. (1768). *Prodromo di un'Opera da Imprimersi sopra le Riproduzioni animali.* Modena. Translated by or for M. Maty as: Abbé Spallanzani (1769). *An Essay on Animal Reproductions.* London: Becket & de Hondt.

Spallanzani, L. (1784). *Dissertations Relative to the Natural History of Animals and Vegetables*, 2 vols, translation. London: Murray.

Spallanzani, L. (1803). *Tracts on the Natural History of Animals and Vegetables,*

2nd edn. 2 vols, translated by J. G. Dalyell, Esq., Advocate. Edinburgh: Creech.

Tardent, P. (1963). Regeneration in the hydrozoa. *Biological Reviews* **38**, 293–333.

Vasil, I. K. & Vasil, V. (1972). Totipotency and embryogenesis in plant cell tissue cultures. *In Vitro* **8**, 117–27.

Vorontsova, M. A. & Liosner, L. D. (1960). *Asexual Propagation and Regeneration*. 489pp. London: Pergamon.

Wallace, H. (1981). *Vertebrate Limb Regeneration*. 276 pp. Chichester: Wiley.

Wallace, H., Maden, M. & Wallace, B. M. (1974). Participation of cartilage grafts in amphibian limb regeneration. *Journal of Embryology and Experimental Morphology* **32**, 391–404.

Yamada, T. (1977). *Control Mechanisms in Cell-Type Conversion in Newt Lens Regeneration*. 126 pp. Basel: Karger.

# Gradients, position and pattern: a history

L. WOLPERT

Department of Anatomy and Biology as Applied to Medicine, The Middlesex
Hospital Medical School, London, UK

On the assumption that the concept of positional information is a useful one, I wish to review the origin of the concept, and particularly its relationship to that of gradients. In particular I want to consider why the link between gradients and that of positional information took so long to establish. A possible explanation is that the paradigm up to the 1950s which dominated thinking about gradients was that of energy metabolism. Only when the paradigm shifted to information transfer in biochemistry, with the coding problem in relation to DNA and protein clearly defined, did the concept of positional information emerge.

In 1888 Roux reported that when one of the blastomeres of the frog egg was killed at the 2-cell stage, the other blastomere formed a half-embryo. This was taken as support of the Roux–Weismann theory that nuclear division distributes different 'structures' to the cells. In 1891, Driesch set out to repeat the experiment on sea-urchin embryos. 'But things turned out as they are bound to do and not as I had expected; there was a typically whole gastrula on my dish the next morning, differing only by its small size from a normal one' (Driesch, 1908). Thus, contrary to expectation, Driesch found that a half embryo could give rise to a small, but normal, whole embryo. This was the first demonstration in a developing system of what came to be known as regulation. It was experiments of this type that lead Driesch to draw the fundamental distinction between prospective fate and possible fate; the possible fate being much wider in range than the prospective fate.

It was similar experiments that led Driesch to suggest also that 'the prospective fate of any blastula cell is a function of its position in the whole.' The 'whole' could be related to any three axes

347

drawn through the normal undisturbed egg, on the hypothesis that there exists a primary polarity and bilaterality of the germ; the axes which determine this sort of symmetry may, of course, conveniently, be taken as co-ordinates; but that is not necessary. Driesch (1899) regarded the early sea-urchin embryo as a 'harmonious equipotential system'. This was in spite of reports as early as 1895 by Zoja that animal halves did not gastrulate. Hörstadius (1972) comments that 'he did not dare to draw the correct conclusion against the authority of Driesch.' Boveri too, as we shall see, reported that animal halves did not gastrulate.

Driesch formalised his concept of position and the harmonious equipotential system in the following way. On the assumption that the system – the early sea-urchin embryo – will develop normally, no matter which part is removed, he argues that the prospective fate ($p.v.$) of any element $X$ of the embryo is a function of the size of the system ($S$) and its relative distance from one of the boundaries ($l$) and a factor $E$. Thus

$$p.v. \ (X) = f(S, \ l, \ E)$$

$E$ is that factor at work in every case which is not a variable and can be thought of as the prospective potency, with special regard to the proportions embraced by it. The situation is also illustrated in Fig. 1.

Driesch then rejected '... all sorts of chemical morphogenetic theories put forward to explain the problem of localization.' There cannot be 'a machine nor any sort of causality based upon correlation underlying the differentiation of harmonious-equipotential systems. For a machine, typical with regard to the three chief dimensions of space cannot remain itself if you remove parts of it or if you rearrange its parts at will.' (Driesch, 1908). Thus Driesch was driven to the concept of Entelechy for $E$ and developed a theory of Vitalism. This virtually removed his influence on thinking about pattern problems. We now know that Driesch was wrong on two grounds. The first, experimental, is that the sea-urchin as both Zoja and Boveri showed is not a harmonious equipotential system. While meridional halves develop into normal embryos, animal and vegetal halves behave quite differently. Animal halves lack a gut. Secondly, it is quite easy to devise models – machines – which will give the required properties for assigning position in space when parts are removed. This is shown by, for example, solutions to the French Flag problem (Wolpert, 1968) and by the behaviour of reaction diffusion systems (Gierer, 1981). I

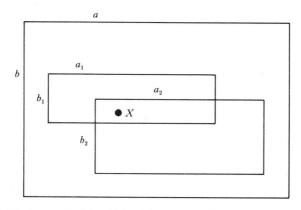

Fig. 1. Diagram to illustrate Driesch's concept of a harmonious equipotential system. Consider the sheet of cells with boundaries $a$, $b$ that undergoes development into a variety of structures such that the region $X$ develops into a particular structure. Its fate, and that of all the other points in the sheet, have a prospective fate that is specified by its position within the sheet. If parts of the sheet, for example, the rectangle bounded by $a_1$, $b_1$, and $a_2$, $b_2$ are isolated then each of these rectangles will develop the same pattern of structures as the sheet $ab$. However, in each case, $X$ will develop according to its position within the whole: that is, its position with respect to the boundaries $a_1 b_1$ and $a_2 b_2$. Thus the prospective value of $X$ will be different in each case.

shall return again to embryos and the sea-urchin in relation to gradients but must first consider the concepts that were emerging in relation to regeneration, particularly polarity.

The concept of polarity was first established in relation to regeneration. While earlier workers such as Trembley, Spallanzani and Bonnet knew that in general the end of the piece of an animal from which a head has been cut off is the one which develops a head, Allman in 1864 was the first to call this polarity (reviewed, Morgan, 1901). Driesch had drawn an analogy between the polar structure of the egg and a magnet: on this basis fragments should behave similarly, particularly if they were part of a 'harmonious equipotential field'. But Boveri (1901) pointed out that fragments of sea-urchin eggs behaved differently and this could only be explained by a stratification of the egg. This stratification is, according to Spemann's (1938) reading of Boveri, conceived of as a gradient. The more one proceeds from the vegetative to the animal pole, the less vegetative do the strata become. At a definite level they are insufficiently vegetative to gastrulate. 'Boveri ... has even stated the idea that the most vegetative part of each egg fragment, starting from a definite minimum of vegetative quality, might be a "region of preference" from which the rest of the germ would be determined in its development.' Thus Boveri (1901) is usually credited with being the first to have identified polarity with gradients. However, our reading of this paper does not easily confirm this view. While Boveri talks extensively about regulation and polarity there is no clear statement about gradients.

By contrast, Morgan (1905*a,b*; 1906) is quite explicit in relating gradients to polarity. The first statement comes in a paper on the regeneration of heteromorphic tails in pieces of planaria (Morgan, 1905*a*). He is puzzled as to why the length of the piece is so important.

It is very significant, I think, to find that in planarians the shortness of the piece is a factor that enters into the problem as to the character of the new part. I have suggested tentatively that this means that in *Planaria maculata* the tendency is stronger for the new structure to become a head than a tail, and that when the influence of polarity is removed a head appears on each end of short cross-pieces. In other worms, as in *Planaria simplicissima*, the tendency in certain posterior regions to produce a tail is stronger than that to produce a head, and two tails appear when the polarity is reduced or removed. Why should the length of the piece be so important a factor? Can it be that there is a greater difference, chemical or physical, between the two ends of a longer piece, so that a stronger polarity is present? In short pieces, from this point of view, the ends being near together are so much alike that the polarity is correspondingly reduced, and, under these conditions, the specification of the material of the old part is not sufficiently strong to determine the nature of the new part. These and many other equally obscure questions remain for future investigation to explain.

In the same year, studying the regeneration of *Tubularia*, Morgan (1905*b*) examines the nature of the structures developed. Considering both the nature of the structure developed and the timing, he writes 'We may assume that the gradation of the material is of such a kind that the hydranth forming material decreases from the apical towards the basal end.' He goes on

I have assumed that the stem of *Tubularia* is not homogeneous, but that from the hydranth to the base there is a graded difference and this gives the order or stratification of the material. The stimulus of the water acting on the free end arouses the formative changes which act in the direction of this existing material order. How does such a view differ from the old assumption of a 'polarity' in the material? In my view there is no such directive force residing in the material as the term polarity suggests, but the polarity is only a name for the gradation of the material and on this as a basis the formative changes are carried out.

Here we have a very clear statement of the relation between gradients and polarity.

Morgan, curiously did not really pursue his ideas on polarity and gradients. Instead they were taken up by Child (1911). Studying regeneration in *Planaria*, he concluded

All of these facts indicate that a graded difference of some sort in the dynamic processes exists along the axis, and together with the results concerning size of pieces, the experiments show, first, that a certain minimal portion of this graded difference is necessary for the regulation of a piece into a whole, and second, that the rate of gradation along the axis differs in different regions of the body. In other words, the minimal length of pieces which are capable of forming a whole is different in different regions of the body. In a later paper, however, I shall show that these factors are by no means constant, but depend, at least in large degree, upon the rate of metabolism or of certain metabolic processes.

It is this dynamic gradation along the axis, together with the complex of correlative conditions associated with it, which I regard as constituting physiological polarity. According to this idea polarity is not a condition of molecular orientation, but is essentially a dynamic gradient in one direction or different gradients in opposite directions along an axis, together with the conditions, and particularly the physiologic correlation between different parts along the axis, which must result from the existence of such a gradient or gradients.'

This seems very similar to Morgan's views, but not to Child.

According to the above view then, a certain minimal fraction of the axial gradient or gradients is necessary in every case for the formation of a whole. But it is necessary to call attention not merely to the existence of the axial gradient, but to the correlative factor in polarity. Morgan's hypothesis of the gradation of substances possesses certain features in common with my idea of a metabolic gradient, but the assumed gradation of substances is more stable and requires the assumption of migration; the most unsatisfactory feature of this hypothesis, however, is its failure to recognise the correlative factor in polarity. The gradation of substances alone cannot account for the fact that the same cells give rise under certain conditions to a head, under certain others to a tail, but when we consider that a gradation of any kind along the axis may give rise to a variety of correlative conditions, the phenomena become less puzzling, and we see that the directive or apparently directive feature of organic polarity is in reality a matter of physiologic correlation.

I find it difficult to understand what Child is driving at and in what substantial way his views about polarity differ from those of Morgan. Nevertheless, it was he who went on to promote gradient theory, and not Morgan. His early views are summarised in his book *Individuality in Organisms* published in 1915 and one of its remarkable features is that all the references, with one exception, are to his own work. He makes clear that his ideas on biological organisation come in part from an analogy he draws between the organism and the state. Order, he argues, requires authority, and in the organism this is provided by a dominant region which sets up a metabolic gradient.

While Child was, at this stage, primarily interested in gradients in relation to regeneration, clear statements about gradients in eggs were made by Boveri (1910) and Runnström (1914). Boveri (1910) put forward the possibility of gradient in relation to the development of the egg of *Ascaris*. Runnström was more explicit, basing his ideas on his studies on sea urchin bilateral symmetry and the effect of different ions on it. He quotes neither Child nor Morgan.

During development certain power is created which transforms into 'qualities'. As the basis of these powers we may think of a particular specific chemical material (see Lundegärdhis, 1914). The chemical material is directed into certain directions. These directions follow the axis of the embryo or more correctly the expression of the axes is shown in the direction of the chemical. The material is localised in a way that different parts of the embryo have different concentrations. It creates a concentration gradient. The phenomenon of polarity is an expression of the presence of this concentration gradient.

Both the ideas of Child and the Swedish school of Runnström and Hörstadius were considerably elaborated upon (Hörstadius, 1972; Child, 1941). One of the most important experiments illustrating the quantitative nature of the animal–vegetal gradients in sea-urchins was that in which Hörstadius (1935) showed that a greater number of micromeres are required to be added to a fragment from the animal pole, in order for normal development to occur, than for an animal fragment taken from slightly more towards the vegetal pole.

By focusing on gradient concepts I have neglected a major area of research – induction and the concept of the embryonic field. This is so large a topic it can only briefly be touched on here. The first demonstration of induction, that is the ability of one piece of tissue to induce a developmental change in an adjacent tissue, was probably that for the lens by Lewis in 1904 (see Spemann, 1938). As early as 1901, Herbst had suggested, on the basis of malformations observed in the head and eyes of amphibians, that the development of the lens was dependent on the optic cup. In the same year, Spemann pointed out how this was to be tested: either by removal of the optic cup or by grafting the optic cup to a different site. Lewis grafted the developing optic cup under the epidermis and found that a lens formed above the transplanted optic vesicle. At about the same time E. N. Browne (1909) showed that the head region of hydra could, when grafted into the body of another animal, induce the formation of a new axis. However, it was Spemann's work in the 1920s and 1930s on induction in amphibian development that created a revolution in development (see Spemann, 1938). His worked showed, for example, that the nervous system of the frog would not develop unless induced to do so by the underlying mesoderm. More dramatically, he showed that one part of the embryo, the dorsal lip of the blastopore, could when appropriately grafted induce an almost complete second axis. This region became known as the organiser.

Spemann was the first to introduce the field concept into embryology. In 1921 he remarked that it appeared that the organiser established a 'field of organization', borrowing the concept from physics. Others too, took up the concept but its meaning remained rather obscure (Waddington, 1956). However, it has relevance to gradient theories as Huxley & de Beer (1934) considered that within fields 'various processes concerned with morphogenesis appear to be quantitatively graded, so that the most suitable name for them is field-gradient systems, or simply gradient-fields'.

Perhaps the clearest view of gradients some thirty years after their conception is the brilliant synthesis of Huxley & de Beer (1934). Largely on the basis of the behaviour of regenerating systems, they put forward a number of rules in relation to field-gradient systems. The most relevant for this paper are as follows, (i) The origin of polarity is to be sought in external factors though in some cases the system is already polarised by virtue of possessing parts of the original gradient system. (ii) In regeneration the apical region, or head, is the first to be formed and its formation is autonomous. (iii) The newly formed apical region influences adjacent regions (this is the dominant region of Child). Its influence is to establish a field. (iv) The development of all other regions is dependent on influences that proceed from more apical levels. (v) At least one of the influences

exerted by the more apical regions on lower levels is that of inhibition. The presence of the apical region inhibits lower levels from
regeneration. (vi) If a portion of tissue comes to lie outside the field
dominated by an existing apical region, a new apical region will
arise. (vii) The frequency or absence of regeneration and the type of
structure regenerated depend on the level of the cut surface in the
gradient field and the form and steepness of the gradient established.
As can be seen, these rules begin to integrate the gradient concept
with the properties of the apical region which has both organiser and
inhibitory activities.

In his Silliman Lectures, Spemann (1938) gave a quite detailed
critical analysis of what he called 'The Gradient Theory'. One of the
main problems that Spemann had with gradient theory (quoting
Bautzmann, 1929) is the difficulty in understanding how qualitative
differences may arise from quantitative ones in the different regions.
'... the gradient, notwithstanding the different steepness of its
single stretches must be conceived of as continuous, whereas the
series of formation whose differentiation would be determined by
that gradient, notwithstanding a general decrease in mass, is absolutely discontinuous.'

A more quantitative approach was introduced by Dalcq & Pasteels
(see Dalcq, 1938). Considering early amphibian development, they
introduced two factors – the yolk gradient $V$ and a cortical factor $C$
which they termed the dorso-ventral field which decreased in
intensity from the grey crescent – it too was graded. They then
defined the product $CV$ as the morphogenetic potential and most
importantly they made use of the concept of a threshold. The parts of
the embryos thus acquired properties depending on whether $CV$ was
above or below a threshold. This was a major step in making gradients more quantitative even though how they were set up, or
regulated, or changed, was ignored. Also, they give credit for the
threshold concept to Huxley (1935) whom they quote as saying 'such
threshold phenomena are probably of great importance in development and are essential to the understanding of epigenesis' though
Huxley says no more than that. They made use of the $CV$ ratio
particularly in relation to sea-urchin development. They also began
to talk about the diffusion of 'organosine' as the basis of induction.

While there was a useful set of rules relating to gradients and
dominance, it was striking how little attention had been given to the
theoretical side of the problem, particularly the construction of

working models. In this connection, Child's (1941) synthesis is rather disappointing. While recognising that the high region of a gradient is 'primarily the chief dominant region' he did not see the logical step, namely that regulation required the establishment of a similar highpoint at the cut surface. His ideas on the physiological basis of gradient were naive. For example, he argued that 'if decrease in concentration of amount of a certain substance or substance–complex occurs in one direction along a gradient, there must be increase in concentration or amount of another substance or substances unless there is a decrease in volume.' He was, like his contemporaries, obsessed with metabolism and it was only when this paradigm was replaced with one based on information transfer that new ideas emerged.

Child assumed that gradients were different and changes in form or growth involved changing the gradient. 'The metabolism of different gradients in the same individual may vary widely...' 'If gradient characteristics, length, steepness, etc., are concerned in determining localization of particular differentiations, it is evident that genetically determined changes in these characteristics may alter localizations of organ systems and organs relatively to others.' He thus interprets d'Arcy Thompson's (1915) co-ordinate transformation in terms of alterations in gradient patterns. While recognising concentration gradients – though not their physicochemical implication – he preferred metabolic gradients. 'If a single metabolic gradient may be associated with two opposed substance gradients, it, rather than the substance gradients, is the effective factor in developmental pattern... There is no theoretical difficulty as regards origin of specific or qualitative differentiation at different levels of a primarily qualitative gradient.' He even recognises that these gradients can alter enzyme activity and even gene action. However, he does not take up the question of threshold. It is important to realise that Child does not equate gradients with position. Though he does recognise that the gradient system can be viewed as a physiological coordinate system, he did not pursue the idea.

In contrast to the rather vague ideas of Child, Spiegelman (1945) and Rose (1952) put forward very specific models. Spiegelman (1945) pointed out that one of the features of regeneration is that the potentiality for forming a particular part of the pattern is present in a larger area than actually forms the part. He thus argued for a 'principle of limited realization' which he suggested must involve two

distinct mechanisms. The first involves the suppression of the realisation of potentialities and the second for providing for differences
between parts of the system. If, for example, two parts of the system
are capable of forming a particular region, and only one does, then it
is essential to have both a difference between these two parts, and the
suppression of the development of one part by the other. He stressed
that a difference as manifested by a gradient in some property, would
not, in itself, provide an adequate mechanism.

Rose's model (1952) was even more explicit. In fact it was the
most explicit model of its time. He suggested that the genesis of a
pattern could be a consequence of a hierarchy of self-limiting reactions, together with the spread of inhibition from one differentiating region to others further down the gradient. The gradient
provided the basis of the rate of differentiation, the reactions proceeding fastest at the highpoint of the gradient. The reaction that
predominates is that closest to the top of the hierarchy which has not
yet been inhibited. Each reaction is self-limiting by virtue of the
inhibitor it produces (for some criticisms of the model, see Wolpert,
1968). It was the first gradient model to generate a pattern.

The gradient concept gained considerable support from Sander's
(1959, 1960) experiments on the axial body pattern in the leafhopper
*Euscelis* which were interpreted in a semiquantitative way in terms
of gradients and thresholds, and the results on the reversal of segment sequence were particularly striking and provided very strong
evidence for gradients. They also emphasised the importance of
boundary regions.

The first very clear statement of the concept of positional information was Hildegard Stumpf's (1966) paper entitled 'Mechanism by
which cells estimate their location in the body.' She starts by
pointing out that all cells in the body have the same genetic information and thus those that form a particular organ need 'some special
information that others do not get.' Earlier, Locke (1959) had
carried out a series of very elegant studies on the ripple patterns
in *Rhodnius* cuticle which showed very persuasively that there was a
gradient along the main axis. She interpreted this as a concentration
gradient in a substance which was made at one margin of the segment
and destroyed at the other. This was the first clear statement of how a
concentration gradient could be set up. Using the abdominal segments of *Galleria*, which are divided axially into three parts, she then
did experiments 'to determine whether the postulated concentration

gradient is the means by which the partition is determined.' The experiment showed that this was the case. She thus concluded 'The concentration gradient, the existence of which is confirmed by these results, obviously has two functions. (i) To orient the scales by its direction; (ii) to supply the cells by its absolute values (or ranges of concentration) with the necessary information about their distance from the segment margins and to induce the corresponding circular structures.' Here, at last we are back with Driesch's problem.

It must be noted that Lawrence (1966), quite independently of Stumpf, put forward a gradient model for the insect epidermis. His model was based on observations on the orientation of hairs and bristles in *Oncopeltus*. For this he put forward a gradient model based on the properties of a sand gradient: the chemical analogue being active transport against a diffusion gradient.

It was again quite independently that I (Wolpert, 1969) put forward the concept of positional information which was so clearly spelled out by Stumpf. I had arrived at it from consideration of the French Flag problem (Wolpert, 1968). I had been working on the morphogenesis of sea-urchin embryo with Trygve Gustafson in Sweden in the early 1960s. While we were concentrating our attention on cell movement and the cellular forces responsible for bringing about changes in shape during, for example, gastrulation, I could not but be aware of the strong Swedish tradition on sea-urchin regulation with its emphasis on animal and vegetal gradients. The thinking was still in terms of the gradients providing one region with a metabolic rate advantage over adjacent regions. I just could not understand how these gradients could specify pattern or how they were regulated. I was particularly impressed that the proportions of the main regions of embryo along the main axis – mesoderm, endoderm, ectoderm – were constant over an 8-fold difference in size of embryo. In about 1964 I started to work on hydra regulation with Gerry Webster. Hydra seemed to be a good model system for thinking about spatial organisation (Webster & Wolpert, 1966). Again, the properties of regenerating hydra were constant over a wide size range. In order to formulate the problem of pattern formation rigorously, I invented the French Flag Problem. Part of the motivation was to emphasise the importance of pattern formation which, on the whole – Waddington was a notable exception – was grossly neglected. Embryologists were, it seemed, obsessed with induction, which provided little insight into spatial organisation and merely provided a co-ordinating mechanism.

The French Flag Problem addressed how a line of cells, each of which was capable of differentiating into red, blue or white, could be organised so that it looked like a French Flag: that is one third blue, one third white, and one third red. Part of the problem was that the pattern should be invariant with size; it was for just that reason I had chosen a flag. I presented the problem at the first of the Theoretical Biology Conferences that Waddington organised in Bellagio in 1966. There are a number of solutions to the problem and I was looking for one that was simple and universal. Universal because I was sure that there would be basic mechanisms of pattern formation. At that time I was collaborating with Mary Williams and Michael Apter, and the latter was keen on an obvious solution: number the cells from each end and the cells can then compute which third of the line they are in. For some time I resisted this solution as being too complex, but then suddenly realised that it provided a marvellous basis for a large number of pattern problems. I was particularly struck that it could account for the genetic mosaic experiments of Curt Stern (1968). It seemed to provide a sensible way of making use of the fact that all cells contained a complete genetic program. I was also influenced by information theory and had written a critique of its application to biology (Apter & Wolpert, 1965). The idea of positional information was presented at the Third Theoretical Biology meeting in Bellagio at Easter 1968 where it was sympathetically received. But at Woods Hole in the same year, it was met with particular hostility, and only Sydney Brenner, who was thinking along similar lines, was encouraging, though Howard Schneiderman showed great interest.

The framework of positional information had the following features. (1) It emphasised the nature of the coordinate system and what this implied: boundaries, vectors, scalars. (2) It raised the possibility of the idea of a universal coordinate system. (3) It discussed regeneration in terms of re-establishing boundaries and positional values. (4) It made an attempt at a quantitative approach – coordinate systems were small (less than 1 mm) and took hours to set up. (5) It emphasised interpretation as the major process which was quite separate from specifying position. These ideas were in marked contrast to thinking about pattern formation at that time as can be seen in Waddington (1962). Developments since then, such as the formulation of the polar coordinate model for epimorphic regulation (French, Bryant & Bryant, 1976) cannot be considered here (but see Wolpert & Stein, 1984).

It was quite outside the tradition just described that Alan Turing published a remarkably original paper in 1952. Turing's approach to pattern formation was to try and set up a prepattern using re-action–diffusion mechanisms. As such it has nothing to do with the tradition that thought in terms of gradients.

However, its relevance here is that it provided a mechanism whereby gradients could be organised by chemical reactions, an approach that is exemplified by the Gierer–Meinhardt model (Gierer, 1981). The whole of Turing's abstract is worth quoting in full. Note that he coined the term 'morphogen'.

It is suggested that a system of chemical substances, called morphogens, reacting together and diffusing through a tissue, is adequate to account for the main phenomena of morphogenesis. Such a system, although it may originally be quite homogeneous, may later develop a pattern or structure due to an instability of the homogeneous equilibrium, which is triggered off by random disturbances. Such reaction–diffusion systems are considered in some detail in the case of an isolated ring of cells, a mathematically con-venient, though biologically unusual system. The investigation is chiefly concerned with the onset of instability. It is found that there are six essentially different forms which this may take. In the most interesting form, stationary waves appear on the ring. It is suggested that this might account, for instance, for the tentacle pattern on *Hydra* and for whorled leaves. A system of reactions and diffusion on a sphere is also considered. Such a system appears to account for gastrulation. Another reaction system in two dimensions gives rise to patterns reminiscent of dappling. It is also suggested that stationary waves in two dimensions could account for the phenomena of phyllotaxis.

Why had it taken so long to bring together the concepts of gradi-ents and position, with all the implications that this carries, particu-larly since Driesch had been quite explicit about coordinate systems? It is possibly that, because of Driesch's vitalism, no one took his ideas very seriously. But much more important was the paradigm within which developmental biologists worked. Child, we have seen, was obsessed with metabolism and this makes thinking about concepts such as positional information very difficult. Metab-olism is about energy and rates; position is about scalars, vectors and boundaries. An analogy may be drawn with the history of the gene concept. As late as 1947 Muller considered that the action of genes was to affect energy flow within the cells (see Mayr, 1982). The transition from thinking about metabolism to thinking about in-formation flow came with the understanding of the structure of

proteins as linear sequences and the recognition of the coding problem in the 1950s. This history is documented by Freeman Judson
(1979) and Mayr (1982). It is hard to realise now how big a change
this was, and an adequate analysis of the process is not yet available.
Thus, Mayr is surprised that he cannot find a reference to information transfer earlier than 1947. He now regards the idea as obvious;
but that is with the wisdom of hindsight. Similarly one can now
think of the concept of positional information as obvious but though
Driesch put it forward at the beginning of the century it was only in
the 1960s that it emerged again.

I am indebted to Amata Hornbruch for translations of German papers and to Maureen
Maloney for preparing this manuscript.

# References

Apter, M. J. & Wolpert, L. (1965). Cybernetics and development. I. Information
and theory. *J. Theoret. Biol.* **8**, 244–57.
Bautzmann, H. (1929). Über Induktion durch vordere und hintere Chorda in
verschiedenen Regionen des Wirtes. *Roux Archives* **119**, 1–46.
Boveri, T. (1901). Über die Polarität Des Seeigeleies. *Verk. phys-med. Ges.
Würzburg, N. F. 31*, **34**, 145–70.
Boveri, T. (1910). Die Potenzen der Ascaris-Blastomeren bei abgeänderter Furchung Zugleich ein Beitrag zur Frage qualitative rungleicher Chromoso-
menteilung. In *Festschrift Z. 60 Geburtstag von R. Hertwig*, vol. 3, pp. 131–220.
Jena: Fischer.
Browne, E. N. (1909). The production of new hydranths by the insertion of small
grafts. *J. exp. Zool.* **7**, 1–23.
Child, C. M. (1911). Studies on the dynamics of morphogenesis and inheritance in
experimental reproduction. I. The axial gradient in *Planaria dorotocephala* as a
limiting factor in regeneration. *J. exp. Zool.* **10**, 265–320.
Child, C. M. (1915). *Individuality in Organisms*. Chicago: University of Chicago
Press.
Child, C. M. (1941), *Patterns and Problems of Development*. Chicago: University
of Chicago Press.
Dalcq, A. M. (1938). *Form and Causality in Early Development*. Cambridge: Cam-
bridge University Press.
Driesch, H. (1899). Die Lokalisation morphogenetischer Vorgänge. Ein Beweis
vitalistischen Geschehens. *Archives Entwicklungsmechanik* **8**, 35–111.
Driesch, H. (1908). *The Science and Philosophy of the Organism*. London: Adam &
Charles Black.
French, V., Bryant, P. J. & Bryant, S. V. (1976). Pattern regulation in epimorphic
fields. *Science* **193**, 969–81.
Gierer, A. (1981). Some physical, mathematical and evolutionary aspects of bio-
logical pattern formation. *Phil. Trans. Roy. Soc.* B **295**, 425–617.

Herbst, C. (1901). *Formative Reize in der Tierischen Ontogenese*. Leipzig.

Hörstadius, S. (1935). Über die Determination in Verlaufe der Eiachse bei Seeigeln. *Pubbl. Stn. zool. Napoli* **14**, 251.

Hörstadius, S. (1972). *Experimental Embryology of Echinoderms*. Oxford: Clarendon Press.

Huxley, J. S. (1935). The field concept in biology. *Trudy Do Dinamike Razvitiya* **10**, 269–89.

Huxley, J. S. & de Beer, G. R. (1934). The elements of experimental embryology. Cambridge: Cambridge University Press.

Judson, H. F. (1979). *The Eighth Day of Creation*. London: Jonathan Cape.

Lawrence, P. A. (1966). Gradients in the insect segment: the orientation of hairs in the milkweed bug *Oncopeltus fasciatus*. *J. exp. Biol.* **44**, 602–20.

Lewis, W. H. (1904). Experimental studies on the development of the eye in Amphibia. I. On the origin of the lens. *Rana palustris*. *Am. J. Anat.* **3**, 505–36.

Locke, M. (1959). The cuticular pattern in an insect *Rhodnius prolixus*. *J. exp. Biol.* **36**, 459–77.

Mayr, E. (1982). *The Growth of Biological Thought*. Cambridge, Mass.: Harvard University Press.

Morgan, T. H. (1901). *Regeneration*. New York: Macmillan.

Morgan, T. H. (1905*a*). Regeneration of heteromorphic tails in posterior pieces of *Planaria simplicissima*. *J. exp. Zool.* **1**, 385–93.

Morgan, T. H. (1905*b*). An attempt to analyse the phenomena of polarity in *Tubularia*. *J. Exp. Zool.* **1**, 589–91.

Morgan, T. H. (1906). "Polarity" considered as a phenomenon of gradation of materials. *J. exp. Zool.* **2**, 495.

Rose, S. M. (1952). A hierarchy of self-limiting reactions as the basis of cellular differentiation and growth control. *Amer. Nat.* **86**, 337.

Runnström, J. (1914). Analytische Studien über Seeigelentwicklung I. *Arch. Entwick.* **40**, 526–64.

Sander, K. (1959). Analyse des ooplasmatischen Reaktionsystem von *Euscelis plebejus* FALL (CICADINA) durch Isolieren und Kombinieren von Keimteilen. I. Mitt. Differenzie-rungsleistungen vorderer und hinderer Eiteile. *Wilhelm Roux Arch. Entwicklungsmech. Org.* **151**, 430–97.

Sander, K. (1960). Analyse des ooplasmatischen Reaktionssystem von *Euscelis plejebus* FALL (CICADINA) durch Isolieren und Kombinieren von Keimteilen. II. Mitt. Die Differenzie-rungsleistungen nach Verlegen von Hinterzolmaterial. *Wilhelm Roux Arch. Entwicklungsmech. Org.* **151**, 660–707.

Spemann, H. (1938). *Embryonic Development and Induction*. Yale: Yale University Press

Spiegelman, S. (1945). Physiological competition as a regulatory mechanism in morphogenesis. *Q. Rev. Biol.* **20**, 121–46.

Stern, C. (1968). *Genetic mosaics and other essays*. Cambridge, Mass.: Harvard University Press.

Stumpf, H. (1966). Mechanism by which cells estimate their location within the body. *Nature* **212**, 430–1.

Thompson, D. A. (1915). Morphology and mathematics. *Trans. Roy. Soc. Edinburgh* **50**, 857–95.

Turing, A. M. (1952). The chemical basis of morphogenesis. *Phil. Trans. Roy. Soc.* B **237**, 37–84.

Waddington, C. H. (1956). *Principles of Embryology.* London: Allen & Unwin.

Waddington, C. H. (1962). *New Patterns in Genetics and Development.* New York: Columbia University Press.

Webster, G. & Wolpert, L. (1966). Studies on pattern regulation in *Hydra.* I. Regional differences in time required for hypostomal determination. *J. Embryol. exp. Morph.* **16**, 91–104.

Wolpert, L. (1968). The French Flag Problem: a contribution to the discussion on pattern development and regulation. In, *Towards a Theoretical Biology,* ed. C. H. Waddington, pp. 125–33. Edinburgh University Press.

Wolpert, L. (1969). Positional information and the spatial pattern of cellular differentiation. *J. Theoret. Biol.* **25**, 1–47.

Wolpert, L. & Stein, W. D. (1984). Positional information and pattern formation. In *Pattern Formation,* ed. G. M. Malacinski and S. V. Bryant, pp. 3–21. New York: Macmillan.

# The role of genes in ontogenesis – evolving concepts from 1883 to 1983 as perceived by an insect embryologist

KLAUS SANDER

Institut für Biologie I (Zoologie), Albert-Ludwigs-Universität, Albertstrasse 21a,
D-7800 Freiburg i.Br., West Germany

## (1) Inheritance and development – inseparable aspects of life

Biologists trained in the first half of this century came to know genetics and developmental biology, or rather 'experimental embryology', as separate disciplines with little in common but well defined (and defended) boundaries and some links to cytology. This strict separation of disciplines – one studying the transmission, the other the expression of heritable traits – may have contributed to scientific progress for some time, but it is by no means a requirement imposed by Nature herself. The separation was indeed preceded by a unified view of transmission and expression. This unified view is clearly reflected in a question which August Weismann, a key figure of this period, asked one hundred years ago: 'How is a single germ cell capable of reproducing the entire body with all its detail?' (Weismann, 1883).

The speculative answer to this question, the 'Keimplasmatheorie' worked out by Weismann during periods of failing eye sight, was known to every biologist at the turn of the century, and is mentioned by a surprising number of textbooks even today, probably because it was the first (and visionary) attempt to provide a molecular link between heredity and ontogenesis, especially spatial diversification, in animals (Weismann, 1892). Expressed in modern terms, Weismann's explanation for the orderly development of the embryo is based on a three-dimensional spatial pattern of determinants preformed in the zygote nucleus. This 'neopreformationist' view of ontogenesis was shared by Wilhelm Roux, who initiated experimental embryology and coined the term 'Entwicklungsmechanik' for it, the

363

Kantian term 'mechanics' simply denoting causality (see e.g. Roux, 1923, p. 146).

## (2) Genetics and experimental embryology: the post-Weismannian schism of disciplines

As is well known, the Weismann–Roux hypothesis was refuted very soon on the basis of experimental studies provoked by it, which led to a different, more 'epigenetic' view of ontogenesis. The 'analytical' approach to ontogenesis propagated by Driesch and the rediscovery (in 1900) of Mendel's laws sparked a schism between the disciplines of 'developmental physiology' – so named by Driesch – and Mendelian or transmission genetics (formal genetics) which lasted for several decades. The difference in outlook separating the adherents of these disciplines is aptly characterised by a sketch that Oscar Schotte used to draw on the board when lecturing on the topic (Fig. 1).[1]

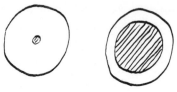

Fig. 1. The cell as perceived by the embryologist (left) and the geneticist (right) – the nucleus is hatched. Free rendering of a blackboard sketch by Oscar Schotte in the 1950s.

Driesch's specific and lasting contribution to the embryologist's view of the cell and ontogenesis concerns the cytoplasm. In contrast to Weismann, he saw (and clearly stated) that the cytoplasm must direct certain nuclear activities in development. 'The nucleus in our view is an important hereditary substance, but it is not the exclusive organ of "inheritance"; without the specifically structured egg cytoplasm, it could not accomplish development' (Driesch, 1894, p. 124). Like some of his contemporaries, notably Oscar Hertwig and E. B. Wilson, Driesch was convinced that the nucleus 'maintains its totality' in each and every cell (1894, pp. 47–8, 81). In order to account for the different functions directed or performed by the identical nuclei lodged in different cells of the embryo, Driesch proposed that the nucleus consists of enzyme-like substances, and that each substance directs, without getting lost, one of the specific steps or 'elementary processes' inferred to occur during ontogenesis (1894, pp. 88, 179). He claimed that the nuclear functions are under control of stimuli derived from (or transmitted through) the cytoplasm, and that the specific enzymatic activities thus initiated change the cytoplasm in turn (1894, p. 88).

However, Driesch was not satisfied with reductionist 'explanations' for ontogenesis, and in the end turned to vitalism using the riddles of morphogenesis as his key argument (Driesch, 1909, p. 143). Driesch justly claims that the simple stimuli by which one may alter development experimentally – for instance the separation of blastomeres (Driesch, 1892) – cannot account for the complexity of the response: the spark triggering an explosion also does not explain why thermal energy is released and whence it comes. Today, we entrust synergetics (Haken, 1977) (section 7) and stored genetic information with ultimately taking care of the complexity of developmental responses. Driesch preferred to invoke vitalistic forces for this purpose. His choice could be excused as guided by creed where suitable data were lacking (and to some degree still are). However, the argument accompanying Fig. 14 in the Gifford Lectures (refutation of the 'machine-theory' of morphogenesis, see Fig. 1 in Wolpert's article in this book) is a far cry from Driesch's earlier well reasoned views and, like the title of the lectures from which it was taken (Driesch, 1909), signals that its author had traded science for philosophy. He no longer took notice of mundane data and interpretations, more specifically of Boveri's well founded claim 'that the nucleus, to which we now may ascribe any degree of structural complexity, behaves exactly as postulated for a "machine" by Driesch in his considerations' (Boveri, 1902, p. 84).

Driesch's vitalistic inclinations recruited adversaries rather than proselytes among the scientific community. His friend and co-author of earlier days (1895), Thomas Hunt Morgan, wrote when looking back on the rise of genetics:

If another branch of zoology that was actively cultivated at the end of the last century had realized its ambitions, it might have been possible to-day to bridge the gap between gene and character, but despite its high-sounding name of Entwicklungsmechanik nothing that was really quantitative or mechanistic was forthcoming. Instead, philosophical platitudes were invoked rather than experimentally determined factors. Then, too, experimental embryology ran for a while after false gods that landed it finally in a maze of metaphysical subtleties. It is unfortunate, therefore, that from this source we can not add, to the three contributory lines of research which led to the rise of genetics, a fourth and greatly needed contribution to bridge an unfortunate gap.

(Morgan 1932)

Whether the geneticists of the day were really intent on bridging that gap was doubted in some quarters. Richard Goldschmidt, for one, found Morgan guilty of neglecting problems (and publications) related to the ontogenetic aspects of inheritance,[2] and complained that the progress that had been made in understanding Mendelian transmission 'led many geneticists to the erroneous idea that this meant having found also a theory of inheritance, and therefore they were lacking any interest in work reaching beyond the investigation of the mechanisms of transmission' (Goldschmidt, 1927, translation by the present author).

On the embryological side of the spiritual gap (and on the other side of the ocean), the dominating counterpart of Morgan was Hans Spemann. When Morgan (in his Nobel lecture of 1933) again proclaimed that experimental embryology 'took a metaphysical turn', he expressly excluded Hans Spemann, but regretted at the same time that 'the evidence from the organizer has not yet helped to solve the more fundamental relation between genes and differentiation' (Morgan, 1935). This was and is certainly true, and might follow from the fact that amphibians are fairly unsuitable for genetic work. However, this technical handicap alone cannot explain the really remarkable restraint of Spemann and his students concerning the role and even the name of genes. In the German edition of the Silliman Lectures, his magnum opus, Spemann (1936) mentions the term 'Gen' not once in nearly 300 pages, while 'Genom' and 'Genotyp' appear in a few statements quoted from other authors.[3] The non-technical German equivalents for these terms (Erbfaktoren, Erbanlage, Erbmasse, Erbschatz) are employed once or twice each. The same abstinency pervades the paper of Spemann & Schotté (1932) describing reactions triggered by a foreign archenteron in the ectoderm of amphibian chimaeras – reactions which most strikingly testify to the role of the genotype in development but are here soberly labelled as 'herkunftsgemäß' (i.e. 'adequate to origin').

This reluctance to employ genetic concepts did not follow from failure to notice that developing cells react according to the prescriptions contained in their genome. Clearly (albeit casually) Spemann states in two places that when forming lens or mouth structures the induced ectoderm acts 'gemäß dem Erbschatz der Art' (Spemann, 1936, pp. 231, 237); even before the experiment was made he hinted at its potential for analysing 'die Art der Aktivierung der Erbfaktoren' (Spemann, 1924, p. 292). If nonetheless he refrained in his book from any general discussion of the role which genes could play in development, that might be due to his fear of following Weismann's lead into the territory of speculation, or to a tendency to maintain independence from the thoughts of others, especially Morgan and his school, Wilson[4] and Goldschmidt[5]; in addition, the 'atomistic' character of the gene theory could well have been uncongenial to him (Hamburger, 1980).

Whatever motivations may have caused Spemann's reluctance to stress the role of the genes, there was in his day good sense in doubting that gene activity can account for the whole of embryogenesis, as believed by many a geneticist. Identifying a gene required mutations, but most mutations recognised in the first half of our century did not alter the basic traits of the body which originate in early embryogenesis. Claims that early embryogenesis must be directed by genes were therefore based on conviction rather than scientific proof, especially since even Theodor Boveri, in his classical paper providing 'the first exact insights into the role of the nucleus in ontogenesis', entrusted the ooplasm with providing 'the most general basic form, the frame in which then all specific traits are being filled in by the nucleus' (Boveri, 1902, pp. 82, 83, translations by the present author). It was but 'normal science' (Kuhn, 1962) when most contemporary embryologists refused to share this conviction and instead believed that the 'embryo in the rough' (Loeb, 1919) arises from properties in the egg cytoplasm independent of the Mendelian factors; the latter were held responsible only for subordinate traits such as pigmentation or bristles.[6] It is worthwhile remembering that the single mutant then known to modify early development – the sinistrally coiling mutant of *Lymnea peregra* (Sturtevant, 1923) – left the basic construction of the animal unaltered, changing only its chirality. Thus the relevance of this mutant for the embryologist's key problem, namely the generation of pattern, remained open to debate (and creed).

One further handicap delaying full recognition of the role which genes play in development was the virtual absence of any knowledge on gene regulation. This absence is neatly documented in the questions which Morgan raised (and essentially left unanswered) in 1935:

If as is generally implied in genetic work (although not often explicitly stated), all of the genes are active all the time, and if the characters of the individual are determined by the genes, then why are not all the cells of the body exactly alike? . . . At every division of the egg, the chromosomes split lengthwise into exactly equivalent halves. Every cell comes to contain the same kind of genes. Why then is it that some cells become muscle cells, some nerve cells, and others remain reproductive cells?

Before being willing to spend time and thoughts on these problems, geneticists and their views were of little use to the embryologist.[7]

Among those who did spend time and thoughts (and did publish the latter!), Richard Goldschmidt was the outstanding figure. In his *Physiologische Theorie der Vererbung* (1927), he is reminiscent of Weismann in his inspirations as well as his errors. We mention only two of his tenets which are relevant for our subsequent discussions: his scheme of 'substrate-induced' differential gene activation (Fig. 2), and his speculation that the nurse cells are responsible for polarity and/or gradients in the insect egg cell (Goldschmidt, 1927, pp. 107–8). In one of his programatic statements, Goldschmidt voiced the feeling of the *avant garde* of his period: 'That they (the genes) are present and acting has been shown by the theory of (Mendelian) factors. How they act, is the next step to be cleared on the way towards a theory of heredity' (Goldschmidt, 1927, p. 8; translation by the present author). However, such advances by geneticists towards the domain of development were viewed with mixed feelings from the embryological side: (The geneticists') 'previous progress has been amazing, and it is not from a feeling of futile labours but rather from being aware of their paramount powers of appropriation that geneticists now are on the look-out for new connexions. They have cast their eye on us, on Entwicklungsmechanik...' (Spemann, 1924, p. 293; translation by the present author). How strikingly this sentence presages the feelings of many an embryologist forty to fifty years later, when molecular biologists set out to solve all the problems of ontogenesis!

Aloof from these hagglings stood Edmund B. Wilson and his 'Cell'. Its first two editions (1896, 1900) antedated the schism and, if heeded by the opponent parties, might have suppressed it from the beginnings. After reviewing and weighing the relevant publications, those of Driesch, Oscar Hertwig, Boveri and many others besides his own, Wilson writes (1896):

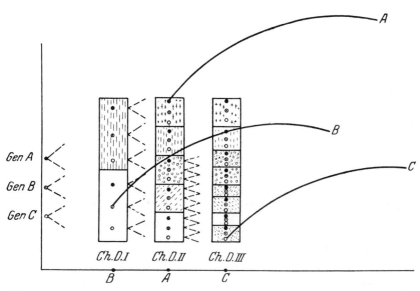

Fig. 2. 'Activation of genes' in various regions of the oocyte as envisioned by Richard Goldschmidt in 1927. The genes are thought of as catalysts being present and ready to act at any time. 'Activation' ensues when an appropriate substrate makes its appearance which enables the catalyst to function. If the oocyte were to contain the substrate for gene *B* in its lower half only (as in box *Ch. D. I*), gene *B* could act only there (see curve probably symbolising the concentration of the reaction product over time). Interaction within the oocyte between gene-catalysed products and/or the influx of additional substances from auxiliary cells will provide further substrates and thus permit the action of further genes (symbolised by *A* and *C*). The result is visualised as an increasingly complex stratified pattern. Note that although genes are involved the cytoplasm is assigned a crucial role in differentiation. From Goldschmidt (1927).

'... we reach the following conception. The primary determining cause of development lies in the nucleus, which operates by setting up a continuous series of specific metabolic changes in the cytoplasm. This process begins during ovarian growth, establishing the external form of the egg, its primary polarity, and the distribution of substances within it. The cytoplasmic differentiations thus set up form as it were a framework within which the subsequent operations take place, in a more or less fixed course, and which itself becomes ever more complex as development goes forward' (p. 320).

'In accepting this view we admit that the cytoplasm of the egg is, in a measure, the substratum of inheritance, but it is so only by virtue of its relation to the nucleus, which is, so to speak, the ultimate court of appeal. The nucleus cannot operate without a cytoplasmic field in which its peculiar powers may come into play; but this field is created and moulded by itself' (i.e. the nucleus) (p. 327).

Thirty years later, in the third edition (1928), Wilson puts it all in a nutshell: 'Heredity is effected by the transmission of a nuclear preformation which in the course of development finds expression in a process of cytoplasmic epigenesis' (p. 1112). If European and particularly German embryologists in the first half of this century are ever to be blamed for a disservice to their discipline, it is for their general failure to introduce their students to Wilson's magnificent book. Was its outlook too broad or its style too conciliatory?

# (3) Early lines of approach to gene action in insect development: the sequential aspect

Insects among all higher animals provide the most readily available material for genetic studies, and their early embryos are much more accessible than those of mammals, the 'runners-up' in view of genetic knowledge. The developmental genetics of birds and mammals is of greater practical importance and can boast many historical highlights, but insects seem to have played a dominating role in the evolution of basic concepts of gene action in ontogenesis. This judgement is apt to be biased by the author's long-standing interest in insect development, but he takes this bias as an excuse for restricting the subsequent sections mainly to work carried out on insects. How the relative merits and chances of studies on insects and vertebrates were judged at the end of the 'schism' can be gathered from the reviews of Kühn (1952) and Hadorn (1952).

All papers on ontogenetic gene action in insects published in the early decades of our century share one feature: it is the abstinency from any direct attack on what was considered the 'essence of development' by visionaries like Driesch and Goldschmidt, namely the generation of ordered spatial patterns or, stated differently, the increase in ordered spatial complexity. Rather than tackling gene expression coordinated in two or three dimensions of space, as required for spatial patterning, the investigators of that period were satisfied with studying one-dimensional or linear problems. Their topics were the distribution of different gene activities on the time scale of development, and the sequence of gene effects required for the expression of a specific trait. These problems will be considered in this section and the next, respectively.

Timing in ontogenesis was made a topic by Valentin Haecker (1912) in a primarily descriptive approach which he termed 'phenogenetics' (Haecker, 1918). Haecker set out from the visible differences between heritable traits or variants of the same phenotypical character, and proposed to trace back both traits from their fully expressed state to that earlier stage in development when they cannot be distinguished as yet. This approach was intended to uncover the apparent bifurcation point or 'phenocritical phase' where the two variants visibly start diverging in their development. Haecker proposed working back from there with refined methods towards the 'real' bifurcation points, i.e. the first physiological or molecular differences between the variants which might well occur as far back as gametogenesis. The latter part of the program was never carried out by him or under his influence but the terms 'phenogenetics' and 'phenocritical phase', and the idea of bifurcation have persisted in some way or other, even to Conrad H. Waddington's 'epigenetic landscape' (Waddington, 1940). Haecker's comments on contemporary phenogenetic efforts in Morgan-dominated America (Haecker, 1926) make interesting reading. They much resemble Spemann's attitude of 'We were here first', voiced a year earlier in his paper on 'Vererbung und Entwicklungsmechanik' (see section 2). To quote Haecker:

'In recent years some American geneticists, too, have taken interest in the links between development and inheritance. This is a border field which by us has been tilled for quite some time . . . The American researchers have not taken notice of this'.

Haecker then discusses books of H. E. Walter (1922) and T. H. Morgan, C. B. Bridges & A. H. Sturtevant (1925), and an article by O. Riddle (1924) dealing with the 'no-mans-land of somatogenesis' (Walter's term for phenogenetics). But in closing he returns to his introductory theme:

In the main camps of research on heredity, where at present nearly everyone strives at extending Mendelian analysis, the developmental foundations of heredity are held in esteem by very few researchers indeed, and Riddle has acquired great merits simply by pointing out this shortcoming, especially in America where breeding analysis and genetical cytology are flowering to such an extent'.                    (Translations by the present author).

The first approach to ontogenetic gene action in *Drosophila* was to establish 'effective' or sensitive periods during which a controlled environmental factor could alter the ultimate phenotype. In other insects, the influence of environmental factors such as temperature on externally visible traits had been analysed already before the advent of genetics, as a rule under the aspect of evolutionary mechanisms (Weismann, 1875, 1894); for instance, Merrifield (as quoted by Driver, 1927) had established by 1891 that in moth pupae the 'temperature effective period' for geometrical markings of the adult occurs earlier than the period for adult colouration.

In *Drosophila*, it was various mutants rather than the wild type that were at first subjected to modifying treatment. Driver (1931) lists a number of earlier papers establishing periods when the expression of mutant characters could be altered by external factors, and he advertises such studies as an approach towards 'the way in which the gene leads to the production of the character'. However, some pitfalls of this approach soon became apparent. For instance, Curt Stern showed in 1927 that low temperature exerts its 'normalising' effect on the curled wings of the mutant *gaunty* much later than the wild type gene could possibly do: it delays hardening of the cuticle and thereby provides the newly ecclosed fly with the time it requires for straightening its visibly crooked abnormal wing rudiments. Nonetheless, such temperature-sensitive mutants (as they are called today) when interpreted judiciously are valuable tools for research on gene action in development, and they are quite invaluable for maintaining conditionally lethal stocks in the laboratory (see reviews by Suzuki, 1970; Suzuki *et al.*, 1976).

Inducing developmental modifications in wild type *Drosophila* became fashionable a few years later and reached its climax in a paper of Karl Henke and his collaborators (Henke, Fink & Ma, 1941) (Fig. 3). One lasting result of this approach was a class of modifications apparently mimicking the abnormal phenotypes of certain mutants, called 'phenocopies' by Goldschmidt (1935). Several phenocopies are evident in Fig. 3, for instance modification no. 18 which mimics weakly expressed *bithorax* or *tetraptera* alleles. Note that the sensitive phase for this modification occurs during early embryogenesis, about the time of blastoderm formation, a fact which was confirmed some years later with ether as the phenocopying agent (Gloor, 1947). Henke *et al.* (1941, pp. 285–6) concluded that the heat pulse which produces modification no. 18 might interfere with the initial events differentiating between the 2nd and the 3rd segments of the thorax, and that the mutation copied by the heat pulse might exert a single detrimental effect very early in development (see section 6). In other

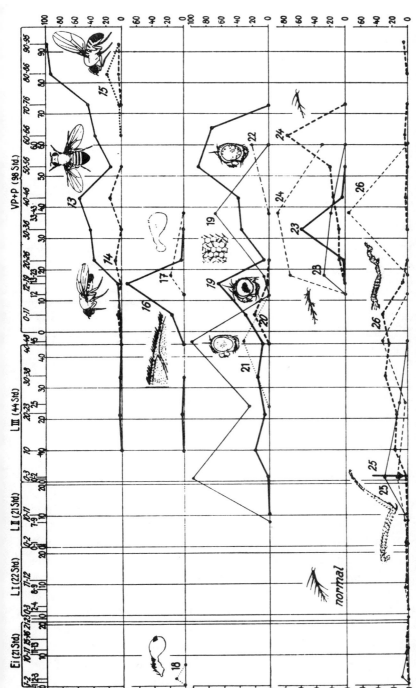

Fig. 3. The phenocritical phases or sensitive periods for various modifications of external morphology in the fruitfly *Drosophila*, established by Karl Henke and his collaborators using pulses of elevated temperature. The curves indicate percentages of animals affected, serial numbers identify the different modifications. Note the early sensitive period for modification no. 18. This modification is characterised by wing bristles appearing on the haltere (phenocopy of a *bithorax* mutation, see Fig. 7). From Henke *et al.* (1941).

modifications the sensitive period was found to occur later than the phenocritical phase of the mutant copied, and therefore the external stimulus could not possibly have copied the first step altering development in the mutant (Henke *et al.*, 1941).

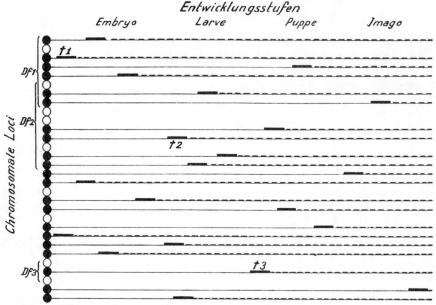

Fig. 4. The stepwise onset of gene action in ontogenesis, as illustrated by Ernst Hadorn in 1949 on the basis of lethal factors. The discs at the left rim mark gene loci on a chromosome, black discs indicating that the respective gene function is vital, i.e. indispensible for development. Elimination of part of the loci by a deficiency (*Df* 1, 2 or 3) causes a lethal crisis (+) at the time when the earliest 'vital' gene deleted (or its products) should have started acting. From Hadorn (1948).

A third approach along similar lines of thought in the 1930s was based on the timing of developmental crises in individuals carrying a lethal factor. It yielded complementary results (Fig. 4) but derived additional incentives from cytogenetical data, particularly from the giant chromosomes identified about that time. Analysing the effects of small overlapping deletions, Donald F. Poulson (1937, 1943) demonstrated that the absence of the *Notch* (*facet*) locus on the X-chromosome of *Drosophila* is followed by a specific anomaly during early embryogenesis. *Notch* embryos convert an excessive fraction of the ectoderm into neural tissue, at the expense of the epidermis (Campos-Ortega, 1983). Deficiencies in another region of the X-chromosome (the *white* region) led to internal anomalies but the epidermis (and most other parts) developed to the hatching level. Poulson (1943) concluded from his studies that different genes on the X-chromosome have quite different and often clearly separable functions in development. As a test for this view, he suggested that embryos of specific mutants should be supplied with the substances they might be lacking due to their genetic defects. This suggestion was not followed up until some decades later when injections of egg cytoplasm (Garen & Gehring, 1972) or of defined molecules (pyrimidine nucleosides; Okada, Kleinman & Schneiderman, 1974) were shown to 'rescue' lethally deficient embryos derived from mutant mothers. However, comparable experiments 'restoring' wild type colouration in eye colour mutants were under way already at the time of Poulson's work (see next section).

The essence of the studies reported in the present section is that genes at different loci start acting at different stages of development – as envisioned earlier by Goldschmidt (Fig. 2). The puffing patterns of giant chromosomes provided visible proof a few years later, once puffs had been recognised as signs of primary gene activity (Beermann, 1952, 1961; Becker, 1959). They also served to establish the earliest links between hormonal control of metamorphosis and gene activation (Clever & Karlson, 1960), a topic beyond the scope of this article.

## (4) Gene-dependent reaction chains and the development of insect pigments

The first biochemical approach linking gene action and ontogenesis, yet again in linear rather than spatial fashion, was initiated by studies on gynanders or gynandromorphs. Such individuals represent a mosaic of male and female territories. They were found in *Drosophila* soon after this animal had been 'domesticated' in the first decade of our century. In a monumental study, Morgan and his group (1919) showed that in gynandromorphs 'male and female parts and their sex-linked characters are strictly self-determining, each developing according to its own aspiration. No matter how large or how small a region may be, it is not interfered with by the aspirations of its neighbors, nor is it overruled by the action of gonads' (Morgan & Bridges, 1919).

However, Alfred H. Sturtevant (1920) observed that the light colouration of the *vermilion* (v) eye was changed towards the wild type in gynanders which, in some other organs, were carrying the wild type allele of *vermilion*. A similar effect was observed by Ernst Caspari after transplanting wild-type gonads into caterpillars of an eye colour mutant (named *a*) in the flour moth *Ephestia* (Kühn, 1932; Caspari, 1933). The apparently diffusible substances responsible for these effects were called 'gene hormones' by Alfred Kühn (Kühn, Caspari & Plagge, 1935). Parallel efforts to analyse these substances and their effects further were made in the laboratories of Beadle, Ephrussi, and Kühn, respectively. Their investigations, which were reviewed several times and from different standpoints (Ephrussi, 1942; Butenandt, 1952; Beadle, 1958; Tazima, 1964), provided the basis for the concept of gene reaction chains (Genwirkketten). The first gene-dependent product required for the brown eye pigments (ommochromes) to form was identified as kynurenine (Butenandt, Weidel & Becker, 1940), and this turned out to be a chromogen (= structural precursor of the pigment molecule) rather than a hormone or biocatalyst (Kikkawa, 1941). In vertebrates, kynurenine is a product of the enzymatic degradation of tryptophan. By analogy (and because tryptophan accumulates in the $a/a$ mutant of the flour moth), the *vermilion*$^+$ gene (= $a^+$ gene of *Ephestia*) was credited with providing the enzyme for that step in insects, too (Fig. 6). Relations between, in modern terminology, genes and enzymes had been a topic of speculation since the time when Driesch (1894) suspected the nucleus to consist of ferments; in later years there was much debate as to whether the gene could be a catalyst (or producer of catalysts) and at the same time an autocatalyst (see e.g. Morgan, 1926: Goldschmidt, 1928). However, it was insect pigment synthesis (and comparable pathways discovered in *Neurospora*; Beadle, 1958) that convinced the majority of biologists of the 'one gene – one enzyme' hypothesis. More importantly in our context, insect pigment synthesis was considered by some developmental biologists (e.g. Kühn, 1955) as *the* model for gene action in development, since it is stage-specific and thus closely interwoven with ontogenesis.

Moving back to embryonic stages, it is worth noting that eye pigment precursors synthesised under control of the maternal genome can pile up in the egg cell during oogenesis, and may subsequently be used for pigment synthesis in the embryo and even in the early larval stages. This 'maternal inheritance', discovered by Toyama (1912) in silkworm mutants, strengthened at first the cause of those who (following Driesch and Loeb) entrusted the ooplasm with a key role in embryonic development (see section 2). However, Tanaka (1924) recognised it as 'maternal determination', that is, gene expression delayed by one generation. A clear-cut example for such a 'maternal effect' (as it is called today) was provided by Kühn & Plagge (1937). They mated *Ephestia* females heterozygous for the allele *a* (see above) to homozygous *a/a* males. When the offspring hatched, the larval eyes were pigmented in all individuals, but 50% (the *a/a* individuals) lost this pigmentation after some moults. Apparently their maternal store of precursors for pigment synthesis was exhausted by then.

## (5) The prepattern concept: generation or mere expression of spatial complexity?

*Drosophila* mosaics and the flour moth, already introduced in the previous section, played a crucial role also in the first genetic attacks on 'one of the great problems of biology, that of differentiation' (Stern, 1954a); differentiation meaning spatial pattern formation in present-day terms. The wing patterns of various moths and the changes they suffer due to mutation or modifying treatments were a major topic of Goldschmidt's *Physiologische Theorie der Vererbung* (1927), but mainly under the heading of graded reaction speeds as a cause for their variation. Kühn and his collaborators studied in great detail the effects of the mutation *Sy* on the banding pattern in the flour moth wing (Kühn, 1936). However, as this gene-dependent effect could be phenocopied by fairly damaging treatments, the mutant was essentially interpreted as hampering a process of pattern generation defined in embryological rather than genetical terms ('streams of determination'). By contrast, the investigations on *Drosophila* mosaics led to a novel concept aspiring to define the role of genes in pattern generation, that is, in raising the ordered spatial complexity of the developing system. In fruit flies homozygous for the mutation *achaete*, certain large bristles on the 'thorax' (= meso- notum) appear in wrong positions or altogether fail to form. Studying these changes in the bristle pattern, Curt Stern (1954a) set out the hypothesis that they were due to 'different over-all organiz- ation of the specific strains or stresses set up in the folded embryonic tissue' which at metamorphosis could serve as local signals for bristle formation. The differences in bristle position between wild-type and *achaete* flies would thus be 'founded upon different patterns in the pre-differentiated tissues' of the larva (Stern, 1954*b*). But this assumption was overthrown by the analysis of gynanders which, due to an appropriate combination of genotypes, were also a patchwork of *achaete* and non-*achaete* (= wild type) areas. Whenever the spot where a certain bristle ought to appear was occupied by wild type cells (*ac*⁺), the bristle indeed did appear, no matter how much or how closely that spot might be surrounded by mutant epithelium (Fig. 5). From this, Stern concluded that the *achaete* phenotype follows from the failure of *ac*/*ac* cells to respond properly to local cues from a covert, pre-existing pattern not altered by the mutation. He introduced the term 'prepattern' to designate such patterns which, although themselves invisible, provide the localised cues for the individual elements of the visible pattern. At first he used the most general definition possible: 'A prepattern is a descriptive term for any kind of spatial differentness in development' (Stern, 1954*a*, pp. 367–8). In order to be transformed into visible patterns, prepatterns need the response of differentiating (in this case cuticle-forming) cells.

Fig. 5. A key result from which Curt Stern derived the prepattern concept. In these dorsal views of halves of the mesothorax from individual *Drosophila* mosaics the *achaete* patches are shown in black. Mutant epithelium surrounding the site of the posterior dorsocentral bristle (small circle) does not prevent the wild-type cells from forming the bristle. Thus, the mutant apparently does not alter the invisible localised cues triggering bristle formation but rather prevents the mutant cells from reacting adequately to these cues, which results in the mutant phenotype (bristles lacking or formed in wrong location). Adapted from Stern (1954*b*).

Stern's concept, although attracting much attention, failed to gain support from embryologists interested in pattern formation. The main reason for this rejection, in the view of the present writer, is the (secondary?) elaboration of the concept to mean that each element of the visible pattern owes its existence and location to a specific singularity in the initial prepattern. In this form, prepatterns imply a regressus ad infinitum (see e.g. Waddington, 1962; Sondhi, 1963) and provide no explanation whatsoever for the ontogenic increase in spatial complexity.[8] More recently, Lewis Wolpert paid full tribute to Stern (1969, p. 29) but nonetheless called the prepattern concept the antithesis of his own approach (Wolpert, 1971, p. 192) which has proved highly seminal since. The crucial difference, as pointed out by Wolpert, is that a terminal prepattern needs to embody a singularity for each pattern element it has to specify while the positional information concept envisions singularities in the response (interpretation) by the cells placed at the respective positions; it thus circumvents the regressus ad infinitum ascribed to the prepattern concept by most of its critics (see above).

The lack of embryological enthusiasm for this specific version might explain the curious fact[9] that Stern failed to mention the prepattern concept altogether in his contribution to *Analysis of Development*, a monumental synopsis edited by three outstanding embryologists (Willier, Weiss & Hamburger, 1955); instead he took recourse to embryonic fields, although he had drawn a clear distinction between prepatterns and fields the year before (Stern, 1954*a*, p. 367). This seems all the more vexing in view of the fact that Stern's original approach did indeed provide for an increase in spatial complexity, as is evident from his approvingly quoting Driesch:

Development starts with a few ordered manifoldnesses; but the manifoldnesses create, by interactions, new manifoldnesses, and these are able, by acting back upon the original ones, to provoke new differences and so on. With each effect immediately a new cause is provided and the possibility of new specific action. (Stern, 1954*b*, p. 224)

The same thoughts read, perhaps somewhat less succinctly, in Stern's own words as follows:

Development thus may be regarded as a sequence of prepatterns, each one being a realized pattern as compared to its predecessor, and a new prepattern as the basis for its successor. At any stage in the sequence, differential genic response to a prepattern will create differential prepatterns for the next stage. (Stern, 1954*a*, p. 368)

Whatever its exact meaning, this statement clearly implies that mutation will alter pattern generation when mutant cells respond incorrectly to the current prepattern while involved in generating the prepattern for the next stage. Clearcut examples for gene defects

altering early prepatterns are provided by the maternal effect mutants *bicaudal* (Bull, 1966) and *dicephalic* (Lohs-Schardin, 1982),
and by the zygotic *bithorax* mutants which in a given region of the
developing embryo divert all subsequent patterning (see next
section). However, many other patterning mutants apparently
change only the terminal response to the last prepattern, as shown by
Stern for *achaete*.

## (6) Genes and the generation of spatial patterns in insects – a present-day view

Trying to write history without the benefits of hindsight is a risky
undertaking at best, and one which may ruin friendships of long
standing when the deeds and thoughts of living persons are involved.
The subsequent sections should therefore be considered documents
(hardly historical) of this author's limited perception of his field of
research, rather than documents of the field's history.

For the embryologist, the major conceptual advance in the field of
gene action in ontogenesis during the last decade or two concerns his
age-old central riddle – the generation of ordered diversity or spatial
patterns. More specifically, the advance regards the interplay in this
process among genes, and between genes and extrakaryotic
components of the developing or regenerating system. To what
extent this advance resulted from cross-fertilisation between the
embryological and the genetical approach – the schism (section 2) is
still (and most naturally) felt in method if not in creed! – is difficult
to gather by merely reading the respective publications, with
(almost) everyone still quoting mainly from their own schools and
traditions; even interrogating those actively involved would probably not reveal the subconscious effects of ideas picked up here and
there. However, the outlook seems to have changed for the better in
recent years, with papers based on mutant analysis discussing cytoplasmic gradients and experimental embryologists invoking gene
regulation – although usually (and perhaps naturally) without reference to the origins of these alien concepts. Even more significantly,
however, integrative formal models have made their appearance
which treat genes as one of several interacting components in a
dynamic patterning system. This approach to ontogenesis is best
characterised by the term synergetics (Haken, 1977) which, however, is likely to be shunned by the pioneer workers in this field, as

they have set out on their quests under the flags of biomathematics, systems theory, or biocybernetics (see section 7).

As examples of present-day thinking, the subsequent paragraphs will discuss concepts for gene-dependent pattern formation in insects, based on clonal restriction and homeotic mutants, and outline two recent 'synergentic' models aspiring to account for pattern formation in insect development. These current concepts are based on, or adapted to, results from insect work and therefore are likely to be inadequate for other animal groups. But for the latter no concepts of comparable sophistication have come forth so far, especially with justified emphasis on gene action in pattern generation. Work on some other animal groups, e.g. sea-urchins, has greatly furthered present-day thinking on gene regulation during ontogenesis, albeit without much regard to spatial aspects. This work cannot further be discussed here, but as an outstanding example we mention the theoretical considerations of Davidson & Britten (1979) and the research that preceded and followed them (see also Davidson, 1976).

Homeotic mutations cause development locally to take the course germane to a different body region, thus producing a normal body part in an abnormal position. That such mutations are important for understanding gene action in development was clearly realised some decades ago (Goldschmidt, 1938; Hadorn, 1955; Kühn, 1955; Waddington, 1956). However, the conclusion that they pertain to a specific class of regulatory genes was apparently drawn only after 1970 when Antonio Garcia-Bellido coined the term 'selector gene' for this class (Garcia-Bellido, 1975). Selector genes are thought to select between alternative developmental pathways open to individual cell populations, by releasing and coordinating the specific activities of (probably) hundreds of other genes. This function would explain why the mutation of a single gene is capable of transforming the developmental fate of a whole cell population, to the effect that, for instance, part of a 3rd thoracic segment is replaced by homologous structures of the 2nd segment (mutant *bithorax*, see below), or part of the antenna by leg parts (mutant *antennapedia*). A modificatory (rather than mutational) switch from function to non-function in such genes would also provide a genetic basis for the long-standing riddle of heteromorphic regeneration (Loeb, 1891) and for the transdetermination effects (Hadorn, 1966, 1978).

The developmental pathways leading to different parts of the adult insect are taken to be specified not by a single selector gene each, but by the specific combination of several selector genes. A 'combinatorial binary code' specifying separate developmental pathways was first inferred by Stewart Kauffman (1973) (see section 7).

The selector gene concept was prompted by the discovery of 'clonal restriction'. Peter Lawrence (1973) and Garcia-Bellido, Ripoll & Morata (1973) had noted that the clonal descendants of an embryonic or larval insect epidermal cell are restricted in their possible contributions to further development. For instance, the descendants of a cell in the posterior part of a wing imaginal disc of *Drosophila* apparently cannot take part in constructing the anterior part of the wing blade, even if by a technical trick they are made to outnumber the descendants of any other cell in the disc (Fig. 6) or if the cells for the anterior wing part were missing altogether (Garcia-Bellido, 1975). The epidermis of a developing fly was consequently envisioned as a patchwork of discrete cell populations, each programmed to build a certain part of the body surface while being incapable of making contributions to any other part. These cell populations were named 'developmental compartments' by Garcia-Bellido (1975). In genetical terms, the member cells of a compartment are thought to share the same specific combination of active selector genes. The developmental restrictions ascribed to compartment boundaries (Fig. 6) came as a great surprise to experimental embryologists raised to believe in nearly unlimited powers of developmental 'regulation' (= functional replacement of lost cells by others); indeed, more recent data indicate that this restriction may at times be relaxed, for example during regeneration.

The most detailed current concept linking gene action to *embryonic* pattern generation is based on a genetic approach of the utmost thoroughness, carried on over decades by Edward B. Lewis at Pasadena (see Lewis, 1978). He studied a gene complex named after the *bithorax* locus whose earliest known mutants had been phenocopied by Henke and his co-workers (Fig. 3). As envisioned by these authors, the *bx*-locus is indeed involved in differentiating between the 2nd and 3rd segments of the thorax. Its functional failure due to mutation or the effects of heat or ether vapours, respectively, causes a part of the 3rd segment to develop as if it were a 2nd segment; it thus serves selector functions but Lewis does not use this term (which illustrates the problem of schools and terminology outlined in the second paragraph of this section). Adjacent parts of the complex have been found to differentiate between some more posterior

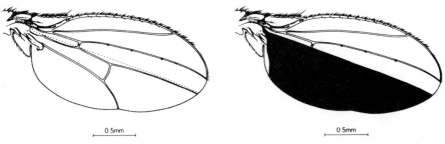

0 5mm   0 5mm

Fig. 6. Clonal restriction in the *Drosophila* wing. Left: normal wing to show wing venation. Right: the black wing area which contains several thousand cells was produced by the clonal descendants of a single cell altered genetically by X-rays during early development. The genetic alteration (based on somatic crossing-over) was twofold: it affected some cuticle-forming property of the descendant cells, thus visible marking the territory they contribute to the wing cuticle (shown black), and it made them divide much faster than the other cells in the wing Anlage (which carry a genetic defect slowing down their division rate). Note that despite its tremendous growth advantage, the mutant clone did not transgress the dashed line shown in the left figure. This line was therefore considered the border between anterior and posterior compartments. It would be similarly 'respected' by any other clone induced during early development or later on, whether posterior or anterior to the line; in the latter case, the line would restrict the clone posteriorly. From Crick & Lawrence (1975).

segments in a similar way, and it is a striking feature that (almost) all of these functions are lined up on the chromosome in the same spatial sequence as the respective segments in the body. Figure 7 outlines the hypothesis by which Lewis (1978) covers these findings. More recently, the molecular structure of the complex locus harbouring these genes has been analysed in considerable detail and with some surprising results (Bender *et al.*, 1983; Hogness *et al.*, 1984).

This mutant analysis of the *bithorax* complex no doubt represents the foremost front of research on gene action in ontogenetic pattern formation. However, the ooplasmic gradient invoked to coordinate this gene action (Fig. 7) is a conventional postulate rather than being based on compelling genetic evidence. On a general level (see Wolpert, this volume), the pioneering ideas of Theodor Boveri (1910) on graded differences in the ooplasm were followed by extensive physiological work (see Child, 1941), but it was apparently Leopold von Ubisch (1953) who first considered in detail the possible relationship between morphogenetic gradients and genes in pattern formation. In insects, gradients as a means of coordinating gene activities had first been proposed on experimental evidence from

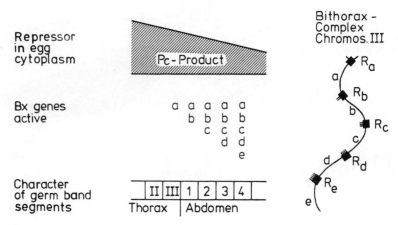

Fig. 7. Genetic determination of different body segments in *Drosophila*. This
synoptic diagram (from Sander, 1984*b*) illustrates the main features proposed by Ed
B. Lewis in 1978. The different genes a . . . e in the bithorax complex (right figure)
are controlled each by a regulatory element ($R_a$ . . . $R_e$, black squares). The affinity
of the regulatory elements to a common repressor (the *Pc* product) is graded, as
symbolised by the number of 'binding site' bars projecting from the black squares.
The repressor is distributed through (part of) the egg cell in gradient fashion,
decreasing from anterior to posterior (top figure). Repressor bound to a regulatory
element prevents the attached locus (a . . . e) from being active. The regulatory
element with the lowest affinity ($R_a$) is binding the repressor only at the highest
concentrations and consequently gene a is permitted to be active over a large region
of the embryo. Activity of the other genes is restricted to increasingly more posterior regions (middle figure) due to the increasing affinities of their regulatory
elements to the repressor; the element with the highest affinity (in this diagram $R_e$)
permits its gene to be active only at the lowest repressor concentrations, i.e. at the
posterior end of the embryo. The character of the individual germ-band segments
(bottom) is determined by the set of bithorax genes locally active (central panel).
After deletion of the entire bithorax complex, all segments shown assume the
character of segment II of the thorax (which does not require any of the genes a . . .
e). Absence of the *Pc* function, on the other hand, causes the same segments to look
like the most posterior segment since the lack of repressor permits all bithorax genes
to be active in all segments. Note that this is a strongly simplified representation (see
text, and Lewis, 1978).

lower insects (Sander, 1960, p. 703). The type of interaction which
led to this postulate was subsequently also found in higher insects
(Herth & Sander, 1973; Schubiger, Moseley & Wood, 1977) and the
gradient notion was also used for explaining the patterning
anomalies caused by maternal effect mutants in *Drosophila* (Nüsslein-Volhard, 1979). The earlier experimental data on pattern formation in insect embryos had been interpreted without any recourse

to gene action or to gradients directing it in space (Seidel, 1955, 1966).

However, *qualitative* differences in the egg cytoplasm which would account for 'certain genes being more affected in one region of the egg, other genes in other regions'[10] had been linked to gene activity much earlier: back in 1927 Goldschmidt had suggested that 'gene activation' would assist in generating a stratified pattern of organ-forming substances in the insect oocyte (Fig. 2). The pattern thus laid down was to include localised determinants for nearly each part of the future body. This mosaic-type view of the insect egg has since been shown to be untenable (see chronological review in Sander, 1984*a*), but Goldschmidt's general notion that the nurse cells determine oocyte polarity is supported strikingly by recent findings: mutant *Drosophila* follicles with a group of nurse cells on either pole instead of only the anterior pole produce anterior traits at both ends of the egg shell, and among the (rare) embryos developing in such eggs the majority are of the 'double-anterior' pattern (mutant *dicephalic*, Lohs-Schardin, 1982).

# (7) Systems theory or synergetics – a decisive advance towards understanding the role of genes in ontogenesis

That genes act in ontogenesis as part of a complex system had been recognised by nearly all writers after Weismann's time. However, many of the early (and later) authors fail to acknowledge expressly the complexity which such systems might imply, or the dynamics of interplay between their components; and those who paid tribute to the incredible degree of complexity involved in biological pattern formation took recourse to vitalism (Driesch, 1909) or to hinting at the complexity of mental functions as an analogy (Spemann, 1936). Consequently, most research was aimed at analysing individual components of the developing system to the greatest possible depth – the molecular structure of the gene or the biochemistry of the primary inducer, for example – but not at analysing the principles by which the system as a whole would perform its marvellous feats. The earliest author expressly taking a different stand appears to have been Ludwig von Bertalanffy (1933). Under the headline 'The System Theory' he wrote in his closing chapter:

The fundamental error of 'classical' mechanism lay in its application of the additive point of view to the interpretation of living organisms. It attempted to analyse the vital process into particular occurrences proceeding in single parts or mechanisms independently of one another. In Weismann's machine theory of development we encounter the classical example of this point of view. Vitalism, on the other hand, while being at one with the machine theory in analysing the vital processes into occurrences running along their separate lines, believed these to be co-ordinated by an immaterial, transcendent entelechy. Neither of these views is justified by the facts. We believe now that the solution of this antithesis in biology is to be sought in an *organismic* or *system theory* of the organism which, on the one hand, in opposition to machine theory, sees the essence of the organism in the harmony and co-ordination of the processes among one another, but, on the other hand, does not interpret this co-ordination as Vitalism does, by means of a mystical entelechy, but through the forces immanent in the living system itself.

This proclamation, however far-sighted, was apparently of little practical consequence until Norbert Wiener (1948) rang in the era of cybernetics and Alan Mathison Turing (1952) modelled the first physicochemical system capable of generating simple spatial patterns. Subsequent work, now aided by powerful computers, on systems which can generate spatial patterns relied mainly on *Hydra* as a model system, but it was insects again that required (and justified) the inclusion of genes among the indispensable components of patterning systems. We will briefly review the routes of approach taken by the pioneers in this field, Stewart Alain Kauffman and Hans Meinhardt. We shall shun the mathematical equations involved, but not without a contemplative comment. These equations – and thus the principles considered the very essence of pattern formation – are as immaterial as Driesch's entelechy: apparently we are incapable (as yet?) of grasping pattern generation without a heavy degree of abstraction. But, needless to add, the difference between these immaterial principles is that mathematical models provide a challenge for physicochemical investigations while entelechy by its very definition would not yield to this approach.

The model of Kauffman (Kauffman, 1973; Kauffman, Shymko & Trabert, 1978) is strongly influenced by data from imaginal discs, regional patches of embryonic epithelia persisting through the larval stages and giving rise, during metamorphosis, to adult structures mainly of the body surface. As adapted to early embryogenesis (Fig.

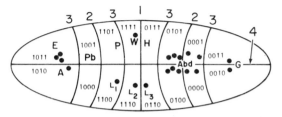

Fig. 8. Stewart Kauffman's model for the generation of compartment boundaries and of binary combinatorial specification of gene expression in the early *Drosophila* embryo. The letters in the figure refer to landmarks in the blastodermal fate map (shown as dots). The combinatorial code address (four digits) specifies different active gene combinations. The compartment boundaries arise in successive generations (marked 1,2,3,4). They correspond to the sites of nodal lines in spatial concentration patterns which can be described as steady-state concentration waves. The successive generations of nodal lines would arise if the 'wavelength' of oscillations in the system decreased during early embryogenesis, e.g. because diffusion constants increased with the increasing numbers of cell borders. At one side of a newly established nodal line, a correlated selector gene would assume the 'on' state (= 1), on the other side the 'off' state (= o); for example, the first digit is thought to differ in the code addresses on either side of the boundary marked 1, and the third digit on either side of the boundaries marked 3. From Kauffman *et al.* (1978).

8), the model is difficult to reconcile with some experimental data (see Sander, 1981) but for the sake of comparison with the embryological concepts of Lewis (Fig. 7) and Meinhardt (Fig. 9) we shall consider Kauffman's approach only at this level. The early embryo is thought to harbour a reaction–diffusion system which generates a sequence of differently shaped chemical patterns before segmentation is completed. Each pattern might represent a steady-state chemical wave, with successive patterns being characterised by successively shorter 'chemical wavelengths'. The geometry of the system (Fig. 8) would then cause the nodal line of the first pattern to appear at location No. 1, the nodal lines of the second pattern to appear at the locations marked 2, and so on. Gene action enters this picture by the provision that the first digit in a binary 'epigenetic code' would take on the value o on one side of the first nodal line, and the value 1 on the other. The second digit would differ likewise on the two sides of nodal lines 2, and so on. Thus, successive binary decisions concerning different selector genes would subdivide the embryo progressively into ever smaller cell populations (compartments) sharing the same developmental fate.

Meinhardt's approach to gene activity in pattern formation

follows quite different lines, and the few contradictions between his model and experimental results appear less fundamental than those implied in Kauffman's approach. Meinhardt assumes coupled chemical reactions to set up gradients of some morphogen(s) in the early embryo which serve to provide positional information (see Wolpert, this volume). He envisions a set of 'structure-determining genes' in which each gene is thought to specify the course of development leading to a single body segment. By a cunning scheme combining autocatalytic activation of these genes, their differential response to local morphogen levels, and mutual gene repression, the system arranges the different gene activities along a spatial axis in the sequence required for the segment patterns to appear (Fig. 9).

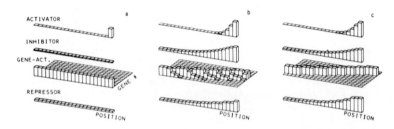

Fig. 9. Hans Meinhardt's model for the position-dependent activation of the members of a series of (selector) genes. A computer simulation of the system's state at three successive stages (a–c) is shown. Gene activity is indicated by columns in a grid combining spatial position (long axis) and serial number of gene (short axis). Interactions between two substances, the activator and the inhibitor, are capable of generating an inhibitor gradient (completed in c) which provides positional information. Increasing inhibitor levels trigger increasingly 'posterior' genes (see middle panel in b; the anteriormost gene is shown at the front). Mutual repression via a common repressor (bottom) provides for 'all or nothing' differences in the activity of a particular gene at adjacent positions (see middle panel in c) which would account for the abrupt transition in selector gene activity associated with compartment borders. From Meinhardt (1982).

It is safe to predict that even at the purely formal level both these models represent at best a rough approximation of what is really happening in gene-dependent pattern formation. However, they provide some impression of the dynamics required for patterning systems to function, particularly when the system is challenged with experimental perturbation (as done for a century by experimental embryologists), and therefore the heuristic value of these models for further research should be considerable. But even without this value they could justly claim a place in this essay, as the most recent episode in our evolving views on gene action in ontogenesis.

The author wishes to record here his debt to W. E. Ankel who first kindled his interest in the History of Biology, to the British Society for Developmental Biology, who invited him to talk on this topic, and to Tim Horder for asking stimulating questions. Ms Margrit Scherer and Gudrun Mahlke provided invaluable assistance in typing the manuscript(s), locating references and reproducing illustrations. I wish to thank them all.

# Endnotes

1. This was kindly related to me by his student Malcolm Steinberg, Princeton University.
2. Goldschmidt (1927, pp. 1–2) wrote on the reception of a preceding series of publications (Goldschmidt, 1917, 1920):

   However, it went as much as unnoticed by the 'inner circle' of geneticists, apart from a few laudable exceptions. There are several reasons for this... A contributory cause, finally, may have been the rise of a school of geneticists (especially in America) to whom biological knowledge apart from Mendelism did not deem necessary, whereby they were entirely content with knowing the work of the schools most closely akin to their own approach. (Translation by the present author)

3. These three terms all were coined by writers using the German language: Gen and Genotyp by Johannsen (1909) and Genom by Winkler (1920) (see Rieger, Michaelis & Green, 1968).
4. Wilson's book, which like Spemann's was dedicated to Theodor Boveri, was presented to Spemann by Dr and Mrs Parmenter of Philadelphia during his 1932 visit (note on the front page of his copy).
5. In his 1924 review, Spemann devoted much of his space to Goldschmidt's concepts.
6. As put for vertebrates by E. Conklin (1919): ... we are vertebrates because our mothers were vertebrates and produced eggs of the vertebrate pattern; but the colour of our skin and hair and eyes, our sex, stature and mental peculiarities were determined by the sperm as well as by the egg from which we came (i.e. by the Mendelian genes). Quoted from Sturtevant (1966). P. Hertwig (1934) quotes W. Johannsen as follows: 'Wir haben mit der Kreuzungsanalyse nur die Kleider geprüft, das Tieferliegende der Organisation ist nicht analysiert'. She ascribes this to the 3rd edition (1926) where however I could not locate it.
7. A relevant statement from Lillie (1927) is quoted in the fine book of Raff & Kaufman (1983, p. 21) published after completion of this review.
8. At the editor's request I recollect here my own reactions to Stern's ideas which became known in Germany at the time when I got interested in embryonic pattern formation (late 1950s). I was not impressed by the prepattern concept because if each pattern element had its antecedent in the prepattern, then pattern generation (the *increase* in ordered spatial diversity) remained unexplained, whereas if the prepattern was structurally simpler than the visible pattern, it seemed just another word for a morphogenetic field. My lack of

enthusiasm may partly also have been due to the fact that the bristle patterns studied by Stern were a far cry from the embryo where all patterning starts – and that was my concern. Developmental genetics at that time was very poorly represented in Germany, with Kühn retired and Henke no longer alive; even Mendelian genetics was poor in most universities, partly because having taught genetics made a professor suspect at the time of denazification; the examiner at my Diplom (= M.Sc.) interrogation, W. E. Ankel, was very pleased when I was able (in 1952!) to explain the X-linked transmission of the *white* gene. It was the stimulating thoughts of L. von Ubisch and C. H. Waddington rather than the painstaking and well documented publications of A. Kühn and K. Henke which finally directed my interests towards the more general topics of ontogenesis, particularly pattern formation, and towards the roles which genes might play in these.

9. This fact was kindly pointed out to me by Dr Philip Grant, Eugene/Oregon.
10. A phrase from Morgan (1935).

# References

Beadle, G. W. (1958). Genes and chemical reactions in *Neurospora*. In *Nobel Lectures in Molecular Biology 1933–1975*, with a foreword by D. Baltimore, pp. 51–65. New York: Elsevier.

Becker, H. J. (1959). Die Puffs der Speicheldrüsenchromosomen von *Drosophila melanogaster*. I. *Chromosoma* **10**, 654–78.

Beermann, W. (1952). Chromomerenkonstanz und spezifische Modifikationen der Chromosomenstruktur in der Entwicklung und Organdifferenzierung von *Chironomus tentans*. *Chromosoma* **5**, 139–98.

Beermann, W. (1961). Ein Balbiani-Ring als Locus einer Speicheldrüsen-Mutation. *Chromosoma* **12**, 1–25.

Bender, W., Akam, M., Karch, F., Beachy, P.A., Pfeifer, M., Spierer, P., Lewis, E. B. & Hogness, D. E. (1983). Molecular genetics of the bithorax complex in *Drosophila melanogaster*. *Science* **221**, 23–9.

Bertalanffy, L. von (1933). *Modern Theories of Development. An Introduction to Theoretical Biology*. Oxford: Oxford University Press.

Boveri, Th. (1901). Über die Polarität des Seeigel-Eies. *Verh. Phys.-Med. Ges. Würzburg*, NF **34**, 145–76.

Boveri, Th. (1902). Über mehrpolige Mitosen als Mittel zur Analyse des Zellkerns. *Verh. Phys.-Med. Ges. Würzburg*, NF **35**, 67–90.

Boveri, Th. (1910). Die Potenzen der *Ascaris*-Blastomeren bei abgeänderter Furchung. *Festschr. für R. Hertwig*, vol. **3**, Jena: Fischer.

Bull, A. L. (1966). *Bicaudal*, a genetic factor which affects the polarity of the embryo in *Drosophila melanogaster*. *J. Exp. Zool.* **161**, 221–41.

Butenandt, A. (1952). Über die Wirkungsweise der Erbfaktoren. *Endeavour* **11**, 188–92.

Butenandt, A., Weidel, W. & Becker, E. (1940). Kynurenin als Augenpigmentbildung auslösendes Agens bei Insekten. *Naturwissenschaften* **28**, 63–4.

Campos-Ortega, J. A. (1983). Topological specificity of phenotype expression of neurogenic mutations in *Drosophila*. *Wilh. Roux's Arch.* **192**, 317–26.

Caspari, E. (1933). Über die Wirkung eines pleiotropen Gens bei der Mehlmotte *Ephestia kühniella* Zeller. *Wilh. Roux' Arch.* **130**, 353–81.

Child, C. M. (1941). *Patterns and problems of development*. Chicago, Ill.: Chicago University Press 1941. pp. vii and 811.

Clever, U. & Karlson, P. (1960). Induktion von Puff-Veränderungen in den Speicheldrüsenchromosomen von *Chironomus tentans* durch Ecdyson. *Exp. Cell Res.* **20**, 623.

Crick, F. H. C. & Lawrence, P. A. (1975). Compartments and polyclones in insect development. *Science* **189**, 340–7.

Davidson, E. H. (1976). *Gene Activity in Early Development*, 2nd edn. New York: Academic Press.

Davidson, E. H. & Britten, R. J. (1979). Regulation of gene expression: possible role of repetitive sequences. *Science* **204**, 1052–9.

Driesch, H. (1892). Entwicklungsmechanische Studien. I. Der Werth der beiden ersten Furchungszellen in der Echinodermentwicklung. Experimentelle Erzeugung von Theil- und Doppelbildungen. *Z. Wiss. Zool.* **53**, 160–84.

Driesch, H. (1894). *Analytische Theorie der organischen Entwicklung*. Leipzig: Verlag Wilhelm Engelmann.

Driesch, H. (1909). *Philosophie des Organischen*. Gifford Vorlesungen, Aberdeen 1907–1908. Leipzig: Verlag Wilhelm Engelmann.

Driesch, H. & Morgan, T. H. (1895). Zur Analyse der ersten Entwicklungsstadien des Ctenophoreneies. *Wilh. Roux' Arch.* **2**, 204–26.

Driver, E. C. (1927). The temperature-effective period – the key to eye facet number in *Drosophila*. *J. Exp. Zool.* **46**, 317–32.

Driver, E. C. (1931). Temperature and gene expression in *Drosophila*. *J. Exp. Zool.* **59**, 1–28.

Ephrussi, B. (1942). Chemistry of 'eye color hormones' of *Drosophila*. *Quart. Rev. Biology.* **17**, 327–38.

Garcia-Bellido, A. (1975). Genetic control of wing disc development in *Drosophila*. In: *Cell Patterning, Ciba Foundation Symp* (New Series) **29**, 161–78. North-Holland: Elsevier.

Garcia-Bellido, A., Ripoll, P. & Morata, G. (1973). Developmental compartmentalisation of the wing disc of *Drosophila*. *Nature New Biol.* **245**, 251–3.

Garcia-Bellido, A., Lawrence, P. A. & Morata, G. (1979). Compartments in animal development. *Sci. Amer.* **241**, 102–10.

Garen, A. & Gehring, W. (1972). Repair of the lethal development defect in *deep orange* embryos of *Drosophila* by injection of normal egg cytoplasm. *Proc. Natl. Acad. Sci. USA* **69**, 2982–5.

Gloor, H. (1947). Phänokopie-Versuche mit Aether an *Drosophila*. *Rev. suisse Zool.* **54**, 637–713.

Goldschmidt, R. B. (1917). A preliminary report on some genetic experiments concerning evolution. *American Naturalist* **52**, 28–50 (1918).

Goldschmidt, R. B. (1920). Die quantitativen Grundlagen von Vererbung und Artbildung. *Vortr. und Aufsätze zur Entwicklungsmech. d.Org.* **24**, 1–163.

Goldschmidt, R. (1927). *Physiologische Theorie der Vererbung*. Berlin: Springer.

Goldschmidt, R. (1928). *Einführung in die Vererbungswissenschaft*. Berlin: Springer.

*K. Sander*

Goldschmidt, R. (1935). Gen und Außeneigenschaft (Untersuchungen an *Drosophila*). I. *Z. Abstammungslehre* **69**, 38–69.

Goldschmidt, R. (1938). *Physiological Genetics*. New York: McGraw-Hill.

Hadorn, E. (1948). Genetische und entwicklungsphysiologische Probleme der Insektenontogenese. *Folia Biotheoretica* **3**, 109–26.

Hadorn, E. (1952). Genetik und Entwicklungsphysiologie. *Naturwissenschaften*, **40**: 85–91.

Hadorn, E. (1955). *Letalfaktoren in ihrer Bedeutung für Erbpathologie und genphysiologie der Entwicklung*. Stuttgart: Georg Thieme Verlag.

Hadorn, E. (1966). Konstanz, Wechsel und Typus der Determination und Differenzierung in Zellen aus männlichen Genitalanlagen von *Drosophila melanogaster* nach Dauerkultur *in vivo*. *Dev. Biol.* **13**, 424–509.

Hadorn, E. (1978). Transdetermination. In *The Genetics and Biology of Drosophila* (ed. M. Ashburner & T. R. F. Wright), vol. 2c, pp. 556–617. London: Academic Press.

Haecker, V. (1912). Untersuchungen über Elementareigenschaften. *Zeitschr. indukt. Abst. Vererb.* **8**, 36–47.

Haecker, V. (1918). *Entwicklungsgeschichtliche Eigenschaftsanalyse (Phaenogenetik)*. Jena: G. Fischer Verlag.

Haecker, V. (1926). Phänogenetisch gerichtete Bestrebungen in Amerika. *Zeitschr. indukt. Abst. Vererb.* **41**, 232–8.

Haken, H. (1977). *Synergetics. An Introduction. Nonequilibrium Phase Transitions in Physics, Chemistry and Biology*. Berlin: Springer-Verlag.

Hamburger, V. (1980). Evolutionary theory in Germany: A comment. In *The Evolutionary Synthesis: Perspectives on the Unification of Biology* (ed. E. Mayr & W. B. Provine), pp. 303–8. Cambridge, Mass.: Harvard University Press.

Henke, K., v. Fink, E. & Ma, S.-Y. (1941). Über sensible Perioden für die Auslösung von Hitzemodifikationen bei *Drosophila* und die Beziehungen zwischen Modifikationen und Mutationen. *Z. Vererbungslehre* **70**, 267–316.

Herth, W. & Sander, K. (1973). Mode and timing of body pattern formation (regionalization) in the early embryonic development of cyclorrhaphic dipterans (*Protophormia, Drosophila*). *Wilh. Roux' Arch.* **172**, 1–27.

Hertwig, P. (1934). Probleme der Vererbungslehre. *Naturwissenschaften* **22**, 425–30.

Hogness, D. S., Beachy, P. A., Gavis, E., Harte, P., Lipshitz, H., Paro, R., Peattie, D. & Sakonju, S. (1984). Regulation and products of the bithorax complex in *Drosophila melanogaster*. *J. Embryol. Exp. Morphol.* **82** (Suppl.), 134.

Johannsen, W. (1926). *Elemente der exakten Erblichkeitslehre* (3 Aufl.). Jena: Gustav Fischer.

Kauffman, St. A. (1973). Control circuits for determination and transdetermination. Bistable control circuits like those in bacteriophage lambda may function in *Drosophila* development. *Science* **181**, 310–18.

Kauffman, St. A., Shymko, R. M. & Trabert, K. (1978). Control of sequential compartment formation in *Drosophila*. *Science* **199**, 259–70.

Kikkawa, H. (1941). Mechanism of pigment formation in *Bombyx* and *Drosophila*. *Genetics* **26**, 587–607.

Kühn, A. (1952). Entwicklung und Problematik der Genetik. Naturwissenschaften **40**: 65–9.

Kühn, A. (1932). Entwicklungsphysiologische Wirkungen einiger Gene von *Ephestia kühniella*. *Naturwissenschaften*. **20**, 974–7.

Kühn, A. (1936). Versuche über die Wirkungsweise der Erbanlagen. *Naturwissenschaften*. **24**, 1–10.

Kühn, A. (1955). *Vorlesungen über Entwicklungsphysiologie*. Heidelberg: Springer-Verlag.

Kühn, A., Caspari, E. & Plagge, E. (1935). Über hormonale Genwirkungen bei *Ephestia kühniella* Z. *Nachr. Ges. Wiss. Göttingen, Nachr. a. d. Biol. N.F.* **2**, 1.

Kühn, A. & Plagge, E. (1937). Prädetermination der Raupenaugenpigmentierung bei *Ephestia kühniella* Z. durch den Genotypus der Mutter und durch arteigene und artfremde Implantate. *Biol. Zbl.* **57**, 113–26.

Kuhn, Th.S. (1962). *The Structure of Scientific Revolutions*. University Press of Chicago.

Lawrence, P. A. (1973). A clonal analysis of segment development in *Oncopeltus* (Hemiptera). *J. Embryol. Exp. Morphol.* **30**, 681–99.

Lewis, E. B. (1978). A gene complex controlling segmentation in *Drosophila*. *Nature* **276**, 565–70.

Lillie, F. R. (1927). The gene and the ontogenetic process. *Science* **66**, 361–8.

Loeb, J. (1891). *Untersuchungen zur physiologischen Morphologie der Tiere. I. Über Heteromorphose*. Würzburg 1891: Georg Hertz, 80 pp.

Loeb, J. (1919). The physiological basis of polarity in regeneration. I. *J. Gen. Physiol.* **1**, 337–62.

Lohs-Schardin, M. (1982). *Dicephalic* – a *Drosophila* mutant affecting polarity in follicle organization and embryonic patterning. *Wilh. Roux's Arch.* **191**, 28–36.

Meinhardt, H. (1982). *Models of biological pattern formation*. London: Academic Press.

Morgan, T. H. (1926). Genetics and the physiology of development. *Amer. Naturalist* **60**, 489–515.

Morgan, T. H. (1932). The rise of genetics. *Science* **76**, 261–7 and 285–8.

Morgan, T. H. (1935). The relation of genetics to physiology and medicine (Nobel lecture). Stockholm: Kungl. Boktryckeriet, P.A. Norstedt & Söner.

Morgan, T. H. & Bridges, C. B. (1919). The origin of gynandromorphs. In *Contribution to the Genetics of Drosophila melanogaster*. Carnegie Inst. of Washington.

Morgan, T. H., Bridges, C. B. & Sturtevant, A. H. (1925). The Genetics of Drosophila. *Bibliographia genetica* **II**, pp. 1–162.

Nüsslein-Volhard, Ch. (1979). Maternal effect mutations that alter the spatial coordinates of the embryo of *Drosophila melanogaster*. In *Determinants of Spatial Organization*, ed. St. Subtelny & I. R. Konigsberg, 37th *Symp. Soc. Dev. Biol.* 185–211. New York & London: Academic Press.

Okada, M., Kleinman, I. A. & Schneiderman, H. A. (1974). Repair of a genetically-caused defect in oogenesis in *Drosophila melanogaster* by transplantation of cytoplasm from wild-type eggs and by injection of pyrimidine nucleosides. *Dev. Biol.* **37**, 55–62.

Poulson, D. F. (1937). Chromosomal deficiencies and the embryonic development of *Drosophila melanogaster*. *Proc. Natl. Acad. Sci. USA* **23**, 133–7.

Poulson, D. F. (1943). Induced chromosome deficiencies. *Yale Scient. Mag.* **17**: 3–5.

Raff, R. A. & Kaufman, T. C. (1983). *Embryos, Genes and Evolution. The Developmental–Genetic Basis of Evolutionary Change.* New York: Macmillan Publ. Co.

Riddle, O. (1924). Any hereditary character and the kinds of things we need to know about it. *Amer. Naturalist* **58**, 410–25.

Rieger, R., Michaelis, A. & Green, M. M. (1968). *A glossary of genetics and cytogenetics.* Berlin, Heidelberg, New York: Springer-Verlag.

Roux, W. (1923). Wilhelm Roux. In *Die Medizin der Gegenwart in Selbstdarstellungen*, pp. 141–206. Leipzig: Meiner.

Sander, K. (1960). Analyse des ooplasmatischen Reaktionssystems von *Euscelis plebejus* FALL. (Cicadina) durch Isolieren und Kombinieren von Keimteilen. II. *Wilh. Roux' Arch.* **151**, 660–707.

Sander, K. (1981). Pattern generation and pattern conservation in insect ontogenesis – problems, data and models. In *Progress in Developmental Biology*, ed. H. W. Sauer, *Fortschritte der Zoologie*, **26**, 101–19. Stuttgart: G. Fischer.

Sander, K. (1984*a*). Embryonic pattern formation in insects: basic concepts and their experimental foundations. In *Primers in Developmental Biology*, ed. G. M. Malacinsky, pp. 245–68.

Sander, K. (1984*b*). Extrakaryotic determinants, a link between oogenesis and embryonic pattern formation in insects. *Proceedings, XIX Annual Meeting of Arthropodan Embryological Society of Japan, Yamada, May 28–29, 1983.*

Schubiger, G., Moseley, R. C. & Wood, W. J. (1977). Interaction of different egg parts in determination of various body regions in *Drosophila melanogaster*. *Proc. Natl. Acad. Sci. USA* **74**, 2050–3.

Seidel, F. (1955). Geschichtliche Linien und Problematik der Entwicklungsphysiologie. *Naturwiss.* **42**, 275–86.

Seidel, F. (1966). Das Eisystem der Insekten und die Dynamik seiner Aktivierung. *Zool. Anz. Suppl.* **29**, 166–87.

Sondhi, K. C. (1963). The biological foundations of animal patterns. *Quart. Rev. Biol.* **38**, 289–327.

Spemann, H. (1924). Vererbung und Entwicklungsmechanik. *Z. Indukt. Abstammungs- und Vererbungslehre* **33**, 272–94.

Spemann, H. (1936). *Experimentelle Beiträge zu einer Theorie der Entwicklung.* Berlin: Springer.

Spemann, H. & Schotté, O. (1932). Über xenoplastische Transplantation als Mittel zur Analyse der embryonalen Induktion. *Naturwiss.* **20**, 463–7.

Stern, C. (1927). Der Einfluß der Temperatur auf die Ausbildung einer Flügelmutation bei *Drosophila melanogaster*. *Biol. Zbl.* **47**, 361–9.

Stern, C. (1954*a*). Two or three bristles. *Am. Scient.* **42**, 213–47.

Stern, C. (1954*b*). Genes and developmental patterns. *Caryologia* **6**, 355–69.

Sturtevant, A. H. (1920). The *vermilion* gene and gynandromorphism. *Proc. Soc. Exp. Biol. and Med.* **17**, 70–1.

Sturtevant, A. H. (1923). Inheritance of direction of coiling in *Limnaea*. *Science* **58**, 269–70.

Sturtevant, A. H. (1966). *A History of Genetics.* New York: Harper and Row.

Suzuki, D. T. (1970). Temperature-sensitive mutations in *Drosophila melanogaster*. *Science* **170**, 695–706.

Suzuki, D. T., Kaufman, T., Falk, D. & the U.B. C. *Drosophila* research group (1976). Conditionally expressed mutations in *Drosophila melanogaster*. In *The Genetics and Biology of Drosophila*, ed. M. Ashburner & E. Novitski, vol. **1a**, pp. 208–63. London: Academic Press.

Tanaka, Y. (1924). Maternal inheritance in *Bombyx mori* L. *Genetics* **9**, 479–86.

Tazima, Y. (1964). *The Genetics of the Silkworm*. London: Logos Press.

Toyama, K. (1912). On certain characteristics of the silkworm which are apparently non-Mendelian. *Biol. Zbl.* **32**, 593–607.

Turing, A. (1952). The chemical basis of morphogenesis. *Phil. Trans. Royal Soc. B.* **237**, 37–72.

von Ubisch, L. (1953). *Entwicklungsprobleme*. Jena: Gustav Fischer Verlag.

Waddington, C. H. (1940). *Organizers and Genes*. Cambridge: Cambridge University Press.

Waddington, C. H. (1956). *Principles of Embryology*. London: George Allen and Unwin. New York: Macmillan.

Waddington, C. H. (1962). *New Patterns in Genetics and Development*. New York: Columbia University Press.

Walter, H. E. (1922). *Genetics*. New York: Macmillan.

Weismann, A. (1875). Studien zur Deszendenztheorie I. Über den Saisondimorphismus der Schmetterlinge. Leipzig: Verlag Wilhelm Engelmann.

Weismann, A. (1883). *Über die Vererbung*, Jena: Gustav Fischer Verlag. (Reprinted 1892 in *Aufsätze über die Vererbung und verwandte biologische Fragen*. Jena: Gustav Fischer Verlag.)

Weismann, A. (1892). *Das Keimplasma. Eine Theorie der Vererbung*. Jena: Gustav Fischer Verlag. (English edition: Roennefeldt, London 1893).

Weismann, A. (1894). *Äussere Einflüsse als Entwicklungsreize*. Jena: Verlag G. Fischer.

Wiener, N. (1948). *Cybernetics*. New York: John Wiley.

Willier, B. H., Weiss, P. A. & Hamburger, V. (1955). Analysis of development. Philadelphia: W. B. Saunders.

Wilson, E. B. (1896/1900). *The Cell in Development and Inheritance*, 1st and 2nd edn. New York: The Macmillan Company.

Wilson, E. B. (1925, reprint 1928). *The Cell in Development and Heredity*, 3rd edn. New York: The Macmillan Company.

Wolpert, L. (1969). Positional information and the spatial pattern of cellular differentiation. *J. Theor. Biol.* **25**, 1–47.

Wolpert, L. (1971). Positional information and pattern formation. *Curr. Top. Dev. Biol.* **6**, 183–224.

# Genome function in sea-urchin embryos: fundamental insights of Th. Boveri reflected in recent molecular discoveries

ERIC H. DAVIDSON

Division of Biology, California Institute of Technology, Pasadena, CA 91125, USA

The central problem in the molecular biology of early development can be stated as the mechanism by which expression of the genome results in the creation of a differentiated embryo. This problem emerged about a century ago, as a consequence of the cellular chromosome theory of hereditary determinants. However, the role of the genome in embryogenesis remained an obscure and contentious issue for several decades. There was no general acceptance of the view that what we now call genes are located in the chromosomes of every cell, and pervasive disagreement persisted over the functions to be attributed to such genes. A long-standing and sometimes bitter controversy raged as well over the extent to which embryogenesis results from a progression of developmental events directed *de novo* by the embryo, or, on the other hand, is to be regarded as the expression of a preformed maternal program. The resolution of some of these confusions by Theodor Boveri, Edmund B. Wilson, and the school of cellular developmental biologists associated with them constituted a major advance. Their insights provided the orientation that modern developmental biologists assume, in their endeavours to understand the role of the genome in early development.

The year 1983 is the centennial of the date of appearance of the classic paper by Edouard Van Beneden that provided a specific and detailed account of the process by which the genome of an animal is formed through pronuclear fusion (Van Beneden, 1883). Van Beneden's drawings (see Fig. 1) demonstrate that the haploid chromosome sets of the two pronuclei in a fertilised *Ascaris* egg together constitute the diploid zygote nucleus, with two copies of each individual chromosome, and also show clearly the replication of

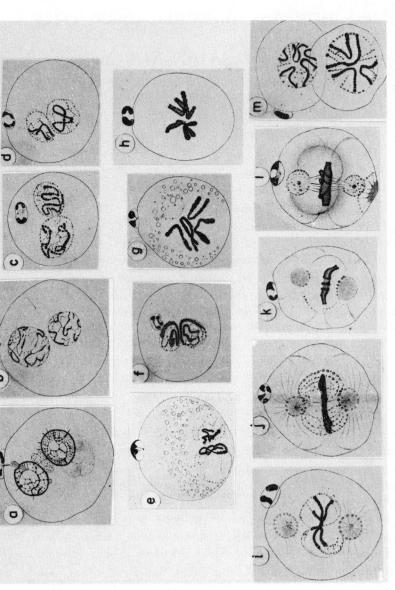

Fig. 1. Successive stages of pronuclear fusion and first cleavage mitosis in *Ascaris* as given by Van Beneden (1883). The chromosomes become visible while the pronuclei are still separate (a)–(e). As fusion occurs the four chromosomes remain clearly identifiable (f) and (g), and can be observed as the first cleavage metaphase plate forms and mitosis is carried out (h)–(m). (From E. Van Beneden (1883), *Arch. Biol.* 4, 265–640.)

the diploid set in the nuclei of the first two blastomeres. Within the next several years it was proposed explicitly in the writings of Weissmann, O. Hertwig, Nägeli, Strasburger and others that the fundamental genetic function of fertilisation is the formation of the genome of each organism from the gametic parental genomes, that the genomic determinants are located in the chromosomes, that these determinants control all the properties of the organism, and that the nuclei of all the somatic cells in the organism contain a complete set of genetic determinants because they contain a complete and equivalent copy of the chromosomes in the zygote nucleus. A balanced and justly famous retrospective account of these imposing theoretical achievements is given by Wilson (1925).

A flurry of more or less successfully conceived experiments on early embryos followed the proposal of the cellular chromosome theory. One achievement of these experiments was certainly the observation that the nuclei of early embryo blastomeres are equivalent and totipotent (Driesch, 1892; Wilson, 1896). However, by the turn of the century no clear consensus had appeared on the developmental function of genes, either in determining the properties of the embryo, or in directing the process of differentiation.

There were several sources of confusion. The most basic and general difficulty was that there were few convincing examples of the kinds of specific organismal properties actually controlled by individual genes, a problem that of course was alleviated only after the rediscovery of Mendelism and the subsequent development of experimental genetics. In addition, as Moore (1983) has described in an enlightening recent essay on the early development of T. H. Morgan's concepts of heredity, before 1910 it was not generally accepted that meiotic reduction provides the cytological basis for the genetic behaviour of Mendelian determinants. Thus the cellular chromosome theory of heredity remained a doubtful proposition, even after what now appears as the thoroughly convincing treatment of this problem by Sutton (1903) and Boveri (1904). With specific regard to the problem of gene action in early development, embryologists were impressed with species hybrid experiments, which tended to show that morphogenesis initially follows the course characteristic of the maternal species, and that paternal or hybrid characters can be discerned only in later embryos. An interpretation that was not easily dismissed was that the *general form* of the embryo is determined by the egg cytoplasm – 'the embryo in the rough,' as

Loeb (1916) put it – while the genes in the blastomere nuclei affect specific, detailed characters appearing only at advanced stages. For example the shape of the spicules in sea-urchin embryos had long been appreciated as a species-specific characteristic, and spicules displaying intermediate form had been noticed in sea-urchin species hybrids (for reviews, see Morgan, 1927, and Tennent, 1922). Even scientists such as Morgan, who later became an ardent proponent of the primacy of the gene in regard to all aspects of development, expressed serious doubts that embryogenesis is controlled by chromosomal determinants, rather than being the result of an epigenetic 'physico-chemical reaction' occurring in the egg cytoplasm. Morgan (1910) wrote:

I myself have felt the same disinclination to reduce the problem of development to the action of specific particles in chromosomes. In my own case this hesitation is due ... to a feeling that it is unsafe or unwise to reduce the problem of heredity and development to a single element in the cell, when we have every evidence that in embryonic development the responsive action of the cytoplasm is the real seat of the changes going on at this time ... The feeling against the view that ascribes everything to the chromosomes has been increased also by the assumption that unit characters in heredity are preformed, ... indivisible units (located) in the chromosomes.

The mind of Theodor Boveri provided a conceptual force that was crucial in organising the contemporary assemblage of inconclusive experiments and conflicting arguments. From his own researches, and perhaps most important, his generalised interpretations, derive the paradigms that underlie modern inquiries into the genomic basis of embryogenesis. Boveri concluded that genes control *all* the properties of the embryo, early and late, including the processes of embryonic differentiation. He realised that 'maternal' properties, such as egg polarity, are ultimately created as a result of the function of genes during oogenesis (Boveri, 1918). He showed that early development requires the activity of a complement of diverse genetic determinants in the embryo chromosomes, and explicitly pointed out that these determinants direct the differentiation of the specific structures of the embryo. A particularly significant paper was Boveri's 1902 study of the effects of multipolar mitosis in sea-urchin embryos (Boveri, 1902). Earlier experimentalists had discovered that dispermic eggs produce tetrapolar and tripolar mitotic figures, and that these eggs always fail to develop properly. Boveri's analysis demonstrated that in dispermic eggs the individual chromosome sets inherited by given blastomeress are aneuploid. Particular chromosomes are always missing or present in greater than diploid number in some blastomeres. Among other experiments, Boveri isolated blastomeres receiving such genomes; and demonstrated that the developmental potential of these cells is invariably incomplete, while in contrast properly organised, whole plutei could be obtained from isolated blastomeres of normal embryos. Boveri (1902) concluded as follows:

From our special case we arrive at the general significance of the findings described . . . We give to the cell a nucleus with some parts lacking and follow the effects of this defect. We have found that such a nucleus might suffice for some of the processes comprising the ontogenic events, but not for others, e.g. that it transmits competence for . . . formation of the gut, but does not transmit to the cells destined for the formation of the skeleton the qualities necessary for that, or vice versa. From this we must conclude that only a precise combination of chromosomes, probably only the totality of those which are contained within each pronucleus, represent the entire nature of the form of the organism . . . (in) the embryonic development of which unfold the qualities of the nuclei . . .

A quarter of a century later Wilson was still constrained to rely mainly on a detailed account of this very experiment for direct evidence that chromosomal genes are directly required for embryogenesis. Wilson was of course able to bolster this argument, by general reference to what was by then known of the nature of Mendelian determinants. However, in 1925 Boveri's pessimistic conclusion of 1902 still appeared justified:

The most important thing in the physiological constitution of the nucleus is absolutely inaccessible to an analysis by the present methods of physiological chemistry.

Here he refers to the nature of the specific chromosomal determinants expressed in developing embryos.

Methods have improved rather dramatically in the 80 years since Boveri wrote the above passage. The conclusions that grew from Boveri's conceptual struggles now occupy a central position. It is a remarkable fact, however, that not until recently has convincing molecular evidence that can be considered a demonstration of his ideas regarding gene function in early development been obtained. As late as 1965, there were only a collection of maternal developmental arrest mutants, a few genetic loci known to act in early embryos, some partially mysterious drug inhibition experiments, and a great many species hybrid and enucleation experiments, none analysed at the molecular level. The first experiments directly indicating the existence of maternal mRNA had just been carried out, by Brachet, Ficg & Tencer (1963) and Denny & Tyler (1964). However, there were still lacking both general molecular evidence and specific examples that could demonstrate the role of the genome in early development that Boveri had promulgated (for review see Davidson, 1968). By 1975 the state of such evidence was vastly improved (Davidson, 1976). Current knowledge provides many specific examples, and quantitative molecular analyses, that now support directly the school of thought to which Boveri's researches contributed so basically. In the remainder of this paper I shall summarise briefly some of the results of modern experimentation, with particular reference to the same material that provided Boveri with many of his prescient insights, the sea-urchin embryo.

We now know that at fertilisation the sea-urchin egg contains a hugely diverse store of proteins and RNAs resulting from the prior transcription of the oocyte genomes. Among these proteins are several known to be required specifically for developmental processes. These include both enzymes, such as RNA polymerase (for review see Davidson, 1976), and structural elements, examples of which are maternal actin, which is required for a variety of structural and contractile functions in the egg cortex (Vacquier, 1981), and the tubulins, which are utilised in embryo cilia and mitotic spindles (for review see Davidson 1976; Alexandraki & Ruderman, 1981). The egg also contains messenger RNAs representing at least $10^4$ transcription units, all expressed during oogenesis. These messages are largely destined for use in the embryo, and many of them are translated as well during oogenesis (Hough-Evans, Ernst, Britten & Davidson, 1979).

A quantitative result of general importance that has been surprising to many of us, is the finding that about 90% of the genes expressed in the embryo, even as late as the pluteus stage, are the *same genes* as were utilised during oogenesis in the preparation of the maternal mRNA. In a sense this is the modern denouement of the early twentieth century arguments about the role of the egg cytoplasm in determining 'maternal' characteristics of the embryo. However, neither classical considerations nor the many recent observations of specific gene activations in highly differentiated cells prepare one's intuition for the observation that the large majority of functioning genes, even in *late* embryos, are also expressed in the undifferentiated oocyte. Instead it seemed likely that as the biological complexity of the embryo increases during early development, so would the library of gene products being translated increase, and that later embryos would express largely genes that are developmentally activated. This is flatly wrong, for messages of all levels of prevalence in the embryo (Davidson, Hough-Evans & Britten, 1982; Mauron, Kedes, Hough-Evans & Davidson, 1982). There are clearly fundamental principles yet unknown, regarding the utilisation and assembly of complex sets of presynthesised gene products during development. The genomic transcription of genes coding for these gene products evidently begins to occur long prior to their utilisation in the differentiating embryo.

The 10% of embryo sequences that are activated during development provide dramatic evidence for specific gene functions required during embryogenesis. Among the best known cases are the genes for embryo histones. Active transcription of these genes after the 16-cell stage results in a rapid 10-fold increase in the concentration of histone mRNA in the embryo, from about $10^6$ to $10^7$ molecules of mRNA for each histone species per embryo. These messages are clearly required to provide histones during the period of rapid cell division (Davidson *et al.* 1982; Mauron *et al.* 1982). Furthermore, a different set of genes is needed to synthesise the histone variants that occur in later embryos. The actin genes provide another impressive example. Figure 2 displays direct evidence of ontogenic actin gene activation (Shott, R. J., Lee, J. J., Britten, R. J. & Davidson, E., unpublished observations). The autoradiograph demonstrates that the mRNA transcribed from this particular actin gene appears abruptly at about the 400-cell blastula stage. No detectable message deriving from this gene is stored in the egg. Among other specifically activated genes that are known at the molecular level are several which are expressed only in the ectoderm cells of the embryo (Bruskin *et al.* 1981, 1982). Thus molecular biologists are now rapidly providing the conceptual armature erected by Boveri with a body of detailed knowledge. It will not be long before there are isolated cloned elements of genomic DNA, and their transcripts, that will represent in material form those very genomic 'qualities necessary for the formation of the skeleton', etc., that Boveri imagined in the 1902 passage cited above.

E    16-c    B    G    LG    P

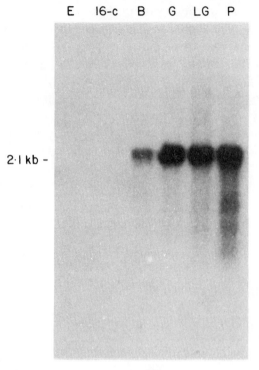

2·1 kb –

CyIIb

Fig. 2. RNA gel blots displaying appearance of actin mRNA in sea-urchin embryogenesis. Equal amounts of total cellular RNA were loaded in adjacent lanes of a gel, resolved according to size by electrophoresis, and blotted to a nitrocellulose filter. The filter was then hybridised with a cloned probe specific for a cytoskeletal actin gene called CyIIb that had been labelled with 32p. No transcript is detectable in RNA of eggs (E) or 16-cell cleavage embryos (16-c). However, the CyIIb transcript, which is 2.1 kb, is present at the 400-cell blastula stage (B), and continues to increase per unit amount of total RNA and per embryo, in gastrula (G), late gastrula (LG), and pluteus (P) stages.

A final remark concerns the analytical approach of modern developmental molecular biology. This is perhaps the most valuable legacy of the best late nineteenth century and early twentieth century cellular developmental biologists. Their prediliction for delving into the *genomic* basis for the developmental process led them inevitably toward a mechanistic concept of the way embryos differentiate. This conference has included some discussion of 'reductionist' as opposed to 'holistic' approaches to embryogenesis, a dichotomy that I regard as tangential or irrelevant. As I have tried to make clear,

modern molecular developmental biology is firmly in the
Boveri–Wilson tradition. In this connection, it is worth quoting
Wilson's (1925) response to Driesch's mystical concept of
'entelechy', but for my purposes applied to attitudes that would
favour less 'reductionist' approaches that would supposedly be
directed at the properties of the embryo as a whole:

This conception has failed wide acceptance partly because it is contrary to
the spirit of modern scientific inquiry, and involves a practical abandonment of the problem; but even more because of new discoveries. The most
important of these was Boveri's experimental demonstration that . . .
*normal development is strictly dependent on the integrity of nuclear organ-
ization* (his italics).

When we do fully understand the mechanisms by which genomic
activity results in three-dimensional morphological differentiation,
we shall indeed possess an *overall* conception of embryonic
epigenesis.

Research for this laboratory was supported by NIH Grant HD-05753.

# References

Alexandraki, D. & Ruderman, J. V. (1981). Sequence heterogeneity, multiplicity
and genomic organization of α-and β-tubulin genes in sea urchins. *Mol. Cell. Biol.*
**1**, 1125–37.

Boveri, T. (1902). Über Mehrpolige Mitosen als Mittel zur Analyse des Zellkerns.
*Verhandlungen der Physikalische – medizinischen Gesellschaft zu Würzburg* **35**,
67–90. Parts of this paper have been translated by Voeller, B. R. (1968). *The
Chromosome Theory of Inheritance*, p. 85. New York: Appleton-Century-Crofts;
and a detailed diagrammatic account is given by Wilson (1925), pp. 921–2.

Boveri, T. (1904). *Ergebnisse über die Konstitution der Chromatischen Substanz
des Zellkerns.* Jena: Verlag von Gustav Fischer.

Boveri, T. (1918). Zwei Fehlerquellen bei Merogonieversuchen und die Entwick-
lungsfähigkeit merogonischer und partiell-merogonischer Seeigelbastarde. *Arch.
Entwm.* **44**, 417–71.

Brachet, J., Ficq, A. & Tencer, R. (1963). Amino acid incorporation into proteins
of nucleate and anucleate fragments of sea urchin eggs: effect of parthenogenetic
activation. *Exp. Cell Res.* **32**, 168–70.

Bruskin, A. M., Tyner, A. L., Wells, D. E., Showman, R. M. & Klein, W. H.
(1981). Accumulation in embryogenesis of five mRNAs enriched in the ectoderm
of the sea urchin pluteus. *Dev. Biol.* **87**, 308–18.

Bruskin, A. M., Bedard, P.-A., Tyner, A. L., Showman, R. M., Brandhorst, B. P.
& Klein, W. H. (1982). A family of proteins accumulating in ectoderm of sea
urchin embryos specified by two related cDNA clones. *Dev. Biol.* **91**, 317–24.

Davidson, E. H. (1968). *Gene Activity in Early Development*, 1st ed. New York: Academic Press.

Davidson, E. H. (1976). *Gene Activity in Early Development*, 2nd ed. New York: Academic Press.

Davidson, E. H., Hough-Evans, B. R. & Britten, R. J. (1982). Molecular biology of the sea urchin embryo. *Science* **217**, 17–26.

Denny, P. C. & Tyler, A. (1964). Activation of protein biosynthesis in non-nucleate fragments of sea urchin eggs. *Biochem. Biophys. Res. Commun.* **14**, 245–9.

Driesch, H. (1892). Entwickelungsmechanisches. *Anat. Anz.* **7**, 584–6.

Hough-Evans, B. R., Ernst, S. G., Britten, R. J. & Davidson, E. H. (1979). RNA complexity of developing sea urchin oocytes. *Dev. Biol.* **69**, 225–36.

Loeb, J. (1916). *The Organism as a Whole*, p. 151. New York: G. P. Putnam's Sons. Others who took this view were Conklin, Jenkinson, and in earlier writings, Boveri. See Wilson (1925), p. 1108.

Mauron, A., Kedes, L., Hough-Evans, B. R. & Davidson, E. H. (1982). Accumulation of individual histone mRNAs during embryogenesis of the sea urchin *Strongylocentrotus purpuratus*. *Dev. Biol.* **94**, 425–34.

Moore, J. A. (1983). Thomas Hunt Morgan – the geneticist. In *Symposium on the Place of T. H. Morgan in American Biology*. *Am. Zool.* **23**, 855–65.

Morgan, T. H. (1910). Chromosomes and heredity. *Am. Nat.* **44**, 449–96.

Morgan, T. H. (1927). *Experimental Embryology*. New York: Columbia University Press.

Sutton, W. S. (1903). The chromosomes in heredity. *Biol. Bull.* **4**, 231–51.

Tennent, D. H. (1922). Studies on the hybridization of echinoids, *Cidaris tribuloides*. *Carnegie Inst. Wash. Publ.* **312**, 1–42.

Vacquier, V. D. (1981). Dynamic changes of the egg cortex. *Dev. Biol.* **84**, 1–26.

Van Beneden, E. (1883). Recherches sur la maturation de l'oeuf et la fécondation et la division cellulaire. *Arch. Biol.* **4**, 265–640.

Wilson, E. B. (1896). On cleavage and mosaic-work. *Arch. Entwm.* **3**, 19–26.

Wilson, E. B. (1925). *The Cell in Development and Heredity*, 3rd ed. New York: Macmillan.

# Reductionism and holism in biology

NEIL W. TENNANT

Department of Philosophy, University of Stirling, Scotland, UK

# I

The modern science of embryology is a fascinating one for the philosopher of science, lying as it does between the 'hard core' sciences, like physics or chemistry, and the 'softer' or less strongly predictive ones, like evolutionary biology and psychology. Embryology is a discipline which highlights various problems with which the philosopher of science is engaged:

- the problem of whether there is an ultimate level of reality (such as that described by fundamental particle physics) which – in a sense to be clarified – *determines* all other levels;
- the problem of whether scientific laws are deterministic and strongly predictive, or at best statistical in character;
- the problem of whether science is a mere sophisticated continuation of commonsense or whether it involves radical departures from our everyday modes of thought and conceptual scheme;
- the problem of whether some entities (such as electrons) must forever be regarded as theoretical or whether they might one day be rendered observational by progress in instrumentation and investigative techniques;
- the problem of whether observational vocabulary, even for middle-sized objects of moderate duration is 'theory laden', so that the observational/theoretical dichotomy is not licit;
- the problem of holism *vs* methodological individualism, i.e. whether complex systems may have emergent properties that cannot be predicted from the properties of their constituent parts; and finally, intimately connected with this problem of emergent properties,
- the problem of whether higher level theories, such as biology,

psychology and sociology, could, in principle, be *reduced* to a chosen lower level theory such as fundamental particle physics. The first problem on this list is known as the problem of physical determinationism; the last one is known as the problem of reductionism. I intend to examine embryology with these problems in mind, in the hope that I might engage the interest of embryologists in philosophical and methodological problems that are, unfortunately in my view, all too often the exclusive concern of professional philosophers of science.

My main focus will be on the first and last problems on the list. I shall explain the significance of a result in mathematical logic that bears importantly on the connection between these two problems. The result is Beth's Theorem on definability. It has the form of an implication

if A then B

in which it is arguable that A could be interpreted as the thesis of physical determinationism and B could be interpreted as the claim that all theories are reducible to physics.

Beth's Theorem is therefore of special significance for the anti-vitalist embryologist who wishes to maintain *logically* privileged autonomy for his discipline. If the interpretation that I have intimated of Beth's Theorem can be sustained, then the anti-vitalist embryologist would have to accept the consequence that his discipline is, *in principle*, reducible to physics.

Of course, whether or not it is reducible *in practice* is quite another question. Perhaps the *practical* impossibility of such reduction – despite its logical possibility – is sufficient to ensure the embryologist his theoretical autonomy and security of employment even under the harshest of selective regimes!

Although my examples and illustrations of general points are thus directed to embryologists, it is worth noting that what I have to say would apply, in principle, to any other 'higher-order' science whose relationship to physics is similarly in question. I shall share the prevailing assumption among modern scientists that if any theory has a claim to be describing the determining level, it is first and foremost physics. Curiously enough, the logical investigations to be described would apply even if the determination were – outrageously – the other way round: even if, say, a theory of social individuals and group minds were taken as describing the determining level, on the bizarre metaphysical conviction that the 'level of reality' whereof it spoke – namely collective unconsciousness, interanimation etc. – determined what was the case even at the level of fundamental particles.

# II

Embryologists can number one of the most accomplished methodologists and philosophers of science of modern times among their company. It was J. H. Woodger who translated the great works of the mathematical logician Alfred Tarski into English, and he was one of the first scientists to absorb the significance of Popper's falsificationist philosophy of the empirical sciences and communicate it effectively to his scientific colleagues. Woodger recommended in 1948 that embryologists engage in a program of conceptual and logical analysis and organisation of their data and theoretical hypotheses. This would require attention to the *language* of embryology:

Language is one tool which is common to all the sciences and without which no science would be possible. And yet very little attention is paid to it compared with the care and research which are lavished upon microscopes and all other scientific instruments . . . From time to time, of course, both in genetics and embryology, muddles of a linguistic origin have become sufficiently acute to demand attention . . . It is difficult to persuade anyone of this who has not felt it himself. It is like persuading a man who does not feel toothache to go to the dentist. The only science which has seriously studied its own language is mathematics and the outcome of these studies has important bearings on the other sciences.

Woodger complained that the data gathering of the day was badly in need of higher-order explanatory hypotheses. Higher-order hypotheses would explain lower-order generalisations by logically implying them (and also the data that they in turn implied); and they would be empirically testable by going beyond the data to make new predictions that were in principle falsifiable by observation and experiment. Embryology had an impressive record of observations of both normal and pathological growth. Much of the pathological growth is at the investigative instigation of the embryologist: grafting, transplantation between species, suturing, irradiating, amputating, homogenising. What end is to justify these means? Woodger's answer was: new explanatory hypotheses. These would, it was hoped and expected, open up new fields of (more humane?) experimentation and provide further impetus to *link* embryology with neighbouring disciplines – expecially biochemistry – and thus to consolidate and broaden its empirical range. But of especial interest, from the point of view of a mathematical logician, is that Woodger was at pains to emphasise how important it was to arrive at a clear understanding of language as a theoretical tool. He urged upon his fellow embryologists the task of classifying their observational and theoretical terms (or predicates) with a view to discovering which of these were basic and indefinable when formalising their data and laws and hypotheses. In this way the logical structure of the discipline would be better understood; and the newly clarified logical structure would suggest theoretical ways forward from the existing data base and current explanatory hypotheses.

It is easy to scoff at the utility – or futility – of pursuing such a program. It may look like nothing more than formalisation for formalisation's sake. A glance at the results of such attempts for theoretical physics itself (Suppes, McKinsey & Sugar, 1953; Suppes & Rubin, 1954; Sneed, 1971) should give the enterprising formalising biologist pause: for biology promises to offer greater complication in the project of formalisation than does physics. Even before its formalisation physics was already a highly mathematical discipline with but a few fundamental notions to be orchestrated in a set of axioms that would allow the derivation of all its laws. But on the other hand the results of such an investigation in the case of embryology could enable philosophical assessment of the claim, say, that modern gradient theory is just as 'metaphysical' and untestable as was Driesch's theory of entelechies. One can command a clear view of the matter and be in a position to answer such a philosophical charge, only by understanding the logical connections between theoretical talk of progress zones, thresholds, gradients and positional values, and observational results in the laboratory and in the wild.

The time may be ripe for a re-assessment of the recommendations Woodger made in 1948. His program was not prosecuted with the vigour it deserved: probably because there were not the skilled, trained analytical minds with sufficient interest in the subject to do the hard work he called for. One mild exception I have found to this claim is Mary Williams' work, aimed at formalising the theory of natural selection. It is wrong on important details: for example, she regards overproduction in a world of limited resources as logically *necessary* for evolution to take place. And it is written in an over-numerical idiom ill-suited to the description of the *qualitative* relationships at issue. But despite these drawbacks, it was interesting work, and ought to be improved upon to the benefit of both evolutionary biologists and philosophers of science.

I shall not offer, with regard to embryology, anything like a definitive 'formalisation' of its theoretical claims. There is enough variety in its terminology for one to be content, at this stage, with only the most tentative classification or categorisation. It is instructive to approach the writings of some leading contemporary embryologists with this task of logical classification in mind. The happy indications are that Woodger's promissory note is being fulfilled by his scientific successors. Wolpert's work (Wolpert, 1978) provides an example of the kind of methodological awareness that could be sharpened and refined by the kind of analysis Woodger recommended. There we find a conspectus of lower-order empirical and higher-order theoretical claims, and of the logical relations among them; as well as a case not only for linking embryology downwards with those disciplines, such as molecular biology, which can clarify mechanisms, but also for linking it upwards with evolutionary biology, by offering the latter grosser mechanisms or models of morphological change.

I have classified terms from the language that embryologists use into the following categories: biochemical terms; natural kind terms from organismic biology; anatomical terms; phase sortals; event and process terms; topological/morphological terms; and theoretical terms. The best explanations are always by way of example, so several representatives for each category are shown in Table 1.

There may be disagreements over entries in these lists. The reader will observe that I have not attempted to 'subordinate' any of these terms to any others. But obviously 'mouse' is subordinate to 'vertebrate', for example. There are some puzzles that I have not been able to resolve to my own satisfaction, concerning where a given term should go. Is 'cell cycle' an observational term, or a theoretical term? Was I right in regarding 'blastoderm' as an anatomical term – denoting a spatial part of something – or is it really, in the mouths of embryologists, a phase sortal term? Where would 'proliferative zone' go? It sounds a touch more observational than 'progress zone', which, in the context of Wolpert's writings, I regard as most definitely theoretical. Is the notion of 'developmental history' strictly theoretical, on the grounds that it applies to all events in an organism's life up to a given point, and therefore concerns events that are unobservable, or that involve entities that are unobservable? Is 'cytoplasm' an anatomical or biochemical term?

Table 1. *Classification of technical terms*

| Biochemical terms | Organismic kinds | Anatomical terms |
|---|---|---|
| macromolecule | amphibian | organ |
| gene | vertebrate | skeleton |
| protein | insect | dermis |
| enzyme | *Escherichia coli* | cell (200 types) |
| haemoglobin | nematode worm | connective tissue |
| | sea urchin | limb |
| | fruit fly | dorsal mass |
| *Phase sortals* | newt | ventral mass |
| | chick | antenna |
| egg | mouse | gut |
| sperm | guinea pig | wing |
| zygote | chimpanzee | muscle |
| morula | human | nerve |
| blastula | | tendon |
| embryo | | cartilage |
| larva | | bone |
| pupa | | blood |
| | | digit |
| *Theoretical terms* | *Topological/* | epiphyseal growth plate |
| vegetal pole | *morphological* | mesoderm |
| co-ordinate system | proximo–distal axis | notochord |
| boundary region | antero–posterior axis | somite |
| threshold | dorso–ventral axis | apical ectodermal ridge |
| classical gradient | | imaginal discs |
| classical organiser | | blastoderm |
| contractile forces | *Terms for events and* | |
| allometric relationship | *processes* | |
| generative program | | |
| positional information | meiosis | fertilisation |
| response selection | mitosis | cytodifferentiation |
| equivalence (of cells) | pattern formation | change in form |
| autonomy (of growth) | gastrulation | invagination |
| progress zone | cell aggregation | cell migration |
| polarising region | cell-to-cell interaction | regeneration |
| developmental history | morphallaxis | epimorphosis |
| diffusible morphogen | induction | heterochrony |
| | abortion | death |

It is interesting to note also that many terms in the jargon of the professional embryologist, which are not current in everyday speech, can be explained easily to a lay person with no detailed knowledge of histology or biochemistry or organismic biology. Among these, for example, are 'morula', 'blastula', 'notochord' and 'somite'.

# III

Latinate impenetrability is one thing; genuine theoreticity is another. Some of the terms I have categorised as theoretical seem very 'removed' from those under other headings. The contrast with any supposed observational/theoretical demarcation in Newtonian physics is quite marked. There the theoretical term 'point mass' is an abstraction rooted in the everyday notion of physical object. The notion of mass is cognate to that of weight, of which we have immediate experiential grasp. The same can be said also of forces. The picture of the physical world that Newtonian physics offers is a skeletal, austere one, which nevertheless enjoys conceptual congruence with the familiar everyday world. In quantum physics and in relativistic physics this conceptual congruence is ruptured, although we are left with scar tissue in the form of preservation theorems such as Ehrenfest's. Ehrenfest's theorem says, roughly, that in the large the predictions of quantum physics coincide with those of Newtonian physics. But the recondite terms of modern physics – such as 'quark', 'charm', and 'spin' – can only be understood by someone who has gained familiarity with the mathematical framework in which they feature. The theoretical terms of quantum physics are at a great remove from conceptual extrapolations within the reach of a lay person. He can grasp them only by ceasing to be lay, and becoming a competent theorist in that area.

Can the same be said of embryology? If not, that is still no criticism of embryology as a scientific discipline. For one must be mindful of the Aristotelian dictum that the level of precision (and one might add: abstraction, or abstruseness, or theoreticity) that we are trying to achieve should be appropriate for the chosen area. When we try to describe growth and development in manifold species of organism, and to discern regularities of pattern, correlations of measurable quantities, and successions of different types of event, we might – in order to be faithful to the phenomena and not pretend to be able to predict and explain more than we can – have to confine ourselves to a much more 'humdrum' vocabulary than does the theoretical physicist. This is a constraint working counter to some of the theoretical directions that, after Woodger, embryologists might have followed. As one seeks new higher-order explanatory hypotheses for embryology, to unify and subsume all that has been observed and hypothesised at lower levels, one must anchor one's new theoretical notions suitably to the observable by various logical links. These links are all important. The Aristotelian caution is that the higher conceptual hatches must be well battened down if we are to show that we can weather experimental tests. The modern notions of gradient, positional value, threshold, and co-ordinate system are cases in point. They must be seen to issue in predictions open to confirmation or refutation by observation and experiment. Woodger himself was keenly aware of this:

... simply to assert that gradients exist in embryos does not help at all. The gradient concept will only be really useful when it enters genuine explanatory hypotheses.

In this respect Wolpert's later work represents progress, but with a curious twist given by Wolpert himself. Not only does he agree implicitly with Woodger's claim that

In order to take further steps in embryology we are *compelled* to invent hypotheses concerning submicroscopical cell structure which will be explanatory of the behaviour of cells during development.

but he sees also linkage upwards, just as much as linkage downwards, as representing further progress. For he offers the prospect of explaining evolutionary change as change in pattern formation; while at the same time offering a 'downward looking' account of what this latter change consists in. The crucial mechanisms for Wolpert will be submicroscopical and biochemical in nature: he suggests there is at least one mechanism by means of which a cell may measure time spent in a 'progress zone', and one other by means of which a cell measures its position, within the organism and according to an organically intrinsic coordinate system, by monitoring levels of a 'diffusible morphogen'. In this regard, says Wolpert (1978, p. 164).

those of us who work on such problems are in the situation of genetics long before DNA was identified as the genetic material: ...

But now comes the twist:

*we have rules governing the phenomenology* but the molecular basis of the phenomena is completely unknown. (My emphasis)

Wolpert may be unfair to himself in saying this. For even pending a molecular mechanism for cell measurement of positional value and of time spent in a progress zone, he has taken an important theoretical leap by importing the last two notions. He may have got more than just the phenomenology right by postulating these notions-in-need-of-a-mechanism. The notion of *gene* was one in need of a mechanism for several decades, yet it helped geneticists during that time to do more than just *describe* the phenomena. The Mendelian laws helped theorists to see the phenomena in a new light, and to search for results that would confirm or deny the existence of a particulate mechanism of heredity. Similarly, in the embryological case, I for one was struck after my first acquaintance with the notion of positional value by the question 'What empirical tests would show that this was the right way to organise the results of our observations?' and had the immediate subsequent gratification of reading such passages as

This model suggests that if there are no long-range interactions between mesenchymal cells, a progress zone should continue to develop autonomously when it is excised and grafted to another site ... Our experimental grafts conform rather well to the theory.

The experiments prompted by the model afford a good example of what Woodger had in mind when he spoke of key hypotheses setting in motion new lines of investigation. Progress in this regard will only be consolidated, however, when we have successfully 'battened down' the new theoretical terms that Wolpert has introduced. Here are some of his key 'upward linking' generalisations that may help to do so (Wolpert & Stein, 1982):

In the evolution of vertebrates the histological cell types have probably not changed much either in quality or quantity ... The difference between (man) and (chimpanzee) may be attributed to pattern formation. (ibid, p. 332)

(I have bracketted their terms to highlight the possibility of other substitutions.)

... the genome provides a generative programme not a descriptive one. There are no genes describing the arm, only genes involved in specifying the processes for making it. (ibid, p. 333).

It is far more difficult to generate new functional proteins that would characterize a new cell-type than to generate new plans for rearranging existing cell types. (idem)

... the basic cellular processes have probably not changed their nature during evolution of multicellular organisms. Differences in form result not from the differences in these cellular activities, but from their spatial and temporal organization. (idem)

... the same set of positional values can be used to generate quite different patterns. This means that there could be a universal coordinate system which is used again and again, both within the same embryo as well as in other embryos. The main change in evolution would thus be in interpretation. (ibid, p. 334)

Non-equivalence enriches the repertoire of evolution, letting small parts of the body change independently of the rest. (ibid, p. 338)

These are representative of the claims 'linking upward' with evolutionary biology and using the new theoretical notions introduced by Wolpert. Like all claims of evolutionary biology itself, they cannot be expected to yield firm predictions. Their function is rather to point to ways of seeing the evidence, of understanding how the pieces of a jigsaw puzzle fit together. This is characteristic of all 'inference to the best explanation'. What, then, of 'downward' linkages involving the new theoretical terms? We have the following:

Differences in positional value can make cells nonequivalent even though they differentiate into a similar cell type. The principle of nonequivalence says that cells of the same differentiation class may have intrinsically different internal states, such as positional value. (ibid, p. 334).

... the pattern of the muscle and tendons uses the same positional field ... (ibid, p. 337)

... positional information is initially specified in a two-dimensional cell sheet, the mesoderm in vertebrates, and ... when this mesoderm comes to underlie the ectoderm, positional information in the ectoderm is specified by transfer of positional values from mesoderm to ectoderm. (ibid, p. 338)

... pattern formation can be viewed as a two-step process: first the cells are assigned positional information and then they interpret that information according to their genetic programme. (Wolpert, 1978, p. 154)

... positional information ... is the same in the antenna as in the leg: it is the interpretation that is different. (idem)

One general feature of positional fields is that they are always small and another feature is that the time required to establish them is long. (ibid, p. 156)

However ... a gradient is established, it can be interpreted by cells if their genetic programme is specified in terms of thresholds: if above a certain concentration the cells differentiate as one type and below it they differentiate as another type. (idem)

... the positional values imparted to the cartilage lead to different growth programs in different regions ... (ibid, p. 158)

Unlike cartilage cells, muscle cells are 'equivalent'. (ibid, p. 161)

... gradients can control the earliest patterning in a developing insect egg. (ibid, p. 162)

Both the posterior cytoplasm of the egg and the zone of polarizing activity of the wing bud appear to act as boundary regions that provide a positional signal. (idem)

Morphallaxis can now be understood as the establishment of a new boundary region at the cut surface and the specification of new positional values with respect to that boundary. In epimorphosis ... (new) positional values are generated in the new tissue. (ibid, p. 164)

... intercalary, or interpolated, regeneration takes place whenever discordant positional values come to the adjacent to each other: new positional values are generated in the growing tissue until the discordance is eliminated. (idem)

This second list of illustrative claims concerning positional value, gradients etc. seems to have genuine empirical import. Whatever mechanism may one day be proposed as underlying Wolpert's theoretical notions, one will be able to return to the claims above and see which ones were *wrong* as descriptions of reality *even pending* specification of the mechanisms. Convinced one day as to the nature of the mechanism to whose existence these claims pointed, we shall be able to re-assess some of them as over-hasty or only approximately true. We surely cannot dismiss *all* of them as on a par with the well known 'dormitive virtues' explanation of why a certain drug can put one to sleep. Wolpert is closer (as in his own estimation) to something more like the gene concept than he is to dormitive virtue. It is hard to read all the claims above as involving only allegedly theoretical cogs that whir but do not engage: as involving something analogous to claiming that when one has described line A being parallel to line B by saying that the direction of A is identical to the direction of B, one has thereby hit upon a new theoretical notion – direction – without which geometry as a science cannot progress. Wolpert's theoretical notions appear not to be of this kind. *Pace* Wolpert, he is not just doing phenomenology (by which I understand the description of what everyone agrees appears to be the case, rather than the more rarified philosophical doctrine of Husserl or the sense-data theorists). He is, rather, going importantly beyond the description of appearances, and even beyond the statement of objective regularities in things and events observed, by importing theoretical notions linked both upwards and downwards to neighbouring disciplines. These at once constrain the search for a mechanism and open up new fields of application: good science by anyone's lights.

Wolpert, like Woodger, believes that the 'lower' levels determine the 'upper' levels. Woodger had written

... we have reached a stage where the *changes* in cells are to be explained, and we can only explain change in a thing by hypotheses that speak about its parts.

And Wolpert wishes to

view development in terms of a generative programme contained within the fertilized egg's DNA.

Thus one can describe each of them as a *physical determinationist*; but whether either would wish to be described as a *reductionist*, is another matter. In what follows I shall explain the difference between these positions, and reflect on arguments for the autonomy of embryology as a scientific discipline.

# IV

Quine (1960) describes physics as limning the ultimate traits of reality. It is commonly held that biological traits are not ultimate. The development of an embryo may be constrained, and ultimately determined by, the underlying physical processes studied in, say, thermodynamics and quantum mechanics. But embryological development as such, so one such view further holds, is not one of the 'ultimate' processes in reality. Laws governing embryological processes – should any exist – do not possess the 'ultimate' character of the laws of physics: not, that is, if they deal directly with specifically embryological concepts such as invagination, gastrulation and the like.

Wherein lies this 'ultimacy' of the laws of physics? And if laws of embryological development are essentially supplementary to the laws of physics, might it nevertheless be essential to supplement the latter with the former? These are the two main questions I shall attempt to answer. More pithily, they can be posed as follows:

What level determines what others?

Can all laws of the levels described be reduced to the laws of the determining level?

Note now the generality of the question schemata. For the purposes of illustration, I have chosen physics and embryology. But, as noted above, the question concerns determination and reduction in general: determination as a relation between different levels of reality, and reduction as a relation between different levels of theoretical description. Physics is the commonest choice of determining theory, concerned with the supposedly 'ultimate' level of reality. Various other 'higher level' sciences (in our example, embryology) may then be contrasted with physics, as being concerned with supposedly 'higher', 'derivative' or 'less basic' levels of reality. In our example, we could have replaced embryology with chemistry, psychology or sociology in order to generate the questions of determination and reduction. The last two choices, however, could be regarded as slightly far-fetched or strained: physics lies 'too far below' psychology and sociology. One can think of the 'levels' of reality, and of corresponding scientific theories, as falling in a rough partial order as given in the following diagram (Fig. 1), with theories arguably

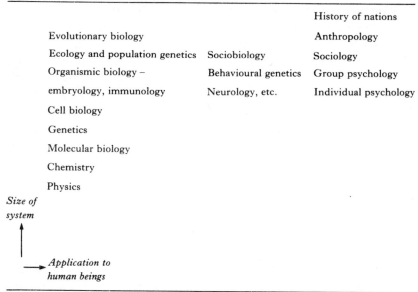

Fig. 1.

becoming less and less scientific as it peters out at the history of nations (compare Riedl, 1979). The diagram reflects both our concern to order systems by containment or size, and our special interest in ourselves. The determination–reduction question is most engagingly posed with respect to close neighbours in the list. Thus we can generate such problems of the past and present as:

the problem of whether evolution is incompatible with thermodynamics (via the pair 'physics–evolutionary biology');

the problem of methodological individualism (via the pair '(individual) human psychology – sociology'); and

the mind–body problem (via the pair 'human neurology–human psychology').

If one believes in the determining strength of a lower level with regard to its next highest neighbour in the list, then by transitivity one would conclude that the relation of determination straddles the list from one end to the other. Likewise, if one believes that one can reduce each higher theory to the one immediately below it, then by transitivity one would conclude also that the relation of reduction straddles the list from top to bottom. Determination concerns *levels of reality* (as described by their theories); reduction concerns *theories* (as they describe their levels). Determination and reduction most plausibly hold, however, with choices of closely neighbouring levels and scientific disciplines within something like the scheme above.

The foregoing talk of levels acquiesces in a convenient intellectual fiction in order to set up the problem of reduction in its stark essentials. But the various levels are never completely insulated from each other, via corresponding theories with no terms in common, like layers of an intellectual onion. Scientists in any discipline borrow from and trespass into others, and must do so as the scope of their 'how' and 'why' questions widen. Their background concepts, laws and theories straddle all available levels. When talking about cell division, there is talk of thermodynamics too; when explaining gross morphological change, one profitably invokes findings of molecular biology; when treating schizophrenia, one would be advised to heed what pharmacology and neurology have to offer. We have seen above how at least one embryologist – Wolpert – envisages bridge laws linking embryology upwards with evolutionary theories of morphological change, as well as laws linking it downwards with molecular biology and biochemistry, once the mechanism of diffusible morphogen (among possibly others) has been specified. So the tapestry of science is a criss-cross affair. The 'levels' interanimate via multifarious bridge laws, including (*pace* Davidson, 1970) laws connecting mental with neurological phenomena.

Quine (1960) has advanced a 'network model' of language and theory, according to which those statements most sensitively attuned to sensory experience lie at the periphery, and fundamental scientific laws and laws of logic at the heart, of an interconnected network of sentences. In his account there is no mention of the sorts of levels that we have been discussing. If one allowed for them in Quine's metaphor, one would obtain what I shall call a 'banded network'. Most vividly, think of Miss Havisham's wedding cake shrouded in dusty webs joining layer to layer. Each layer of the cake on its own is like a Quinean network, with an interior/periphery distinction to be made, on the basis of how liable statements are to revision in the light of experience. The icing consists of observation reports with so-called 'stimulus meaning'. The marzipan consists of Quine's 'standing sentences'. The more fruity interior consists of those higher-order explanatory hypotheses, and those laws of mathematics involved in the discipline (such as embryology) corresponding to that layer. Thus each layer corresponds to one of the main scientific disciplines mentioned above, bringing with it, as just indicated, a rough observational/theoretical distinction as well. Now the webs joining adjacent layers of the cake represent theoretical connections – 'bridge laws' – between neighbouring theoretical levels. And as Needham said about D'Arcy Thompson's theory of growth and form

(u)ntil we can find the links between the superimposed levels of organisation, there remains a certain meaninglessness about the genetics of size and shape or the mathematics of spirals and polyhedra. A unified science of life must inevitably seek to know how one level is connected with the others. (Needham, 1900)

# V

What is reduction? It is common for practitioners in one branch of science to hold out the possibility of reducing the concepts and laws in another to those of the former. For example, (some) physicists think that they can reduce the concepts and laws of chemistry, possibly even biology, to physics. Since Crick and Watson's discovery of the genetic code, evolutionary biology has started to simmer in the reducing heat given off by molecular biology below. Genes have been defined as double-helical structures, whose constituents are of known chemical structure. A biochemical theory of replication, coupled with a biochemical theory of cellular pattern formation, might one day displace evolutionary biology by reformulating all the main claims of the latter in biochemical terms. Yet higher, at another level in the diagram above, materialistically minded philosophers of mind have sought to establish mentality as mere epiphenomenal fog generated by the real processes 'below', namely the neurological ones. Pain-states, for example, might one day be *defined* theoretically as states of the central nervous system: perhaps concentrations of known chemicals are responsible for lowered synaptic thresholds in certain patterns (who knows? – I merely conjecture for the sake of argument). Pain-avoidance behaviour might then be predictable from knowledge of what characteristic stimuli to the motor system would ensue when that physical basis for pain obtains. So the neurological-cum-biochemical theory might one day displace our ordinary predictive apparatus that involves talk of sensation and bodily actions (or reactions).

Let us now move away from these particular examples, away from the inadequate metaphors and contrived reductions with which I have given them. Let us ask, quite generally, what it *means* when one says one can reduce one theory to another. What it means for the logician is this: *one can take the terms expressing concepts of the theory to be reduced and provide explicit definitions of them using only terms for concepts in the reducing theory. Then, using these definitions, and one's reducing theory, one can derive within the reducing theory the laws of the theory to be reduced.*

A fine example of reduction comes from the foundations of mathematics. Von Neumann (1923) gave an analysis of the notion of *natural number*, or in general *ordinal number*, in terms of *set*. The reducing theory in this case was set theory – and the theory to be reduced was arithmetic. Von Neumann defined 0 as the empty set, the set which contains no members (the set of all $x$ such that $x$ is not identical to $x$). He then defined each ordinal as the set that consisted of all preceding ordinals. So, for example, the number 1 was the set whose sole member was the empty set; 2 was the set whose sole members were the empty set and 1, which in turn was the set whose sole member was the empty set; and so on. With this ingenious reduction of the objects of arithmetic, he was able to replace arithmetical talk by set theoretical talk. And by reducing, or reformulating, the axioms of *arithmetic* via these definitions, he was able to derive those axioms as theorems of *set theory*. It is in this sense, the one just defined in a general way, that arithmetic is now regarded as reducible to set theory.

# VI

What is determinationism? *Physical* determinationism is the view that the physical facts determine all the facts. The principle of physical determinationism has been nicely expressed within a mereological framework by Hellman & Thompson (1975). Space–time and the distribution of matter therein is taken as the basic substrate of existence. Then bundles of it, and bundles of bundles, and so on, are available as the only entities one may talk about. (The rest is not there even to be passed over in silence!) The principle of physical exhaustion is then just the thesis that there are no objects, or entities to be talked about by any science, which are not contained in this cumulative hierarchy erected on matter-in-space-time. It rules out vitalism: there are no entelechies, or *vires vitales*, in addition to the basic physicochemical processes in the organism. There are no special mental entities in addition to brains; all mental events are physical events. I take the principle of physical exhaustion to be a working hypothesis of modern science. Just as no mathematician bothers to assert that numbers exist, but proceeds to prove interesting results about them, so too does the modern scientist proceed to formulate and test theories on the implicit assumption that there is, indeed, nothing but the physical.

The thesis of *physical determinationism* goes one step further than the principle of physical exhaustion. There are many 'ways of looking' at the physical systems just granted exhaustive tenure. Some are genes that replicate; some are embryos that invaginate; some are human beings that talk to each other; some are nation states that trade and war with one another. We discuss these systems using the language of genetics, embryological development, everyday psychology and sociology respectively. We ascertain and express genetic, embryological, psychological and sociological *facts* in doing so. Now the principle of physical determinationism simply says that these latter 'higher level' facts about these various (physical) systems, systems identified by 'higher level' concepts (such as 'gene', 'embryo', 'human being', 'nation state'), are nevertheless *fully determined* by the 'low level' *physical* facts concerning them.

Another way of putting the thesis is to say that the 'higher level' in question (say, the mental) is *supervenient upon* the physical. The thesis of supervenience in the philosophy of mind, for example, maintains that there is no change in a mental state without a corresponding change in physical state. Put another way, the physical state of an organism – including perhaps part of its environment, and not just its brain – uniquely determines what its mental state is. If the mental properties vary – say upon satiation or persuasion – then underlying the changed desire or changed belief there must be some alteration of physical state: a change, perhaps, in the pattern of neuronal excitation, or a change in the physical environment. The physical story fixes the mental story: the physical facts determine the mental facts, and indeed *all* the facts there are. This is *not* to say that physical processes are *deterministic*: physical determinationism, as just explained, is compatible with anti-determinism within physics.

Nor is this thesis of *physical* determinationism a way of saying that the world – *pace* Wittgenstein (1922) – *is* all that is *physically* the case. It is rather to say that the world – after Wittgenstein, all that is the case – is *determined* by all that is *physically* the case. One can be a physical determinationist and still grant the existence of, say, mental and social facts; and grant the licitness of mental and social vocabulary in describing these facts. Indeed, a physical determinationist (or supervenience theorist) can even grant the existence of *irreducibly* mental and social facts, and grant the *indispensability* of the idioms that bring them out as such. In short, a physical determinationist need not be a reductionist. This is the philosophical position that I believe modern embryologists to occupy; and that they would wish to see defended.

Physical scientists of this philosophical persuasion often draw comfort and support for their view from some form of *holism* or from an acknowledgement of *emergent properties*. Indeed, I think it fair to say that holism and emergent properties have even been cited as evidence against determinationism. This, however, I think to be misguided, and I shall consider only how, once granted the thesis of physical determinationism, holism and emergent properties have been applied as brakes in the slide to reductionism. But I shall argue that they have been *wrongly* so applied: and that, insofar as the philosophical position above (determinationism with anti-reductionism) can be defended, it is neither correctly nor most effectively defended by appeal to holism and emergence.

# VII

What is holism? Very roughly, it is the idea of global dependence. One can illustrate the idea from branches of enquiry besides physics and biology. In linguistics and semantics, holism is the doctrine that one cannot grasp the meaning of a word in isolation before one understands the whole language to which the word belongs (for a fuller discussion of which, see Tennant, in press (*a*)). In the philosophy of science, one also speaks of holism with respect to the evidence. This is the view that our theories 'face the tribunal of experience as wholes' – every theory must address itself to *all* the evidence, not just evidence selectively presented or emphasised; and as a corollary, should theoretical predictions be at odds with the data, it can prove to be difficult to pinpoint exactly which statements of the theory should be revised. Another holistic view is expressed by Mach's principle in physics. Mach maintained that the inertial mass of a body was determined by the distribution of matter in the universe as a whole.

But the sense of holism which will engage the biologist most acutely is that of 'the whole being greater than the sum of its parts'. This is a difficult idea to make precise, and I shall not attempt to do so. Suffice it here to say that it will not be enough to maintain that

> Perhaps the most important aspect of holism is that it emphasises relationships. I, myself, have always felt that relationships are not given sufficient weight. (Mayr, 1982)

For the same could be said of computer dating bureaux.

The main task in explicating the idea of a whole being greater than the sum of its parts will, I think, be to isolate a sense in which the *causal powers* of the whole cannot be predicted on the basis of the causal powers of its parts. Causally interactive ensembles will have to be shown by the holist to display regularities of interaction that are not to be obtained by any method of superposition or aggregation of the causal interactions of their constituents. Here the anti-holistic trend in modern embryological writing is worth noting: for example, Wolpert's insistence that cell-to-cell interactions, and small changes in their modes, can effect both the bodily growth of individual organisms and gross evolutionary changes in morphology, respectively. Moreover, the causal interactions of the ensembles could still be described by the holist in strictly physicalist language. The version of holism I am canvassing here is quite compatible with physical determinationism, provided only that the 'determining level' of physical reality is amenable to theoretical description at the corresponding 'lowest level' of physical theorising. This may require certain terms in the language of physical theorising to be taken, *primitively*, as applying to physical ensembles, or wholes, that are quite 'high up' in the cumulative hierarchy of Hellman and Thompson described above. For example, 'cell' or 'organism' might be such a term, and be reckoned to the determining level of theory. But this would already be to avoid the problems posed to the would-be reductionist by the phenomenon of holism. *Of course* one can avoid being unable to reduce, say, biological theory to what is commonly regarded as physics, by simply subsuming the former to the latter and regarding biology-cum-physics as the determining theory! But what this devious move in effect amounts to is an acknowledgement, on the part of the holist physical determinationist, of his inability to reduce biological theory to what is commonly regarded as physics.

The holism that thus stands in the way of reductionism can, of course, have even more far-reaching effects. We have thus far imagined wholes to display new causal powers not predictable from the causal powers of their parts. The causation in question has still been easy to regard as *physical* causation. (Lorenz's well known example of the capacitor comes to mind here, Lorenz, 1973.) But what now if the *emergent properties* of the whole started registering themselves, or making themselves felt, in ways that we were forced to describe in terms that we could not regard as belonging to the language of physical theorising? For some collections of cells are so complicated in their aggregative behaviour that they are dignified as *persons*, as having thoughts, beliefs and desires, and as engaging in *social* interactions. The complexity of these ensembles was produced by evolution, as of course was their very own cognitive apparatus for dealing efficiently with the gross overall effects of the highly complex physical processes going on inside and around them – processes which, according to the physical determinationist, do nevertheless still determine what is happening at what we like to call the mental and social levels.

Holism thus goes in hand with the doctrine of emergent properties. The more drastically different in kind the emergent properties of wholes are from those properties of their parts that we imagine ourselves exhaustively to have characterised, the more grave the problems facing the would-be reductionist. Emergent properties are those that arise (and we can think in a temporal or evolutionary sense when we say this) when constituents come together and join up, to produce a new whole that has dramatically different properties from those of the constituents that went to make up the whole. A standard illustration here is the way sodium and chlorine combine to give common salt, whose sharp taste cannot be predicted from knowledge of the chemical properties of sodium and chlorine alone. Here I can only endorse the following extended remark of Waddington (1981):

If we could observe the behaviour of sodium and chlorine only when each is in isolation, and if we regarded these two substances as made up of atoms, we might be able to discover something about these atoms but not very much. There is no reason why we should expect to become aware of the properties which allow them to combine with each other and form common salt. When this compound is formed, it is not that some new emergent properties appear, it is simply that a new avenue is opened to us for discovering a little more about sodium and chlorine atoms.

Even so, the 'little more' discovered by this 'new avenue' concerns the *psychological* effect (sharp taste) of salt on sentient organisms (us). So even if, with Waddington, we refuse to be unruffled as *physical determinationists* by the emergent properties of common salt, those properties might nevertheless give us pause as *reductionists*. It is the same problem to which Mayr alerts us when he denies that

. . . it is part of emergentism to believe that organisms can only be studied as wholes . . . All (that emergentists) claim is that *explanatory reduction* is incomplete since new, and previously unpredictable, characters emerge at higher levels of complexity in hierarchical systems.

(Mayr, 1982)

Still, the would-be reductionist could refuse to be impressed by this justification of emergentism, the observation that new and previously unpredictable characters emerge at higher levels of complexity in hierarchical systems. For the reductionist could argue as follows:

What the discovery of these so-called emergent properties shows you is that there would have been a very complex linguistic predicate made up of the terms of the reducing theory which is of special interest at this new level. If, indeed, the emergent property *can* be reduced to something before, its novelty lies simply in its unpredictable complexity. Among all the very complex formulae that one might have devised using terms of the reducing theory, it is highly unlikely that *this particular* complex formula (the one that successfully achieves the sought reduction of the emergent property) would have been hit upon, before the fact of emergence, as one peculiarly germane in the envisaged circumstances.

This is a cogent defence by the reductionist. He undermines the emergentist's opposition to reduction by pointing to a logical possibility. This is the possibility that, *should* there be a reduction, the complex predicate that captures the emergent property is likely to be so complex in its construction out of the reducing notions that one is *not* liable to identify it as theoretically relevant (in conveying truths about the world) *before* the emergent property has, so to speak, hit one in the eye.

# VIII

It may only be after a rigorous program of logical analysis, such as was advocated by Woodger, that philosophers of biology will be able to give content to the notion that a given theory treats its objects holistically, or gets to grips with emergent properties, or is irreducible to any lower level theory. The contributions of Frege (1903), Russell & Whitehead (1910–13) and Gödel (1931) eventually provided a definitive answer to the question whether arithmetic was reducible to logic. Likewise Montague (1974) has attempted a rigorous analysis of what it means to say that a theory is *deterministic*. Bealer has similarly tried to show that, upon rigorous analysis, the philosophy of mind known as *functionalism* collapses to plain old physicalist reductionism.

In all this work, various philosophical or intuitive notions are clarified upon analysis of logical and syntactical properties of the theoretical systems themselves. The question whether a given theory is reducible to another is a deep and difficult metamathematical question. So much so, that I think it safe to say that one cannot simply *see* that reduction is impossible on mere acquaintance with the two theories concerned. We need detailed argument to support any intuition that we ought to be content to operate at the higher level, and not seek to reduce it to the lower level. Any strong and immediate intuitions to such effect could be expected to be grounded in logical workings of the theories that are not so complicated as those involved in a detailed metamathematical proof that reduction is impossible: otherwise, whence their strength and immediacy? For this reason, I think that logical analysis might succeed in revealing whatever it is about the internal structure of theories that prompts anti-reductionist convictions. Beyond this, I cannot, at this stage, offer any more detailed suggestions; I can only point out a direction research might follow, and a goal that it might thereby reach. Prescinding, however, from details of particular theories, there is another more powerful and more general method of attack on the problem. It has been used independently to date by Hellman & Thompson (1975) and by Bealer (1978). It invokes Beth's theorem (Beth 1953) which states that if a new concept $Q$ can be defined *implicitly* by means of a theory T using concepts $P_1, \ldots, P_n$, then it can be defined *explicitly* in terms of $P_1, \ldots, P_n$, relative to T. Explanation of these notions is required. First, *implicit definability*:

Suppose one understands the concepts $P_1, \ldots, P_n$. One develops a theory using them, and one can identify what count as $P_i$ in the domain to which the theory addresses itself. Suppose further that a new concept $Q$ is imported, and a grasp of it conveyed by means of a set T of statements involving *both Q and* $P_1, \ldots, P_n$. Suppose finally, on the assumption that T is a true account of what is the case in the domain, and that you have settled what, in the domain, count as $P_i$, that there turns out to be *but one* way of understanding what $Q$ applies to. Then we say that we have *implicitly* defined $Q$ in terms of $P_1, \ldots, P_n$ relative to T.

Secondly, *explicit definability*:

Suppose there is a complex concept $R$ built up from $P_1, \ldots, P_n$ but *not* involving $Q$ and that it follows as a logical consequence of T that $R$ and $Q$ apply to exactly the same things. Then we say that $R$ provides an *explicit* definition of $Q$ relative to the theory T.

Beth's theorem, to repeat, states that if $Q$ is *implicitly* defined in terms of $P_1, \ldots, P_n$ relative to T then there is some such $R$ that *explicitly* defines $Q$ in terms of $P_1, \ldots, P_n$ relative to T. Why is it important in the present context? For the following reason: Suppose one is a physical determinationist working with concepts $P_1, \ldots, P_n$ from the determining level (say physics) and with various concepts from a higher level (say embryology). Let T be one's full story *concerning both levels. The claim that the lower level determines the higher level is exactly the hypothesis of Beth's theorem*, namely that each concept $Q$ is implicitly defined in terms of $P_1, \ldots, P_n$ relative to T. Put another way, we may say that in the statement

> if A then B

of Beth's theorem, the antecedent A can be interpreted as the thesis of physical determinationism. So now the implication that is Beth's theorem guarantees an explicit definition of each higher level concept $Q$ in terms of the lower level concepts $P_1, \ldots, P_n$. The consequent B of Beth's theorem, in other words, is the thesis of reductionism. For now replace in T each occurrence of a higher level concept $Q$ by its defining formula $R(P_1, \ldots, P_n)$. The result is a theory that operates at the lower level and yet covers all the higher level phenomena, albeit by the complex defining formulae $R(P_1, \ldots, P_n)$.

Beth's theorem, then, according to the determinationist, *implies* reductionism. It appears to render uninhabitable the intellectual niche sought by the physical determinationist who, prompted by considerations of holism and emergent properties, was a would-be anti-reductionist. It threatens to demolish the philosophical position that I described above as the one most likely to be adopted by modern embryologists (and biologists generally). In vain would Mayr be able to protest that

The claim that genetics has been reduced to chemistry after the discovery of the structure of DNA, RNA, and certain enzymes cannot be justified. To be sure, the chemical nature of a number of black boxes in the classical genetic theory was filled in, but this did not affect in any way the nature of the theory of transmission genetics. As gratifying as it is to be able to supplement the classical genetic theory by a chemical analysis, this does not in the least reduce genetics to chemistry. The essential concepts of genetics, like gene, genotype, mutation, diploidy, heterozygosity, segregation, recombination, and so on, are not chemical concepts at all and one would look for them in vain in a textbook on chemistry. (Mayr, 1982).

For Mayr has come nowhere near establishing the *logical* impossibility of achieving a chemical definitional reduction of the notions of genetics. Just because they are not to be found in any extant textbook on chemistry does not show that it is not, in principle, possible to devise such a reduction. Beth's theorem would ensure that the reducing formulae $R$ exist. They might be fiendishly complex: but they would be there.

If this possibility cannot be foreclosed then we have to live with it. What value would then remain in the insistence that there will always be a place for the holistic, a place for the emergent, as our minds grapple with reality? Terms such as the embryological ones that I categorised above would probably defy *workable* definition in strictly physicochemical terms. The details of such definitions would be elusive, and the definitions themselves would be tiresomely, if not monstrously, cumbersome. What would such a definition of 'blastula' look like? of 'pupa'? of 'mesoderm'? In these definitions one would have (at the very least) to talk of cells of certain types lying in various topological configurations. Then the reference to cells would have in turn to be replaced by strictly physicochemical terminology: so cells would be defined as certain topological configurations of cytoplasm and nuclear substance within a protein-fibre wall; then the cytoplasm, nuclear substance and wall themselves would in turn be defined as . . . and so on. Then there is the further problem that each species' gastrula has a characteristic number of cell types, and possibly also a characteristic rate at which these differentiate further. So 'gastrula' *tout court* as a non-species-specific term in embryology, would probably have to have built into it a disjunction, across species, of these characteristic cell types. The details are too awesome to contemplate. Nor does the resulting definiens capture the open-textured *meaning* of the word 'gastrula'. This meaning can be mastered pre-theoretically after ostensive training. One who has thus grasped it can characteristically recognise a gastrula of a newly discovered species without having to know anything about its characteristic number of cell types. So the physico-chemical definition of *gastrula* served up by Beth's theorem would not be meaning-preserving. The complex predicate F serving in this role would not be in what the philosopher calls *intensional* agreement with the ordinary term *gastrula*. For in some possible world (perhaps even one compatible with the underlying laws of physics that hold in the actual world) the extensions of the term *gastrula* and the complex predicate F may not coincide. This, however, is not fatal to scientific practice in the actual world, so long as extensional agreement there between the terms is guaranteed. But would such agreement ever be *reached*? The practising reductionist would be referring to an entity in incubation not as 'this gastrula here', but as 'this (instance here of the complex physicochemical predicate) F'; and by the time he had uttered F the wretched thing would probably have grown, multiplied and died! So 'gastrula' itself would in all likelihood be retained as a convenient definitional abbreviation for the complex predicate F; and the actual science of embryology would carry on as if nothing, philosophically or methodologically, had altered it. There are many more examples of problematic terms, drawn from the lexicon of embryology that I roughly categorised above, that would similarly resist definitional reduction save at the unbearable cost of hopeless complication, even should agreement on details ever be attainable. But why is it that we are not capable of such circumlocutory precision, and prodigious logical manipulation, so that scientific practice can be made consonant with reductionist conviction? Or, better, why is it that we do science with a language and with concepts that, in deference to such reductionist convictions, would have to be dispensed with in favour of highly complex definitions in reducing terms, only to be resuscitated as convenient abbreviations for them? For a philosopher to suggest to evolutionary biologists the following answer to such a question is to carry coals to Newcastle. But suggest it I shall: I think we have been naturally selected to carve up reality in various ways: to see interactions in terms of the organismic and the functional. So we are to a large extent handicapped in our deeper theoretical endeavours by the very cognitive heritage that has enabled us to survive. The same view, which I shared with a biologist, was expressed elsewhere as follows (Tennant & von Schilcher, 1984):

As long as it is the brains of human organisms that do science, there will be a special place in science for the organismic and the human: and this is in spite of the long drawn out cosmological fact that the physical brought forth the human and the social. Because ours is a world of purpose quite by chance, our only chance is to see purpose.

Not all is lost, however, on the logical front. In a technical sequel to this paper (Tennant, in press (*b*)) I shall examine various counter-arguments that have been given to the application of Beth's theorem in the manner I have been concerned here to describe. I shall argue that these counter-arguments do *not* succeed in preventing the slide, greased by Beth, from determinationism to reductionism. But I shall *also* argue that there *are* definitive objections which *do* halt that slide. So the determinationist need not be in the position of refusing to agree with the thesis of reductionism merely on *practical* grounds. His refusal could be based also on the absence of any cogent *logical* grounds.

I am grateful to Tim Horder for his very helpful editorial advice and suggestions as to the sort of reading in philosophy of science that would interest embryologists. I am grateful also for the invitation to attend the BSDB conference in Nottingham, which enabled me to learn something about embryology.

# References

Bealer, G. (1978). An inconsistency in functionalism, *Synthese* **38**, pp. 333–72.

Beth, E. (1953). On Padoa's method in the theory of definition, *Konik Ned, Akad. Wetensch Proc. Ser.* A56 (= *Indag Math.* 15), pp. 330–9.

Davidson, D. Mental events, In *Experience and Theory*, ed. L. Foster & J. W. Swanson (Massachusetts, 1970).

Frege, G. *Grundgesetze der Arithmetik*, 2 vols. (Jena, 1893, 1903).

Gödel, K. (1931). Ueber formal unentscheidbare Sätze der Principia Mathematica und verwandter Systeme I, *Monatshefte für Mathematik und Physik* **37**, pp. 173–98.

Hellman, G. & Thompson, F. W. (1975). Physicalism: ontology, determination and reduction, *Journal of Philosophy* **72**, pp. 551–64.

Lorenz, K. *Die Rückseite des Spiegels* (Munich 1973).

Mayr, E. *The Growth of Biological Thought* (Harvard 1982).

Montague, R. Deterministic theories. In *Formal Philosophy* (Yale, 1974), pp. 303–59.

Needham, J. (1900). Biochemical aspects of form and growth, In *Aspects of Form*, ed. L. L. Whyte.

von Neuman, Zur Einführung der transfiniten Zahlen, *Acta litterarum ac Scientiarum Regiae Universitas Hungaricae Francisco-Josephine, Sectio Scientiarum Mathematicarum 1* (1923), pp. 24–33.

Quine, W. V. O. *Word and Object* (MIT, 1960).

Riedl, R. *Biologie der Erkenntnis* (Berlin, 1979).

Russell, B. & Whitehead, A. N. *Principia Mathematica*, 3 vols (Cambridge, 1910, 1912, 1913).

Sneed, J. D. *The Logical Structure of Mathematical Physics* (Reidel, 1981).

Suppes, P., McKinsey, J. C. C. & Sugar, A. C. (1953). Axiomatic foundations of classical particle mechanics, *Journal of Rational Mechanics and Analysis*, vol. **2**, pp. 253–72.

Suppes, P. & Rubin, H. (1954). Transformations and systems of relativistic particle mechanics, *Pacific Journal of Mathematics* **4**, 563–601.

Tennant, N., The philosophical significance of Beth's theorem, Paper delivered to the 7th International Congress of Logic, Methodology and Philosophy of Science, Salzburg 1983, in press.

Tennant, N. Holism, molecularity and truth. In *Festschrift for Michael Dummett*, ed. B. Taylor, (Nijhoff), in press (a).

Tennant, N. W. & von Schilcher, F. *Philosophy, Evolution and Human Nature* (Routledge and Kegan Paul, 1984).

Waddington, C. H. *The Nature of Life*, p. 20. (London: Allen & Unwin, 1981).

Williams, M. J. (1970). Deducing the consequences of evolution: a mathematical model, *Journal of Theoretical Biology* **29**, 343–85.

Wittgenstein, L. *Tractatus Logico-Philosophicus* (London, 1922).

Wolpert, L. (1978). Pattern formation in biological development, *Scientific American* **239**, 154–64.

Wolpert, L. & Stein, W. D. (1982), Evolution and development. In *Learning, Development and Culture*, pp. 331–42. (John Wiley).

Woodger, J. H., Growth, *Symposium of the Society for Experimental Biology* (1948).

# Selected references to resources relating to the history of embryology

## General guides to information sources

Anderson, K. (ed.) (1977). *Current British Directories*, 8th edn. Beckenham, Kent: CBD Research Ltd.

Burkett, J. (1979*a*). *Directory of Scientific Directories. A world guide to scientific directories including medicine, agriculture, engineering, manufacturing and industrial directories*, 3rd edn. London: Francis Hodgson. Longman Group Ltd.

Burkett, J. (1979*b*). *Library and Information Networks in the United Kingdom*. London: ASLIB.

Codlin, E. M. (ed.) (1984). *Aslib Directory of Information Sources in the United Kingdom*, 5th edn. London: Aslib.

Corsi, P. & Weindling, P. (1983). *Information Sources in the History of Science and Medicine*. London: Butterworth Scientific.

*Directories of Science Information Services: International Bibliography (1967)*. The Hague: International Federation for Documentation.

Durbin, P. T. (1980). *A Guide to the Culture of Science, Technology and Medicine*. New York: The Free Press.

Ethridge, J. M., ed. (1980). *The Directory of Directories*. Detroit, Michigan: Information Enterprises.

*European Sources of Scientific & Technical Information (1957–1981)*, 5th edn, 1981, ed. A. P. Harvey & A. Pernet. Harlow, Essex: Francis Hodgson (Longman Group Ltd).

Henderson, G. P., ed. (1981). *Current European Directories*, 2nd edn. Beckenham, Kent: CBD Research Ltd.

Higgens, G. L. (1984). *Printed Reference Material*, 2nd edn. London: Library Association.

Kyed, J. M. & Matarazzo, J. M. (1979). *Scientific, Engineering and Medical Societies Publications in Print 1978–1979*. New York and London: R. R. Bowker.

Legenfelder, H. (1980). *Handbook of International Documentation and Information*, vol. 8 of World Guide to Libraries. Munich, New York, London and Paris: G. Saun.

*A Directory of Information Resources in the United States: Biological Sciences*. (1972). Library of Congress National Referral Center. Washington DC: Government Printing Office.

MacReavy, S. E. (1971). *Guide to Science and Technology in the United Kingdom*. A reference guide to Science and Technology in Great Britain and Northern Ireland. Guernsey: Francis Hodgson.

*Medical Reference Works, 1679–1966: a selected bibliography (1967)* (with subsequent Supplements), Blake, J. & Roos, C., eds. Chicago: Medical Library Association.

*Scientific and Technical Books and Serials in Print (1974–)*. 3rd edn, 1984, 3 vols. New York & London: R. R. Bowker Co.

Sheehy, E. P., Comp. (1976). *Guide to Reference Books*, 9th edn. (Supplements, 1980, 1982). Chicago: The American Library Association.

Walford, A. J., ed. (1966). *Guide to Reference Material*, 2nd edn, 3 vols. London: Library Association.

Wasserman, P. & Herman, E. (eds.) (1975). *Library Bibliographies and Indexes*. A subject guide to resource material available from Libraries, Information Centres, Library Schools and Library Associations in the United States and Canada. Detroit: Gale Research Co.

## Archives and museums

Alton, J. (ed.) (1973–8). The catalogues of archives of scientists compiled by the contemporary scientific archives centre. Oxford: Oxford Microform Publications.

Bearman, D. & Edsall, J. T. (eds.) (1980). *Archival Sources for the History of Biochemistry and Molecular Biology: a Reference Guide and Report*. Boston: American Academy of Arts and Sciences, Philadelphia: American Philosophical Society.

Clapp, J. (1962). *Museum Publications*, in 2 parts. New York: Scarecrow Press.

Contemporary Medical Archives Centre, Consolidated Accessions List (1982), compiled by Sheppard, J. and Hall, L. London: Wellcome Institute for the History of Medicine.

*Directory of Archives and Manuscript Repositories in the United States*. Washington DC: National Historical Publications and Records Commission.

Foster, J. & Sheppard, J. (1984). *British Archives*. London: Macmillan.

Hamer, P. M. (1961). *A Guide to Archives and Manuscripts in the United States*. New Haven: Yale University Press.

Hudson, K. & Nicholls, A. (1975). *The Directory of Museums*. London: Macmillan.

MacLeod, R. M. & Friday, J. R. (1972). *Archives of British Men of Science*. London: Mansell.

Markham, S. F. (1948). *Directory of Museums and Art Galleries of the British Isles*. London: The Museum Association.

Museums of the World. A directory to 17 500 museums in 150 countries including a subject index. (1975) 2nd edn enlarged. Munich: Verlag Dokumentation.

National Historical Publications and Records Commission (1978). *Directory of Archives and Manuscripts Repositories in the United States*. Washington DC: National Archives and Records Service.

*Official Museum Directory 1981*. United States and Canada (1980). Washington & Stokie: American Association of Museums and National Register Publishing Co.

Royal Commission on Historical Manuscripts (1982). *The Manuscript Papers of British Scientists 1600–1940*. London: HMSO.

Smart, J. E. (1978). *Museums in Great Britain with Scientific and Technological Collections*, 2nd edn. London: Science Museum.

*Additional sources:*
Archives Sections of *Journal of the History of Biology* and *Mendel Newsletters* (American Philosophical Society).

## Oral history

Columbia University Oral History Collection, New York.
United Kingdom National Sound Archive, British Library, London.

## Films

Audio-visual Materials for Higher Education: British Universities Film Council, London.

*British National Film Catalogue (1963–1983)*, ed. C. Quinn, 21 vols. *Continues as* British National Film and Video Catalogue (1984–). London: British Film Institute.

Higher Education Film Catalogue (1980). Glasgow: Scottish Central Film Library.

Hinchliffe, J. R. (1972). *J. Biol. Education* **6**, 119–23.

Hinchliffe, J. R. (1975). *J. Biol. Education* **9**, 123–6.

National Union Catalog: *Films and Other Materials for Projection (1973–)*, Annually. *Continuation of* National Union Catalog: *Motion Pictures and Filmstrips (1954–1972)* and *Films (1953)*. Totowa NJ: Rowman & Littlefield.

## Guides to libraries

Ash, L. (1978). *Subject Collections*. A guide to special book collections and subject emphases as reported by University, College, public, and special libraries and museums in the United States and Canada, 5th edn. New York and London: R. R. Bowker.

Bowns, R. B. (1972). *American Library Resources*. A bibliographical guide. Chicago: American Library Association.

Cattell Press, ed. (1979). *American Library Directory*, 32nd edn. New York & London: R. R. Bowker.

Lewanski, R. C. (1978). *Subject Collections in European Libraries*, 2nd edn. London and New York: R. R. Bowker.

Linton, W. D. (ed.) (1982). *Directory of Medical and Health Care Libraries in the United Kingdom*. London: The Library Association.

Roberts, S., Cooper, A. & Gilder, L. (1978). *Research Libraries and Collections in the United Kingdom: A Selective Inventory and Guide*. London: Clive Bingley; Hamden, Conn.: Linnet Books.

Thornton, J. L. & Tully, R. J. (1962). *Scientific books, libraries and collectors*: a study of bibliography and the book trade in relation to science. 2nd edn. London: Library Association.

*World Guide to Special Libraries*. (1983). Munich: Saur.

Young, M. L. & Young, H. C. (eds.) (1981). *Directory of Special Libraries and Information Centers*, 6th edn, 2 vols. Detroit: Gale Research Co.

Young, M. L., Young, H. C. & Kruzas, A. T. (eds.) (1977). *Subject Directory of Special Libraries and Information Centers*. 4th edn, 5 vols. Detroit: Gale Research Co.

## Guides to bookshops and book dealers

*Driffs's Guide to all the Secondhand and Antiquarian Bookshops in Britain*. London: Holborn Books.

Lewis, R. H. (1982). *The Bookbrowser's Guide to Secondhand and Antiquarian Bookshops*, 2nd edn. Newton Abbot: David & Charles.

Sheppard Press Ltd (1982). *European Bookdealers: A Directory of Dealers in Secondhand and Antiquarian Books on the Continent of Europe, 1982–1984*. 5th edn. Seven Hills Books.

Sheppard Press Ltd (1983). *Bookdealers in North America: a Dictionary of Dealers in Secondhand & Antiquarian Books in Canada and the USA, 1983–1985*. 9th edn. Seven Hills Books.

Sheppard Press Ltd (1984). *Dealers in Books: a Dictionary of Dealers in Secondhand and Antiquarian Books in the British Isles, 1984–1986*. 11th edn. Seven Hills Books.

Swets & Zeitlinger, B. V. *Catalogues of periodicals for sale*: Swets' Backsets on Medical Sciences and Related Subjects. Lisse, The Netherlands.

## Dictionaries and encyclopaedias

### Guides

*Bibliography of Interlingual Scientific and Technical Dictionaries*, (1961) 4th edn, with Supplement. Paris: UNESCO.

*Bibliography of Monolingual Scientific and Technical Glossaries* (1955–9) Paris: UNESCO.

Brewer, A. S. ed. (1979). *Dictionaries, Encyclopaedias and other Word-Related Books*, 2nd edn. Detroit: Gale Research Co.

Collison, R. L. (1966). *Encyclopedias: Their History Throughout the Ages*. A bibliographical guide with extensive historical notes to the general encyclopedias issued throughout the world from 350BC to the present day. 2nd edn. New York & London: Hafner.

*International Bibliography of Specialized Dictionaries: Fachwörterbücher und Lexika, ein internationales Verzeichnis* (1979), 6th edn, ed. H. Lengenfelder. München & New York: K. G. Saur.

Walsh, S. P. (1968). *Anglo-American General Encyclopaedias. A Historical Bibliography 1703–1967*. New York and London: R. R. Bowker.
*World Dictionaries in Print. A Guide to General and Subject Dictionaries in World Languages.* (1983). New York: Bowker.

**Dictionaries**

Artschwager, E. (1934). *Dictionary of Biological Equivalents, German/English*. London: Baillière, Tindall and Cox.

Asimov, I. (1974). *Words of Science, and the History Behind Them*, 2nd edn. London: Harrap.

Baldwin, J. M. (1901). *Dictionary of Philosophy and Psychology*, 3 vols. London: Macmillan.

Barnes, F. (1881). *A German–English Dictionary of Words and Terms used in Medicine and its Cognate Sciences*. London: Lewis.

Bynum, W. F., Browne, E. J. & Porter, R. (1981). *Dictionary of the History of Science*. London: Macmillan.

Donath, T. (1969). *Anatomical Dictionary*. Oxford: Pergamon.

Dunman, T. (1889). *A Glossary of Anatomical, Physiological and Biological Terms*. London & Sydney: Okeder & Welsh.

Flood, W. E. (1963). *The Origins of Chemical Names*. London: Oldbourne.

Frederick, K. E. & Frederick, R. G. (1981). *Dictionary of Theoretical Concepts in Biology*. Metuchen, NJ: Scaresom Press Inc.

Gray, P. (1967). *The Dictionary of the Biological Sciences*. New York: Reinhold.

Gray, P. (1970). *The Encyclopedia of the Biological Sciences*, 2nd edn. New York: Van Nostrand.

Hyrtl, J. (1880). *Onomatologia Anatomica*. Wien: Braumüller.

Jackson, B. D. (1953). *A Glossary of Botanic Terms*, 6th edn. New York: Hafner.

Jaeger, E. C. (1953). *A Source Book of Medical Terms*. Springfield, Ill.: Thomas.

Jaeger, E. C. (1955). *A Source Book of Biological Names and Terms*. 3rd edn. (1st edn, 1944). Springfield, Ill.: Thomas.

Jaeger. G. *et al.* (1879–1902). *Encyclopedie der Naturwissenschaften*, 43 vols. Breslau: Eduard Trewendt.

Kelly, E. C. (1948). *Encyclopaedia of Medical Sources*. Baltimore: Williams and Wilkins.

King, R. C. (1974). *A Dictionary of Genetics*, (2nd edn.) Oxford: Oxford University Press.

Lapedes, D. N. (ed.) (1976). *Dictionary of the Life Sciences*. New York: MacGraw-Hill.

Mayerhöfer, J. (ed.) (1959–1970). *Lexikon der Geschichte der Naturwissenschaften*. Wien: Hollinek.

McNicoll, D. H. (1863). *Dictionary of Natural History Terms*. London: Lorrell Reeve.

Medawar, P. B. & Medawar, J. S. (1984). *Aristotle to Zoos: A Philosophical Dictionary of Biology*. Oxford: Oxford University Press.

Melander, A. L. (1937). *Source Book of Biological Terms*. New York: The City College.

Power, H. & Sedgewick, L. W. (1882). *The New Sydenham Society's Lexicon of*

*Medicine and the Allied Sciences* (based on Maynes Lexicon), 5 vols. London: New Sydenham Society.

*Rieger, R., Michaelis, A. & Green, M. M. (1976). *A Glossary of Genetics and Cytogenetics*, 4th edn. (1st edn, 1954). Berlin: Springer.

Roux, W., Correns, C., Fischel, A. & Küster, E. (eds.) (1912). *Terminologie der Entwicklungsmechanik der Tiere und Pflanzen*, Leipzig: Engelmann.

*Skinner, H. A. (1961). *The Origin of Medical Terms*, 2nd edn. Baltimore: Williams and Wilkins.

Steffanides, G. F. (1963). *The Science Thesaurus*, 3rd edn. Boston: Best Printers.

Todd, R. B. (1839–1847). *The Cyclopedia of Anatomy and Physiology*. London: Longman.

*Wain, H. (1958). *The Story Behind the Word*. Springfield, Ill.: Thomas.

Wiener, P. (ed.) (1973–4). *The Dictionary of the History of Ideas* (4 vols.). New York: Scribner.

Ziegler, H. E. & Bresslau, E. (1912). *Zoologisches Worterbuch*. Jena: Fischer.

*For chronology of terminology see:*
*Shorter Oxford English Dictionary* and starred references.

**Encyclopaedias with historically important successive editions**
*Encyclopaedia Britannica* (1768–1982), 15th edn, 1982: *New Encyclopaedia Britannica*, 30 vols. Chicago etc: Encyclopaedia Britannica Corp.
*Encyclopedia Americana* (1829–1971). New York: Americana Corp.

## General guides to bibliography

*A Guide to the World's Abstracting and Indexing Services in Science and Technology* (1963). Report No. 102. Washington DC: National Foundation, Science Abstracting and Indexing Services.

Besterman, T. (1965–6). *A World Bibliography of Bibliographies, and of Bibliographical Catalogues, Calendars, Abstracts, Digests and the like*. 4th edn, 5 vols. Lausanne: Societas Bibliographica.

*Bibliographic Index: A Cumulative Bibliography of Bibliographies (1937–)*. Quarterly, with annual and other cumulations. New York: H. W. Wilson.

Collison, R. L. (1968). *Bibliographies Subject and National*. A guide to their contents, arrangement and use. London: Crosby Lockwood and Son.

*Forthcoming Books* (1966 – ). New York: Bowker.

*Guide to Microforms in Print (1984)*, 2 vols. Westport & London: Meckler Publishing.

*Guide to reprints (1978)*. Kent, Connecticut: Guide to reprints, Inc.

Peddie, R. A. (1933). *Subject Index of Books Published up to and Including 1880* (2nd series, 1935, 3rd series, 1939, new series, 1948). London: Grafton & Co.

Reynolds, M. M. (1975). *A Guide to Theses and Dissertations:* an annotated international bibliography of bibliographies. Detroit: Gale Research Co.

Toomey, A. F. (1977). *A World Bibliography of Bibliographies 1964–1974*. A list of works represented by Library of Congress Printed Cards, a Decennial Supplement to Theodore Besterman, *A world bibliography of bibliographies*, 2 vols. Totowa, NJ: Rowman and Littlefield.

UNESCO, (1969–77). *Bibliographical Series Throughout the World*. Paris: Unesco Bibliographical Handbooks.

## Scientific

Bottle, R. T. & Wyatt, H. V. (eds.) (1971). *Use of Biological Literature*, 2nd edn. London: Butterworth.

Drachman, J. M. (1930). *Studies in the Literature of Natural Science*. New York: Macmillan Co.

Grogan, D. J. (1976). *Science and Technology; an Introduction to the Literature*, 3rd edn. London: Clive Bingley.

Henwood, F. & Thomas, G. (1986). *Science, Technology and Innovation: a Research Bibliography*. Brighton: Wheatsheaf Books.

Malcles, L.-N. (1950–8). Les Sources du Travail Bibliographique (4 vols), vol. 4, Bibliographies specialisées (Sciences exactes et techniques). Geneva: Droz, Lille: Giard.

Morton, L. T. (1977). *Use of Medical Literature*, 2nd edn. London: Butterworth.

Morton, L. T. (1983). *A Medical Bibliography* (Garrison & Morton), 4th edn. (1st edn, 1943). Aldershot: Gower.

National Book League (1958). *Science for All: An Annotated Reading List for the Non-Specialist*. London: Cambridge University Press.

*Pure and Applied Science Books, 1876–1982* (1982), 6 vols. New York: R. R. Bowker.

*Royal Society Catalogue of Scientific Papers 1800–1900* (4 series, 19 vols), (1867–1925). London: Cambridge University Press.

Smith, R. C. & Painter, R. H. (1967). *Guide to the Literature of the Zoological Sciences*, 7th edn. Minneapolis: Burgess Publ. Co.

Taschenberg, O. (1887–1923). *Bibliotheca Zoologica 11*, 8 vols. Leipzig: Engelmann.

Wood, C. A. (1931). *An Introduction to the Literature of Vertebrate Zoology*. London: Oxford University Press.

*Systematic bibliography of nineteenth century embryology:*

Minot, C. S. (1893). A Bibliography of Vertebrate Embryology. *Mem. Boston Soc. Nat. Hist.* **4**, 487–614.

## History of science

*Bibliography of the History of Medicine 1965*. Bethesda: National Library of Medicine.

Blake, J. B. & Roos, C., eds. (1967). *Medical Reference Works 1679–1966*. Chicago: Medical Library Association.

*Current Work in the History of Medicine* (1954–). London: Wellcome Historical Medical Library.

Ferguson, E. S. (1968). *Bibliography of the History of Technology*. Cambridge, Mass.; Society for the History of Technology.

Jayawardene, S. A. (1982). *Reference Books for the Historian of Science*. London: Science Museum Library.

Knight, D. M. (1972). *Natural Science Books in English, 1600–1900*. London: Batsford.

Mayerhofer, J. (1959–70). *Lexikon des Geschichte der Naturwissenschaften*. Biographien, Sachworter und Bibliographien. Vienna: Verlag Bruder Hollinek.

Neu, J. (1979). *Isis Comulative Bibliography 1966–75*. A bibliography of the History of Science formed from Isis Critical Bibliographies 91–100. London: Mansell.

Rider, K. J. (1970). *History of Science and Technology: a Select Bibliography for Students*, 2nd edn. London: The Library Association.
Russo, F. (1954). *Histoire des Sciences et des Techniques; Bibliographie*. Paris: Hermann.
Sarton, G. (1952). *A Guide to the History of Science*. New York: Ronald Press.
Scheele, M. & Natalis, G., eds. (1981). *Biologie-Dokumentation: Bibliographie der deutschen biologischen Zeitschriftenliteratur 1796–1965*, 24 vols. Munchen, New York, London, Paris: K. G. Saur.
Smit, P. (1974). *History of the Life Sciences: an annotated Bibliography*. Amsterdam: Asher.
Whitrow, M., ed. (1972–6). *Isis Cumulative Bibliography: A Bibliography of the History of Science formed from Isis Critical Bibliographies*, 1–90, 1913–1965, 3 vols. London: Mansell.

**National**

*Biblio: Catalogue des ouvrages parus en langue française dans le monde entier (1934–1970)*, continued as *Les livres de l'année- Biblio (Bibliographie de la France) (1971–)* Annual. Paris: Librairie Hachette & Cercle de la Librairie.
*British National Bibliography (1950–)*. Weekly, with cumulations. London: The Council of the British National Bibliography Ltd, (1950–1970); The British Library, Bibliographic Services Division (1971–).
*Cumulative Book Index* (World List of Books in English) (1928–). Annual, with cumulations. New York: H. W. Wilson Co.
*Deutsche Nationalbibliographie (1966–)*. Leipzig: Deutschen Bücherei.
*The English Catalogue of Books (1835–1950)*. Annual, with cumulations. London: Sampson, Low, Son & Marston (1835–1900); The Publishers' Circular Ltd (1901–1950).
'*GV*': *Gesamtverzeichnis des deutschsprachigen Schrifttums, 1700–1910* (1979–1984), 126 vols. München, New York, London and Paris: K. G. Saur.
Howard-Hill, T. H. (1969–1982). *Index to British Literary Bibliography*, 6 vols. Oxford: Clarendon Press; New York: Oxford University Press.

**Library catalogues**

*The British Library General Catalogue of Printed Books to 1975*. (Supersedes British Museum Catalogues of Printed Books) (1979–; in progress), several vols. London: Clive Bingley; London, München, New York & Paris: K. G. Saur.
*British Museum Subject Index, 1881–1940* (1966). London: Pordes.
*Catalog of the Library of the Marine Biological Laboratory and the Woods Hole Oceanographic Institution (1971)*, 12 vols, including Journal Catalog. Boston: G. K. Hall.
*Catalogue of the Library of the British Museum (Natural History) (1903)*, 5 vols. Supplement, 3 vols, 1933. Accessions Supplements, 1934–. London: British Museum (Natural History).
Eales, N. B. (1969, 1975) *The Cole Library of Early Medicine and Zoology*, 2 vols. Reading: University of Reading.
*Index Catalog of the Library of the Surgeon General's Office (U.S. Army) (1880–1949)*. Washington: US Government Printing Office. Continues as:

*Armed Forces Medical Library Catalog 1950–1954 (1955).* Ann Arbor: Edwards and then as *National Library of Medicine Catalog.*
*Library of Congress Catalog (Subject Catalog) (1950–).* Quinquennial cumulations. Washington: Library of Congress.
*National Library of Medicine Catalog (1955–).* Quinquennial vols. Bethesda: National Library of Medicine.
*The National Union Catalog* (Library of Congress Catalogs).
   Pre-1956 Imprints (1968–1981), 754 vols. London: Mansell.
   1956–1967 Imprints (1970–72), 125 vols. Totowa: Rowman & Littlefield.
   1968–1972 Imprints (1973), 104 vols. Ann Arbor: Edwards.
   1973–1977 Imprints (1978), 135 vols. Totowa: Rowman & Littlefield.
   1978–1982 Imprints (1979–83, Annual Cumulations, each of several vols). Washington: Library of Congress.
   1983 Imprints–: available on Microfiche.
Wellcome Institute for the History of Medicine and Related Sciences (1980). *Subject Catalogue of the History of Medicine,* 18 vols. München: Kraus International Publications.

## Bibliography of Reviews

*The Book Review Digest.* (1906–). New York: H. W. Wilson.
*Technical Book Review Index.* (1935–). New York: Special Library Association.
*Bibliography of Medical Reviews.* (1955–1967), in *Index Medicus* (1967–).
*Index to Scientific Reviews in the Sciences.* (1980–). Philadelphia: Institute for Scientific Information.

## Periodicals

*British Library Lending Division: Keyword Index to Serial Titles (1980).* Boston Spa: British Lending Library.
*British Union-Catalogue of Periodicals:* a record of periodicals of the world, from the seventeenth century to the present day, in British libraries (1955), 4 vols. Supplement, 1962. London: Butterworths Scientific Publications.
*Irregular Serials and Annuals: An International Directory (1983),* 8th edn. New York: R. R. Bowker Co.
LeFanu, W. R. (1938). *British Periodicals of Medicine, 1684–1938.* Baltimore: Johns Hopkins Press.
*New Periodical Titles: British Union-Catalogue of Periodicals,* incorporating World List of Scientific Periodicals, *1960–(1964–).* Quarterly, with cumulations. London: Butterworths.
*New Serial Titles (Library of Congress) (1950–).* Monthly, with cumulations. New York & London: R. R. Bowker; Washington: Library of Congress.
Scudder, S. H. (1879). *Catalogue of Scientific Serials 1633–1876.* Cambridge, Mass.: Library of Harvard University.
*Serial Publications in the British Museum (Natural History) Library (1980),* 3rd edn, 3 vols. London: British Museum (Nat. Hist.).
*Union List of Serials (1965),* 3rd edn, 5 vols. New York: H. W. Wilson. (*Continued as* New Serial Titles.)

*Ulrich's International Periodicals Directory (1984)*, 23rd edn, 2 vols. New York: R. R. Bowker Co.

Wall, C., ed. (1971). *Poole's Index to Periodical Literature. Cumulative Author Index*, 1802–1906. Ann Arbor: Piernan Press.

*World List of Scientific Periodicals, 1900–1960* (1963). 4th edn, 3 vols, ed. P. Brown & G. B. Stratton. London: Butterworth.

## Guide to periodicals in embryology

Numbers after an entry indicate specific contents of possible interest.

*Key to journals listing*
1. Statement of Journal Aims in first issue
2. Affiliation to Society
3. Notices of meetings, membership of affiliated Society
4. Index
   a = author
   s = subject
   c = cumulative
5. Obituaries
6. Book reviews
7. Guides to literature
8. Abstracts

References, where given, are to papers dealing with the history of the Journal or affiliated Society.

*Acta Biotheoretica* (1935–): 1.
*Acta Embryologiae Experimentalis* (1957–).
*American Naturalist* (1867–). 1, 2 (American Society of Naturalists), 3 (Supplements: Records of American Society of Naturalists), 4a,s, 6. (+ (early vols.) Natural History calendar; occ. Biographies).
   Conklin, E. G. (1934). *American Naturalist* **68**, 385.
*Anatomy and Embryology* (1974–; continuation of *Zeitschrift für Anatomie und Entwicklungsgeschichte* (1921–1974, part of *Zeitschrift für die Gesamte Anatomie*) and *Arbeiten aus Anatomischen Instituten* (1891–1921, part of *Anatomische Hefte: Referate und Beiträge zur Anatomie und Entwicklungsgeschichte*, founded by Merkel & Bonnet, 1891): 4a.
*L'Année Biologique* (1895–): 1, 2 (Federation des Sociétés des Sciences Naturelles), 3, 4a,s,c, 5,6,7,8.
*Archives de Zoologie Expérimentale et générale* 1872–.
   Founder: Henri de Lacaze-Duthiers, Paris: 1, 2 (Centre Nationale de la Recherche scientifique).
*Biological Bulletin* (1900–). (Marine Biological Laboratory, Woods Hole, Mass. 4a,s,c, 5).
*Contributions to Embryology* (1915–1966). Carnegie Institute, Washington: 4a,c.
*Development, Growth & Differentiation* (1961–, continuation of *Embryologia*, 1950–1960).

*Developmental Biology* (1959–): 1, 2 (Society for Developmental Biology) 3, 4a,s,c, 8.
*Differentiation* (1973–): 1,2 (International Society of Differentiation).
*Growth* (1937–): 1, 4a,s, 5, 6.
*Journal of Embryology and Experimental Morphology* (1953–): 1, 4a, s,c.
   JEEM (1979) **54**, 1–3.
*Journal of Experimental Zoology* (1904–): 2 (American Society of Zoologists), 4a,s.
   Harrison, R. G. (1945). *J. Exp. Zool.* **100**, xi–xxxi.
*Journal of Theoretical Biology* (1961–): 1, 4a,s.
*Physiological Zoology* (1928–): 4a,s, 6.
*Wilhelm Roux' Archives of Developmental Biology* (1975–; Continuation of *Wilhelm Roux' Archiv für Entwicklungsmechanik der Organismen* (1925–1975)., *Archiv für Mikroskopische Anatomie u. Entwicklungsmechanik* (1923–1925), *Archiv für Entwicklungsmechanik der Organismen* (1895–1923). Founder: Wilhelm Roux, Innsbruck, 1894): 1, 2 (European Developmental Biology Organisation) 4a.
*Zoologischer Jahresbericht* (1879–1913). Zoological Station, Naples. (Founder: Anton Dohrn): 1, 4a,s, 5 (death notices), 6,7, 8.

## Longer running series of literature compilations; abstracts; reviews; citation indexes; listings or indexes by subject

*Key to listing*
1. Recurrent compilations of literature arranged/indexed by subject/author.
2. Abstracts.
3. Reviews, surveys, commentaries.
4. Citation analysis by key word/author/address/individual publication/journal
5. Alerting service
6. Title key word index
7. Cumulated indexes over one year or longer periods by subject/author

*Anatomischer Anzeiger* (1886–): 1.
*Anatomische Hefte. Referate und Beiträge zur Anatomie und Entwicklungsgeschichte* (1891–): 1, 2, 3.
*Annual Review of Physiology* (1939–): 3.
*Berichte Biochemie und Biologie* (1969–) supersedes *Biochemisches Centralblatt* (1903–): 1, 2.
*Bibliography of Reproduction* (1963–): 1, 5.
*Biologisches Centralblatt* (1881–): 1, 2, 3.
*Biological Abstracts* (1926–). See also
   *Bioresearch Index* (1965–1979) and,         1, 2, 4, 6, 7.
   *Biological Abstracts RRM* (Reports, Reviews, Meetings (1980–).
*British Museum Subject Index* (1881–): 1, 7.
*Bulletin Signalétique* (1956–): 1.
*Current Awareness in Biological Science* (1982–) supersedes *Physiological Abstracts* (1916): 1, 2, 5.
*Current Contents: Life Sciences* (1959–): 1, 3, 5, 6.
*Dissertation Abstracts* (1952): 2.
*Genetics Abstracts* (1968–): 1, 2.
*Index Catalog of the Library of the Surgeon General's Office* (1880–): 1, 7.
*Index Medicus* (1960–) supersedes *Index Medicus* (*1879*): 1, 7.

*Jahresberichte über die Fortschritte der Anatomie und Physiologie* (1872–1918): 1, 2, 3.
*Library of Congress Catalog: Books: Subjects* (1950–): 1, 7.
*L'Année biologique* (1895–): 1, 2, 3.
*Royal Society Catalogue of Scientific Papers* (1800–1914): 1.
*Science Citation Index* (1961–): 1, 4.
*Zoological Record* (1864–): 1.

### Regular reports and symposia

*Abstracts from American Association of Anatomists* – Supplements to *Anatomical Record* (1906–).
*Annual Reports, Marine Biological Laboratory, Woods Hole*, in *Biological Bulletin* (1900–).
*Annual Meetings of Society for General Physiologists*, in *Journal of General Physiology* (1918–).

*The Collecting Net*
*General Embryological Information Service:* An International Directory of Current Research in Developmental Biology. (1949–1980). Utrecht: Hubrecht Laboratory.
*\*Newsletter of British Society for Development Biology.*
*\*Newsletter of International Society of Development Biologists.*
*Symposia of American Society of Zoologists* in *American Zoologist* (1961–). (Am. Zoologist (1982) **22**, 735–48).
*Symposia of the Society for Developmental Biology* (1965–). New York: Academic Press (continuation of *Symposia of Society for the Study of Development and Growth* (1952–64); and *Symposia on Development and Growth* (1939–1951–supplements to '*Growth*').
*Verhandlungen der Anatomischen Gesellschaft* – Supplements to *Anatomischer Anzeiger* (1886–).
Yearbook Carnegie Institution.
*Zoologisches Anzeigen* (1870–), transactions of *Deutsche Zoologisches Gesellschaft*.

*\*Occasional historical notes.*

*English translations from foreign language journals*
*Biological Abstracts* (1926–) includes abstracts in English of foreign language publications; available from 1969 on BIOSIS Data Base, Biosciences Information Service, Philadelphia.
*World Index of Scientific Translations* (1967–77). European Translations Centre, Delft. Quarterly. Continues in *World Transindex*.
*World Transindex* (1978–). International Translations Centre, Commission of the European Communities and Centre Nationale de la Recherche Scientifique, Delft. Monthly, with Annual Cumulative Indexes.

### Computerised data bases

Hall, J. L. (ed.) (1979). *On-line Bibliographic Data Bases*. London: Association of Special Libraries and Information Bureaux.

National Library of Medicine, Washington, DC: Data Bases *MEDLINE* and *HISTLINE* (*History of Medicine*), from 1966 onwards.

## Personal bibliographies

Arnim, M. (1944). *Internationale Personalbibliographie*, 1800–1943, 2nd edn, 2 vols. Leipzig: Hiersemann.

Arnim, M. & Hodes, F. (1984). *Internationale Personalbibliographie*, 1944–1959, 2nd edn, 2 vols. Stuttgart: Hiersemann.

Leistner, O. (1984). *International Bibliography of Festschriften from the beginnings until 1979*. Osnabrück: Biblio Verlag.

# Biography

## Guides

Batts, J. S. (1976). *British Manuscript Diaries of the Nineteenth Century. An Annotated Listing*. Totowa, NJ: Roman & Littlefield.

*Biographical Books 1950–1980*, (1980). New York: R. R. Bowker Co.

*Biography Index:* a quarterly index to biographical material in books and magazines (1946–). New York: H. W. Wilson.

Fruton, J. S. (1977). *Selected Bibliography of Biographical data for the History of Biochemistry since 1800*, 2nd edn. Philadelphia: American Philosophical Society.

Riches, P. M. (1984). *An analytical Bibliography of Universal Collected Biography*. London: Macmillan.

Scott Barr, E. (1973). *An Index to Biographical Fragments in Unspecialised Scientific Journals*. Alabama: Univ. Alabama Press.

Slocum, R. B. (1967). *Biographical Dictionaries and Related Works*. Detroit: Gale Research Co.

Thornton, J. L. (1970). *A Select Bibliography of Medical Biography*, 2nd edn. London: Library Association.

New York Academy of Medicine Library: *Catalog of Biographies (1960)*. Boston: G. K. Hall.

## Collected biographies

Abbott, D. (1983). *Biographical Dictionary of Scientists*. London: Muller.

*Allgemeine deutsche Biographie* (1871–1912), 56 vols. Leipzig: Duncker & Humblot.

Barnhart, J. H. (1965). *Biographical Notes upon Botanists*, 3 vols. Boston: G. K. Hall.

*Biographical Memoirs of the National Academy of Sciences*, Washington (1877–). Annually.

*Biographie Universelle Ancienne et Moderne*, (Michaud) (1857) 2nd edn, 45 vols. Paris: Desplaces.

*Biographisches Lexikon der hervorragenden Aerzte aller Zeiten und Völker* (*1929–1935*), 2nd enlarged edition, ed. W. Haberling *et al*. (1st edn, 1880, ed. A. Hirsch), 5 vols. Berlin & Wien: Urban & Schwarzenberg.

*Biographisches Lexikon der hervorragenden Aerzte der letzten 50 Jahre* (1932–33), ed. I. Fischer, 2 vols. Berlin & Wien: Urban & Schwarzenberg.

*Biographisches Lexikon hervorragender Aerzte des neunzehnten Jahrhunderts*, (1901), ed. J. Pagel. Berlin & Wien: Urban & Schwarzenberg.

Cattell Press, ed. (1982). *American Men and Women of Science*, 15th edn, 7 vols (1st edn, 1906). New York: R. R. Bowker Co.

Debus, A. G. (1968). *World Who's Who in Science: a biographical dictionary of notable scientists from antiquity to the present*. Chicago: Marquis and Who's Who.

*Dictionary of National Biography (1908–1971)*. Oxford: Oxford University Press.

*Directory of British Scientists* (1966–7). 3rd edn. Boston: G. K. Hall.

Elliott, C. A. (1979). *Biographical Dictionary of American Science*. Westport & London: Greenwood Press.

Gillispie, C. C., ed. (1970–1980). *Dictionary of Scientific Biography, 14 vols*. New York: Scribner.

Howard, A. V. (1951). *Chamber's Dictionary of Scientists*. London & Edinburgh: Chambers.

Hyamson, A. M. (1951). *A Dictionary of Universal Biography*, 2nd edn. London: Routledge & Kegan Paul.

*Internationaler Nekrolog: Verzeichnis verstorbener Personen aus Politik, Wirtschaft, Kultur, Wissenschaft und Gesellschaft*. (1982). Pullach: W. Gorzny.

Ireland, N. O. (1962). *Index to Scientists of the World, from Ancient to Modern Times: Biographies and Portraits*. Boston: F. W. Faxon Co.

Jordan, D. S., ed. (1910). *Leading American Men of Science*. New York: H. Holt & Co.

Mayerhöfer, J. (ed.) (1959–70). *Lexikon der Geschichte der Naturwissenschaften*. Wien: Hollinek.

*McGraw-Hill Modern Men of Science* (1968), 2 vols. New York: McGraw-Hill.

*Neue Deutsche Biographie* (1953–), Berlin: Duncker & Humbolt.

*Nouvelle Biographie Générale* (1853–66), 45 vols. Paris: Firmin Didot Frères.

*Obituary Notices of Fellows of the Royal Society of London* (1901–1954), continued as *Biographical Memoirs of Fellows of The Royal Society of London* (1955–). Annually, London: The Royal Society.

Pelletier, P. A. (1980). *Prominent Scientists. An Index to Collective Biography*. New York: Neal-Schuman Pub. Co.

Plesse, W. & Rux, D. (1977). *Biographien bedeutender Biologen*. Berlin: Volk and Wissen.

Poggendorff, J. C. (1863–1974). *Biographisch-literarisches Handwörterbuch zur Geschichte der exakten Wissenschaften*, several vols. Leipzig: Barth; Berlin: Akademie Verlag.

Roberts, F. C. (1978). *Obituaries from the Times 1951–75*, 3 vols. Reading: Newspaper Archive Developments Ltd.

Talbott, J. H. (1970). *A Biographical History of Medicine*. New York: Grune & Stratton.

Turkevich, J. & Turkevich, L. B. (1968). *Prominent Scientists of Continental Europe*. New York: American Elsevier Publ. Co.

*Who Was Who* (1897–1980), vols. I–VII. London: Adam & Charles Black.

*Who was Who in Science*. (1968). Chicago: Marquis.

Williams, T. I. (1982). *A Biographical Dictionary of Scientists*. London: Adam & Charles Black.

Wintle, J. (ed.) (1981). *Makers of Modern Culture: a Biographical Dictionary*. London: Routledge & Kegan Paul.
Wintle, J., ed. (1982). *Makers of Nineteenth Century Culture (1800–1914)*. London: Routledge & Kegan Paul.

*Additional sources:*
New York Academy of Medicine Library: *Portrait Catalogue* (1960), 5 vols, 3 supplements to 1975. Boston: G. K. Hall; *Index Catalog of the Library of the Surgeon General's Office; Index Medicus*.
*The Times Index*, 1785–1790, (1978), 4 vols. Reading: Newspaper Archive Developments Ltd.
*Palmer's Index to The Times*, 1790–1905, (1905–25), 536 vols. Hampton Wick: Palmer.
*The Annual Index to The Times*, 1906–1913, (1907–1914). London: Wright. Continued as *The Official Index to The Times* (1914–1956), quarterly. London: Bland; continued as *The Times Index* (1957–), quarterly; Annual Cumulations from 1977. London: Times Publishing Co; Reading: Newspaper Archive Developments Ltd.
New York Times: Index (1851–). New York: New York Times Co.

**Individual biographies and autobiographies relevant to embryology**

Ackerknecht, E. H. (1953). *Rudolf Virchow: Doctor, Statesman, Anthropologist*. Madison: University of Wisconsin Press.
Allen, G. E. (1978). *Thomas Hunt Morgan. The Man and His Science*. Princeton, NJ.: Princeton University Press.
Baltzer, F. (1967). *Theodor Boveri; Life and Work of a Great Biologist, 1862–1915* (transl. Dorothea Rudnick), Berkeley: Univ. California Press.
Bateson, B. (1928). *William Bateson, F.R.S., Naturalist*. London: Cambridge University Press.
Bolsche, C. E. W. (1900). *Ernst Haeckel, ein Lebensbild*. Dresden and Leipzig: H. Seeman. [Transl. J. McCabe, 1906. London: T. Fisher Unwin. New edn. (1909) issued for the Rationalist Press Association. London: Watts and Co.]
Corner, G. (1981). *The Seven Ages of a Medical Scientist: an autobiography*. Philadelphia: University of Pennsylvania Press.
Driesch, H. (1951). *Lebenserinnerungen. Aufzeichnungen eines Forschers und Denkers in entscheidender Zeit*. Munich and Basel: Ernst Reinhardt.
Evans, M. A. & H. E. (1970). *William Morton Wheeler, Biologist*. Cambridge, Mass.: Harvard University Press.
Gaupp, E. (1917). *August Weismann, Sein Leben und Sein Werk*. Jena: G. Fischer.
Gegenbaur, K. (1901). *Erlebtes und Erstrebtes*. Leipzig: Englemann.
Goldschmidt, R. B. (1956). *Portraits from Memory*. Seattle: University of Washington Press.
Goldschmidt, R. B. (1960). *In and out of the Ivory Tower*. Seattle: University of Washington Press.
Goldsmith, M. (1980). *Sage, a life of J.D. Bernal*. London: Butterworths.
Hemleben, J. (1964). *Ernst Haeckel in Selbstzeugnissen und Bilddokumenten*. Hamburg: Rowohlt.

Kölliker, A. von (1899). *Erinnerungen aus meinem Lebe*. Leipzig: Englemann.
Ludwig, E. (ed.) (1965). *Wilhelm His der Aeltere. Lebenserinnerungen und ausgewahlte Schriften*. Berne and Stuttgart: Hans Huber.
Manning, K. R. (1983). *Black Apollo of Science, the life of Ernest Everett Just*. New York & Oxford: Oxford University Press.
Mocek, R. (1974). *Wilhelm Roux-Hans Driesch. Zur Geschichte der Entwicklungsphysiologie der Tiere*. Jena: Gustav Fischer.
Raikov, B. E. (1968). *Karl Ernst von Baer 1792–1876. Sein Leben und sein Werk* (Transl. H. von Knorre), Leipzig: Johann Ambrosius Barth.
Roux, W. (1923). Wilhelm Roux in Halle a. S. *In* L. R. Grote (ed.), *Die Medizin der Gegenwart in Selbstdarstellungen*, vol. 2, pp. 141–206. Leipzig: Felix Meiner.
Sabin, F. R. (1934). *Franklin Paine Mall*. Baltimore: Johns Hopkins.
Spemann, H. (1943). *Forschung und Leben*, (ed. F. W. Spemann). Stuttgart: Adolf Spemann.
Stieda, L. (1878). *Karl Ernst von Baer. Eine biographische Skizze*. Braunschweig: Friedrich.
Thompson, R. d'Arcy (1958). *D'Arcy Wentworth Thompson*. London: Oxford University Press.
Twitty, V. C. (1966). *Of Scientists and Salamanders*. San Francisco: Freeman.
Uschmann, G. (1955). *Caspar Friedrich Wolff. Ein pionier der modernen Embryologie*. Leipzig: Urania.
Uschmann, G. (1961). *Ernst Haeckel-Forscher, Kunstler, Mensch*, 3rd enlarged revised edn. Leipzig: Urania.
Weissenberg, R. (1959). *Oscar Hertwig 1849–1922 Leben und Werk eines deutschen Biologen*. Leipzig: Barth.

## Portraits

Lane, W. C. & Browne, N. E. (1906). *A. L. A. Portrait Index*. Washington DC: Library of Congress.
Bradshaw, D. N. & Hahn, C. (1982). *World Photography Sources*. Epping: Bowker.
Burgess, R. (1972). *Catalogue of Portraits of Doctors and Scientists in the Wellcome Institute for the History of Medicine*. London: Wellcome Institute for the History of Medicine.
New York Academy of Medicine: *Portrait Catalogue* (1960), 5 vols. (supplements to 1975). Boston: G. K. Hall.

## Membership of Universities, Institutions, Societies etc.

Bates, R. S. (1965). *Scientific Societies in the United States* (3rd edn). Cambridge, Mass.: M.I.T. Press.
*Encyclopedia Of Associations* (1959–). 19th edn, 1984, 5 vols. Detroit: Gale Research Co.
Gribbin, J. H. (1961). *Scientific and Technical Societies of the United States and Canada*, 7th edn. Washington DC: National Academy of Sciences, National Research Council.

Meenan, A. (1983). *A Directory of Natural History and Related Societies in Britain and Ireland.* London: Brit. Mus. Nat. Hist. Publ. 859.

MINERVA: *Jahrbuch der gelehrten Welt* (1891–1934), 30 vols, Continues in 2 parts: *Abteilung Universitäten und Fachhochschulen* (1933–) and *Abteilung Forschungsinstitute, Observatorien, Bibliotheken, Archive, Museen usw*, (1933–37). Now *Internationales Verzeichnis Wissenschaftlicher Institutionen, Wissenschaftliche Gesellschaften (1938–),* Berlin: Walter de Gruyter.

*National Faculty Directory* (1970–). Published annually. Detroit: Gale Research Co.

*Scientific and Learned Societies of Great Britain.* A handbook compiled from official sources (1964). London: Allen and Unwin.

## Scientific meetings

Bishop, W. J. (1958). *Bibliography of International Congresses of Medical Sciences.* Oxford: Blackwell Scientific Publications.

Gregory, W. (ed.) (1938). *International Congresses & Conferences 1840–1937.* New York: H. W. Wilson.

*Index of Conference Proceedings* (1965–). Monthly, various cumulations. Boston Spa: British Lending Library.

*Index to Scientific and Technical Proceedings* (1978–). Monthly, with annual cumulations. Philadelphia: Institute for Scientific Information.

*Interdok: Directory of Published Proceedings* (1965–). Monthly, annual cumulations. White Plains, NY: Interdok Corp.

## Methods in Embryology

Abderhalden, E. (1925). *Hanbuch der biologischen Arbeitsmethoden.* Berlin: Urban & Schwarzenberg.

Adamstone, F. B. & Shumway, W. (1939). *A Laboratory Manual of Vertebrate Embryology* (2nd edn, 1947; 3rd end, 1954). New York: Wiley.

Costello, D. P., Davidson, M. E., Eggers, A., Fox, M. H. & Henley, C. (1957). *Methods of Obtaining and Handling Marine Eggs and Embryos.* Woods Hole, Mass.: Marine Biological Laboratory.

Galtsoff, P. S., Latz, F. E., Welch, P. S. & Needham, J. G. (1937). *Culture Methods for Invertebrate Animals.* Ithaca, NY: Coinstock Pub. Co.

Hamburger, V. (1960). *A Manual of Experimental Embryology.* Chicago: University of Chicago Press.

Just, E. E. (1939). *Basic Methods for Experiments on Eggs of Marine Animals.* Philadelphia: Blakiston.

Lillie, F. R. & Moore, C. R. (1919). *A Laboratory Outline of Embryology, with special reference to the Chick and the Pig,* 2nd edn. Chicago: University of Chicago Press.

New, D. A. T. (1966). *The Culture of Vertebrate Embryos.* London: Logis Press.

Rugh, R. (1962). *Experimental Embryology; techniques and procedures,* 3rd edn. Minneapolis: Burgess.

Waterman, A. J. (1948). *A Laboratory Manual of Comparative Vertebrate Embryology.* New York: Henry Holt.

Whitman, C. O. (1885). *Methods of Research in Microscopical Anatomy and Embryology.* Boston: Cassino.
Wilt, F. H. & Wessels, N. K., eds. (1968). *Methods in Developmental Biology.* New York: Crowell.
Wischnitzer, S. (1975). *Atlas and Laboratory Guide for Vertebrate Embryology.* New York: McGraw-Hill.

**Normal Tables of Developmental Staging:**

Altman, P. L. & Dittmer, D. S. (1972). *Biology Data Book* Vol. 1. 2nd edn. Bethesda: F.A.S.E.B.
Bellairs, R. (1971). *Developmental Processes in Higher Vertebrates*, Appendix. London: Logos Press.

## Collections of embryos

Blechschmidt Collection, Anatomisches Institut Göttingen, Germany.
California Primate Research Center and Carnegie Embryological Collections, University of California, Davis, California, USA.
Central Embryological Collection, Hubrecht Laboratory, Utrecht, The Netherlands.
Cornell University Department of Anatomy, New York State College of Veterinary Medicine, Cornell University, Ithaca, New York, USA.
Department of Anatomy, Minsk Medical Institute, Minsk, USSR.
Department of Embryology and Histology, Charles University, Plzen, Czechoslovakia.
Department of Evolutionary Morphology and Ecology of Animals, Academy of Sciences, Moskow, USSR.
Hochstetter Collection, Anatomisches Institut der Universitat Wien, Austria.
The Hooker/Humphrey Collection, Department of Anatomy, University of Alabama Medical Center, Birmingham, Alabama, USA.
Mossman Collection, University of Wisconsin Zoological Museum, Madison, Wisconsin, USA.
Patten Collection, University of Michigan Collection, Department of Anatomy, Ann Arbor, Michigan, USA.
University of Washington Collection, Seattle, Washington, USA.

## History of biology

**Chronology**

Aschoff, L. & Diepgen, P. (1945). *Kurze Uebersichtstabelle zur Geschichte der Medizin*, 6th edn. Berlin: Springer.
D'Arcy Power (1923). *Chronologia Medica.* New York: Hoeber.
Darmstaedter, L. (1909). *Handbuch zur Geschichte der Naturwissenschaften und der Technik in chronischer Darstellung*, 2nd edn. Berlin: Springer.
Garrison, F. H. (1929). *An Introduction to the History of Medicine, with Medical Chronology, Suggestions for Study and Bibliographic Data*, 4th edn. Philadelphia & London: W. B. Saunders & Co.

Rothschuh, K. E. (1952). *Entwicklungsgeschichte Physiologische Probleme in Tabellenform*. Munich: Urban und Schwarzenberg.

Schmidt, J. E. (1959). *Medical Discoveries, Who and When*. Springfield: Charles & Thomas.

**General histories**

Allen, G. E. (1975). *Life Science in the Twentieth Century*. New York: J. Wiley.

Bahn, B. (1979). *L'ordre et les Temps*. L'anatomie comparée et l'histoire des vivants aux XIXe Siècle. Paris: Masson et Cie.

Bernhem, B. M. (1948). *The Story of the Johns Hopkins*. New York, Toronto: Whittlesey House.

Bodenheimer, F. S. (1958). *The History of Biology: an introduction*. London: Dawson.

Brauer, L., Mendelssohn-Bartholdy, A. & Meyer, A. (1930). *Forschungs-Institute, ihre Geschichte, Organisation und Ziele*. Hamburg: Hartung.

Castiglioni, A. (1958). *A History of Medicine* (transl, & ed. E. B. Krumbhaar), 2nd edn. New York: A. A. Knopf.

Cohen, I. B. *et al.* (1967). *Landmarks of Science*. New York: Readex Microprint.

Cole, F. J. (1944). *History of Comparative Anatomy. From Aristotle to the Eighteenth Century*. London: Macmillan & Co.

Coleman, W. (1967). *The Interpretation of Animal Form*. New York: Johnson Reprint Corps.

Coleman, W. R. (1971). *Biology in the Nineteenth Century: Problems of Form, Function and Transformation*. New York: John Wiley.

Dampier, Sir W. C. (1929). *A History of Science and its Relation with Philosophy and Religion*. Cambridge: Cambridge University Press.

Dawes, B. (1952). *A Hundred Years of Biology*. London: Gerald Duckworth and Co.

Desmond, A. (1982). *Archetypes and Ancestors*. London: Bland & Briggs.

Diepgen, P. (1949). *Geschichte der Medizin*. Berlin: de Gruyter.

Dubos, R. J. (1976). *The Professor, the Institute, and DNA*. New York: Rockefeller University Press.

Farley, J. (1977). *The Spontaneous Reneration Controversy from Descartes to Oparin*. Baltimore and London: Johns Hopkins University Press.

Gardner, E. J. (1922). *History of Biology*, 3rd edn. Minneapolis: Burgess.

Garrison, F. H. (1929). *An Introduction to the History of Medicine, with Medical Chronology, Suggestions for Study and Bibliographic Data*, 4th edn. Philadelphia & London: W. B. Saunders Co.

Gasking, E. (1970). *The Rise of Experimental Biology*. New York: Random House.

Goldstein, P. (1965). *Triumphs in Biology*. Garden City, New York: Doubleday and Co.

Groeben, C. & Muller, I. (1975). *The Naples Zoological Station at the time of Anton Dohrn*. Naples; Stazione Zoologica di Napoli.

Harvey, A. M. (1974). *Adventures in medical research: a century of discovery at Johns Hopkins*. Baltimore: Johns Hopkins Univ. Press.

Hertwig, O. (1908). *Die Entwicklung der Biologie im Neunzehnten Jahrhundert*, 2nd edn. Jena: Gustav Fischer.

Jahn, I. (1982). *Geschichte der Biologie*. Jena: Fischer.

Lenham, U. (1968). *Origins of Modern Biology*. New York: Columbia University Press.

Lenoir, T. (1982). *The Strategy of Life: Teleology and Mechanics in Nineteenth Century German Biology*. Dordrecht: Reidel.

Lillie, F. R. (1944). *The Woods Hole Marine Biological Laboratory*. Chicago: University of Chicago Press.

Locy, W. A. (1915). *Biology and Its Makers*, 3rd edn. New York: Henry Holt.

Locy, W. A. (1925). *The Growth of Biology*. New York: Henry Holt.

Mason, S. F. (1953). *A History of the Sciences; Main Currents of Scientific Thought*. London: Routledge & Kegan Paul.

Mayr, E. (1982). *The Growth of Biological Thought, Diversity, Evolution and Inheritance*. Cambridge, Mass.: Harvard University Press.

Mazzeo, J. A. (1967). *The Design of Life. Major Themes in the Development of Biological Thought*. New York: Pantheon Books.

Merz, J. T. (1965). *A History of European Thought in the Nineteenth Century*. New York: Dover.

*Nobel Lectures in Physiology and Medicine*, vols. I–III, 1901–1961. Amsterdam: Elsevier.

Nordenskiold, E. (1928). *The History of Biology; a Survey*. New York: Tudor Publ. Co.

Petit, G. & Theodorides, J. (1962). *Histoire de la zoologie* (Histoire de la Pensée, **8**). Paris: Hermann.

Radl, E. (1930). *The History of Biological Theories* (transl. and adapted by E. J. Hatfield). London: Oxford University Press.

Ritterbush, P. C. (1964). *Overtures to Biology: The Speculations of Eighteenth Century Naturalists*. New Haven & London: Yale Univ. Press.

Schiller, J. (1979). *La Notion d'Organisation dans l'Histoire de la Biologie*. Paris: Maloine.

Singer, C. J. (1959). *A Short History of Biology*, 3rd edn. London & New York: Abelard-Schmann.

Sirks, M. J. & Zirkle, C. (1964). *The Evolution of Biology*. New York: Ronald Press.

Taton, R. (1963). *History of Science*, transl. A. J. Pomerans. New York: Basic Books.

Taylor, G. R. (1963). *The Science of Life: a Picture History of Biology*. New York: McGraw-Hill.

Thomson, A. L. (1973–5). *Half a Century of Medical Research*, vol. I, Origins and policy of the Medical Research Council (UK), vol. II, The programme of the Medical Research Council (UK), London: HMSO.

Thomson, J. A. (1899). *The Science of Life; an Outline of the History of Biology and its Recent Advances*. Chicago: H. S. Stone.

Uschmann, G. (1959). *Geschichte der Zoologie und der zoologischen Anstalten in Jena 1779–1914*. Jena: Fischer.

Wendel, G. (1975). *Die Kaiser-Wilhelm-Gesellschaft 1911–1914. Zur Anatomie einer Imperialistischen Forschungsgesellschaft*. Berlin: Akademie-Verlag.

Werski, G. (1978). *The Visible College*. London: Allen Lane.

Winsor, M. P. (1976). *Starfish, Jellyfish and the Order of Life: Issues in 19th Century Science*. New Haven: Yale University Press.

*Additional Source:*
*Historical Abstracts* (1955–).

**History of embryology**

Adelmann, H. B. (1942). *The Embryological Treatises of Hieronymus Fabriccius ab Aquapendente*. A facsimile edition with an introduction, a translation, and a commentary, 2 vols. Ithaca: Cornell University Press.

Adelmann, H. B. (1966). *Marcello Malpighi and the Evolution of Embryology*, 5 vols. Ithaca: Cornell University Press.

Cole, F. J. (1930). *Early Theories of Sexual Generation*. Oxford: Clarendon Press.

Gasking, E. (1967). *Investigations into Generation 1651–1828*. Baltimore: Johns Hopkins Press.

Haraway, D. J. (1976). *Crystals, Fabrics and Fields. Metaphors of organicism in twentieth-century developmental biology*. New Haven and London: Yale University Press.

Harvey, W. *De Generatione Animalium*, transl. G. Whitteridge (1982). Oxford: Blackwells.

Meyer, A. W. (1936). *An Analysis of the De generatione animalium of William Harvey*. London: Oxford University Press; Stanford: Stanford Univ. Press.

Meyer, A. W. (1939). *The Rise of Embryology*. London: Oxford University Press; Stanford University Press.

Meyer, A. W. (1956). *Human generation. Conclusions of Burdach, Dollinger and von Baer*. Stanford: Stanford University Press.

Needham, J. (1959). *A History of Embryology*, 2nd edn. Cambridge: Cambridge University Press.

Oppenheimer, J. M. (1967). *Essay in the History of Embryology and Biology*. Cambridge, Mass.: M.I.T. Press.

Roe, S. A. (1981). *Matter, life and Generation: 18th Century Embryology and the Haller–Wolff Debate*. Cambridge: Cambridge University Press.

Russell, E. S. (1916). *Form and Function*. London: Murray.

Tanner, J. M. (1981). *A History of the Study of Human Growth*. Cambridge: Cambridge University Press.

*Essay reviews:*
J. Hist. Biol. (1970) **3**, 158–81; Q. Rev. Biol. (1975) **50**, 373–87; Q. Rev. Biol. (1976), Special 50th Anniversary Issue.

**History of microscopy and cell theory**

Aschoff, L., Kuster, E. & Schmidt, W. J. (1938). *Hundert Jahre Zellforschung*. Berlin: Gebruder Borntraeger.

Bracegirdle, B. (1978). *A History of Micro Technique*. New York: Cornell University Press.

Bradbury, S. (1967). *The Evolution of the Microscope*. Oxford: Pergamon.

Clay, R. S. & Court, T. H. (1932). *The History of the Microscope*. London: Griffin.

Conn, H. J. (1948). *The History of Staining*, 2nd edn. Geneva, New York: Biotech Publ.

Farley, J. (1982). *Gametes and Spores*. Baltimore & London: Johns Hopkins Univ. Press.

Ford, B. J. (1985). *Single lens: the Story of the Simple Microscope*. London: Heinemann.

Freund, H. & Berg, A. (eds.) (1965). *Geschichte der Microskopie: Leben und Werk grosser Forscher*. Frankfurt: Umschau Verlag.

Gray, F. & Gray, P. (1956). *Annoted Bibliography of Works in Latin Alphabet Languages on Biological Microtechniques*. Dubuque, Iowa: Brown.

Gray, P. (1973). *The Encyclopedia of Microscopy and Microtechnique*. New York: Van Nostrand.

Hughes, A. F. G. (1959). *A History of Cytology*. London: Abelard–Schumann.

Kasten, F. H. & Clark, G. (1983). *History of Staining*. Baltimore: Williams and Wilkins.

Klein, M. (1936). *Histoire des Origines de la Théorie Cellulaire*. Paris: Hermann.

Lee, A. B. (1885). *The Microtomist's Vade-Mecum. A handbook of the methods of Microscopic Anatomy*. London: J. & A. Churchill (11th edn, 1950).

Murray, M. R. & Kopeck, G. (1953). *Bibliography of Tissue Culture Research 1885–1951*. New York: Academic Press.

Tuson, J. (1978). *The Cell Doctrine: Its History and Present Status*, 2nd edn. Philadelphia: Lindsay & Blakiston.

## History of microbiology

Bulloch, W. (1979). *The History of Bacteriology*. New York: Dover.

Doetsch, R. N. (1960). *Microbiology: Historical contributions from 1776–1908*. New Brunswick, NJ: Rutgers Univ. Press.

Lechevalier, H. A. & Solotorovsky, M. (1965). *Three Centuries of Microbiology*. New York: McGraw-Hill.

## History of biochemistry

Fruton, J. S. (1972). *Molecules and Life: Historical Essays on the Interplay of Chemistry and Biology*. New York and London: Wiley Interscience.

Keilin, D. (1966). *The History of Cell Respiration and Cytochrome*. Cambridge: Cambridge University Press.

Kohler, R. E. (1982). *From Medical Chemistry to Biochemistry, the Making of a Biomedical Discipline*. Cambridge: Cambridge University Press.

Needham, J. & Baldwin, E., ed. (1949). *Hopkins and Biochemistry 1861–1947*. Papers concerning Sir Frederick Gowland Hopkins ... with a selection of his addresses and a bibliography of his publications. Cambridge: Heffer.

## History of endocrinology

Medvei, V. C. (1982). *A History of Endocrinology*. Lancaster & Boston: MTP Press.

Rolleston, H. D. (1936). *The Endocrine Organs in Health and Disease*. London: Oxford University Press.

## History of botany

Arber, A. (1950). *The natural philosophy of plant form*. Cambridge: Cambridge University Press.

Delaporte, F. (1982). *Nature's second kingdom*. Cambridge, Mass. & London: M.I.T. Press.

Green, J. R. (1909). *History of Botany 1860–1900, Being a Continuation of Sachs' History of Botany*. Oxford: Clarendon.

Morton, A. G. (1981). *History of Botanical Science*. London: Academic Press.

Sachs, J. von (1889). *History of Botany 1530–1860*, transl. H. G. F. Garnsey & I. B. Balfour). Oxford: Clarendon.

Verdoorn, F. (ed.) (1942). *A Short History of Plant Sciences*. New York: Ronald.

Wardlaw, C. W. (1952). *Phylogeny and Morphogenesis. Contemporary Aspects of Botanical Science*. London: Macmillan.

Wardlaw, C. W. (1968). *Essays on Form in Plants*. Manchester: Manchester University Press.

### History of physiology

Brooks, C. M. and Cranefield, P. F. (1959). *The historical development of physiological thought*. New York: Hafner.

Franklin, K. J. (1969). *Short History of Physiology*. London: Hutchinson.

Goodfield, G. J. (1960). *The Growth of Scientific Physiology*. London: Hutchinson.

Hall, T. S. (1969). *Ideas of life matter. Studies in the History of General Physiology 600 BC–1900 AD*, 2 vols. Chicago and London: University of Chicago Press.

Rothschuh, K. E. (1969). *Physiologie im Werden*. Stuttgart: Fischer.

Rothschuh, K. E. (1972). *Geschichte der Physiologie*, 2nd edn. Berlin: Springer. (Issued in translation as History of Physiology (1973), transl, G. B. Risse, New York: Krieger).

Wightman, W. P. D. (1956). *The Emergence of General Physiology*. Belfast: Queen's University of Belfast.

### History of obstetrics and gynaecology

Findley, P. (1939). *Priest of Lucina; the story of obstetrics*. Boston: Little, Brown.

Graham, H. (pseud.) [i.e. I. H. Flack] (1950). *Eternal Eve*. London: Heinemann Medical Books.

Himes, N. E. (1936). *Medical History of Contraception*. (re-issued, 1963). New York: Gamut Press.

Leonardo, R. A. (1944). *History of Gynaecology*. New York: Froben Press.

Ricci, J. V. (1945). *One Hundred Years of Gynaecology 1800–1900*. Philadelphia: Blakiston.

Ricci, J. V. (1950). *The Genealogy of Gynaecology*, 2nd edn. Philadelphia: Blakiston.

Speert, H. (1958). *Obstetric and Gynaecologic Milestones: Essays in Eponymy*. New York: Macmillan.

### History of genetics

Beadle, G. W. (1963). *Genetics and Modern Biology*. Philadelphia: American Philosophical Society.

Carlson, E. A. (1966). *The Gene: a Critical History*. Philadelphia: Saunders.

Delage, Yves (1903). *L'Hérédité et les Grandes Problèmes de la Biologie générale*, 2nd edn, Paris: Reinwald and Sleicher.

Dunn, L. C., ed. (1951). *Genetics in the 20th Century; Essays on the Progress of Genetics during its First 50 Years*. New York: Macmillan.

Dunn, L. C. (1965). *A short history of genetics*. NY: McGraw Hill.
Hsu, T. C. (1979). *Human and Mammalian Cytogenetics: an historical perspective*. New York: Springer.
Jacob, F. (1973). *The Logic of Life: a History of Heredity*. New York: Pantheon Books.
Judson, J. F. (1979). *The Eighth Day of Creation. Makers of the Revolution in Biology*. London: Jonathan Cape.
Koestler, A. (1971). *The Case of the Midwife Toad*. London: Hutchinson.
Mackenzie, D. A. (1981). *Statistics in Britain 1865–1930: the Social Construction and Scientific Knowledge*. Edinburgh: Edinburgh University Press.
Moore, J. A. (1972). *Heredity and Development*. Oxford: Oxford University Press.
Olby, R. C. (1966). *Origins of Mendelism*. London: Constable.
Olby, R. (1974). *The Path to the Double Helix*. London: Macmillan.
Portugal, F. H. & Cohen, J. S. (1977). *A Century of DNA*. Cambridge, Mass.: & London: M.I.T. Press.
Ravin, A. W. (1965). *The Evolution of Genetics*. NY: Academic Press.
Robinson, G. (1981). *Preludes to Genetics*. Lawrence, Kansas: Coronado Press.
Stubbe, H. (1972). *History of Genetics. From Prehistoric Times to the Rediscovery of Mendel's Laws*. Cambridge, Mass: MIT Press.
Sturtevant, A. H. (1965). *A History of Genetics*. New York: Harper & Row.
Watson, J. D. (1968). *The Double Helix. A Personal Account of the Discovery of the Structure of DNA*. New York: Athenaeum. New critical edn (G. S. Stent) (1981). London: Weidenfeld and Nicolson.
Watson, J. D. & Tooze, J. (1981). *The DNA Story: A Documentary History of Gene-Cloning*. San Francisco: Freeman.

**History of evolution theory**
Bowler, P. J. (1983). *The Eclipse of Darwin; anti-Darwinian Evolution Theories in the Decades Around 1900*. Baltimore: Johns Hopkins University Press.
Bowler, P. J. (1984). *Evolution. The History of an Idea*. Berkeley: University of Calif. Press.
Grene, M. G. (ed.) (1983). *Dimensions of Darwinism, Themes and Counterthemes in 20th Century Evolutionary Theory*. Cambridge: Cambridge University Press.
Gould, S. J. (1977). *Ontogeny and Phylogeny*. Cambridge, Mass.: Belknap Press of Harvard University Press.
Kohn, D. & Mayr, E. (eds.) (1985). *The Darwinian Heritage*. Princeton: Princeton University Press.
Mayr, E. & Provine, W. B. (1980). *The Evolutionary Synthesis*. Cambridge, Mass.: Harvard University Press.
Provine, W. B. (1971). *The Origins of Theoretical Population Genetics*. Chicago: University of Chicago Press.

**Readings**
Brock, T. D. (ed.) (1961). *Milestones in Microbiology*. Englewood Cliff, NJ: Prentice-Hall.
Carlson, E. A., ed. (1967). *Modern Biology; Its Conceptual Foundations*. New York: George Braziller.

Flickinger, R. A. (1966). *Developmental Biology, a Book of Readings*. Dubuque, Iowa: Wm. C. Brown Co.

Fulton, C. & Klein, A. O. (1976). *Explorations in Developmental Biology*. Cambridge Mass.: Harvard University Press.

Fulton, J. F. (1966). *Selected Readings in the History of Physiology*. Springfield, Ill.: Thomas.

Gabriel, M. L. & Fogel, S. (1955). *Great Experiments in Biology*. Englewood Cliffs, NJ: Prentice-Hall.

Hall, T. S., ed. (1970). *A Source Book in Animal Biology*. Cambridge Mass.: Harvard University Press.

Knickerbocker, W. S. (1927). *Classics of Modern Science*. New York: Knopf.

Laetsch, W. M. (1969). *The Biological Perspective; Introductory Readings*. Boston: Little, Brown.

Moore, J. A. (1972). *Readings in Heredity and Development*. Oxford: Oxford University Press.

Peters, J. A. (ed.) (1959). *Classic Papers in Genetics*. Englewood Cliffs, NJ: Prentice-Hall.

Pisuner, A. (1955). *Classics of Biology* (transl. C. M. Stern). London: Pitman.

Stern, C. & Sherwood, E. R. (1966). *The Origin of Genetics: a Mendel Source Book*. San Francisco: W. H. Freeman.

Voeller, B. R. (1968). *The Chromosome Theory of Inheritance. Classic Papers in Development and Heredity*. New York: Appleton-Century-Crofts.

Willier, B. J. & Oppenheimer, J. M. (1964). *Foundations of Experimental Embryology*. Englewood Cliffs, NJ: Prentice-Hall.

## National differences

Bates, R. S. (1945). *Scientific Societies in the United States*. New York: Wiley; London: Chapman & Hall.

Ben-David, J. (1971). *The Scientist's Role in Society. A Comparative Study*. Englewood Cliffs, NJ: Prentice-Hall.

Beyerchen, A. (1977). *Scientists under Hitler: Politics and the Physics Community in the Third Reich*. New Haven: Yale University Press.

Bonner, T. N. (1963). *American Doctors and German Universities: A Chapter in International Intellectual Relations 1870–1914*. Lincoln: University of Nebraska Press.

Bruford, B. W. (1962). *Culture and Society in Classical Weimar*. Cambridge: Cambridge University Press.

Conry, Y. (1974). *L'Introduction du Darwinisme en France au XIXe Siècle*. Paris: Vrin.

Cravens, H. (1978). *The Triumph of Evolution: American Scientists and the Heredity–Environment Controversy, 1900–1941*. Philadelphia: University of Pennsylvania Press.

Daniels, G. (ed.) (1972). *Nineteenth Century American Science*. Evanston: Northwestern University Press.

Finkelstein, L. (ed.) (1953). *Thirteen Americans: Their Spiritual Autobiographies*. New York: Harper.

Geison, G. L. (1978). *Michael Foster and the Cambridge School of Physiology.* Princeton: Princeton University Press.

Glick, T. F. (1972). *The Comparative Reception of Darwinism.* Austin: University of Texas Press.

Greenberg, D. S. (1969). *The Politics of American Science.* Harmandsworth: Penguin.

Gregory, F. (1977). *Scientific Materialism in Nineteenth Century Germany.* Dordrecht and Boston: Reidel.

Haller, M. (1963). *Eugenics: Hereditarian Attitudes in American Thought.* New Brunswick: Rutgers University Press.

Hirsch, W. (1968). *Scientists in American Society.* New York: Random House.

Hughes, A. F. M. (1982). *The American Biologist Through Four Centuries.* Springfield, Ill.: Thomas.

Jaffe, B. (1944). *Men of Science in America. The Role of Science in the Growth of Our Country.* New York: Simon and Schuster.

Kelly, A. (1981). *The Descent of Darwin; the Popularisation of Darwinism in Germany, 1860–1914.* Chapel Hill: University of North Carolina Press.

Ludmerer, K. (1972). *Genetics and American Society. A Historical Appraisal.* Baltimore: Johns Hopkins University Press.

Morrell, J. B. & Thackray, A. (1981). *Gentlemen of Science: The Early years of the British Association for the Advancement of Science.* Oxford: Oxford University Press.

Mullen, P. C. (1964). *The Preconditions and Reception of Darwinian Biology in Germany, 1800–1870.* Berkeley: University of California Press.

Oleson, A. & Voss, J. (eds.) (1979). *The Organization of Knowledge in Modern America, 1870–1920.* Baltimore: Johns Hopkins University Press.

Reingold, N. (ed.) (1966). *Science in Nineteenth Century America.* London: Macmillan.

Rossiter, M. (1975). *Emergence of Agricultural Science: Justus Liebig and the Americans, 1840–1880.* New Haven: Yale University Press.

Russett, C. E. (1976). *Darwin in America. The Intellectual Response 1865–1912.* San Francisco: W. H. Freeman & Co.

Schuster, A. & Shipley, A. E. (1917). *Britain's Heritage of Science.* London: Constable.

Struik, D. J. (1957). *Yankee Science in the Making.* New York: Cameron.

Turner, G. L. E. (ed.) (1976). *The Patronage of Science in the Nineteenth Century.* London: Noordhoff.

Van Tassel, D. & Hall, M. (1966). *Science and Society in the U.S.* Homewood: Dorsey Press.

von Aesch, A. G. (1941). *Natural Science in German Romanticism.* New York: Columbia University Press.

Weinreich, M. (1946). *Hitler's Professors.* New York: Yiddish Scientific Institute.

### History of scientific disciplines

Basseau, H. (ed.) (1983). *L'Explication dans les Sciences de la Vie.* Paris: C.N.R.S.

Eulner, H-H. (1971). *Die Entwicklung der medizinischen Spezialfächer an den Universitäten des deutschen Sprachgebietes*. Stuttgart: Enke.

Graham, L. (1983). *Functions and Use of Disciplinary History*. Dordrecht: Reidel.

Lemaine, G. (1977). *Perspectives on the Emergence of Scientific Disciplines*. Paris: Masson.

Mayr, E. & Provine, W. B. (1980). *The Evolutionary Synthesis*. Cambridge, Mass.: Harvard University Press.

## Current Guides to Sociology of Science

Barnes, B. & Edge, D. (1982). *Science in Context: Readings in the Sociology of Science*. Milton Keynes: Open University.

Knorr-Cetina, K. D. & Mulkay, M. (1983). *Science observed. Perspectives on the Social Study of Science*. London: Sage.

Spiegel-Rosing, I. & Price, D. (1977). *Science, Technology and Society: A cross-disciplinary perspective*. London: Sage.

Ziman, J. (1984). *An Introduction to Science Studies* – the philosophical and social aspects of science and technology. Cambridge: Cambridge University Press.

*Additional Sources:*
*Social Science Citation Index* (1973–). Philadelphia: Institute for Scientific Information.
*Social Sciences and Humanities Index* (1965–), New York: Wilson.
*Social Studies in Science*
*Sociological Review*
*Sociology of Science Yearbook*

## Philosophy of biology & theoretical biology relevant to embryology

Ayala, F. J. & Dobzhansky, T. D. (1974). *Studies in the Philosophy of Biology*. London: Macmillan.

Beckner, M. (1959). *The Biological Way of Thought*. New York: Columbia University Press.

Bertalanffy, L. von (1962). *Modern Theories of Development; an Introduction to Theoretical Biology* (transl. and adapted from the original (1933) German edition by J. H. Woodger). New York: Harper.

Cassirer, E. (1950). *The Problem of Knowledge; Philosophy, Science and History since Hegel*. New Haven: Yale University Press.

DeGrood, D. (1965). *Haeckel's Theory of the Unity of Nature: a Monograph in the History of Philosophy*. Boston: Christopher Publishing House.

Driesch, H. (1908). *The Science and Philosophy of the Organism*. London: Adam and Charles Black.

Driesch, H. A. E. (1914). *The History and Theory of Vitalism* (transl. C. K. Ogden). London: Macmillan.

Gregg, J. R. & Harris, F. T. C. (1964). *Form and Strategy in Science*. Dordrecht: Reidel.

Grene, M. (ed.) (1968). *Approaches to a Philosophical Biology*. New York: Basic Books.

Haldane, J. B. S. (1935). *The Philosophy of a Biologist*. Oxford: Oxford University Press.

Hull, D. (1974). *Philosophy of Biological Science*. Englewood Cliffs: Prentice-Hall.

Koestler, A. & Smythies, A. (1960). *Beyond Reductionism*. London: Hutchinson.

Lillie, R. S. (1945). *General Biology and Philosophy of Organicism*. Chicago: University of Chicago Press.

Lovejoy, A. O. (1966). *The Great Chain of Being; a Study of the History of an Idea*. Cambridge, Mass.: Harvard University Press.

Nagel, E. (1961). *The Structure of Science*. London: Routledge & Kegan Paul.

Needham, J. (1936). *Order and Life*. New Haven: Yale University Press.

Rittenbush, P. C. (1968). *The Art of Organic Forms*. Washington: Smithsonian Institution Press.

Ritter, W. E. (1919). *The Unity of the Organism; or, The Organismal Conception of Life*, 2 vols. Boston: R. G. Badger.

Ruse, M. (1973). *The Philosophy of Biology*. London: Hutchinson.

Russell, E. S. (1930). *The Interpretation of Development and Heredity*. Oxford: Clarendon Press.

von Uexküll, J. J. (1926). *Theoretical Biology*. New York: Harcourt, Brace & Co.

Waddington, C. H., ed. (1968–1972). *Towards a Theoretical Biology*, 4 vols. 1–4. Chicago: Aldine Publishing Co.; Edinburgh: Edinburgh University Press.

Weiss, P. A. (1973). *The Science of Life: the Living System – a System for Living*. New York: Futura.

Whyte, L. L. (1968). *Aspects of Form* (2nd edn). London: Lund Humphries.

Whyte, L. L., Wilson, A. G. & Wilson, D. (1969). *Hierarchical Structures*. New York: American Elsevier.

Woodger, J. H. (1929). *Biological Principles: a Critical Study*. London: K. Paul, Trench, Trubner & Co.; New York: Harcourt Brace & Co.

*Additional sources:*

*Bibliotheca Biotheoretica*, Leiden (1941–).

*The Philosophers Index* (1967–). Bowling Green, Ohio: Bowling Green University Press.

# Index

Agassiz, Alexander, 116
Agassiz, Louis, 109, 116
agricultural breeding programmes
  and genetics–embryology split, 141
  and heredity research, 135–7
Allman, George, James
  hydroid evolution, 17
  polarity, 349
allometry, xvi, 53, 63
alternation of generations, 5
*Alternation of Generations, On The* (1845),
  J. J. Steenstrup, xii, 13
*Amblystoma*, 190, 227
amphibia, 80–1, 153ff, 186ff, 246, 252ff,
  265ff, 332ff
  *see also individual species*
*Amphioxus*, 111
*Animal Biology* (1927), J. B. S. Haldane,
  58–9
analogies, 10, 38
*Anthomedusae*, 14
Aristotle, 3
*Ascaris*, 185, 352
  chromosomes, 187, 214
  pronuclear fusion and genome
    formation, 397, 399
Assheton, Richard, 42, 48–9
Astbury, W. T.
  dynamic relations of molecular
    structures, 298
  protein molecular morphology, 297

*Balanoglossus*, 42
Balbiani, E. G., xiii, 14
Balfour, Francis Maitland, xiii, 7, 9, 14, 27
  ancestral and non-ancestral
    developmental stages, 43
  biogenetic law, 12
  at Cambridge, 37, 41–4
  heredity and variation, 29

  larval and embryonic development,
    43–4
  recapitulation, 40ff
  visit to Kleinenberg, 36
  visits to Stazione Zoologica, 37
  on von Baer, 42
  *A Treatise on Comparative Embryology*
    (1880–1), xiii, 42
Barth, L. G., 230
  primary induction factors, 254, 255, 267,
    268, 269
Bateson, Gregory
  and C. H. Waddington, 311–12
Bateson, William, 42, 44, 46
  and C. H. Waddington, 312
  genetics–embryology split, 132
  particulate inheritance and Mendelism,
    45
  on von Baer, 45
  *Materials for the Study of Variation*
    (1894), xv, 45
Baur, Erwin, xvi, 206
Bautzmann, H., 199, 200, 253
Bechold, H., colloids, 293
Belgium, 245ff
Beneden, Edouard Van, xiii, xiv, 246
Berrill, N. John, 26, 27
  continuity of germ-plasm, 13
Bertalanffy, Ludwig von, xvii
  The System Theory, 385–6
Beth's theorem, 408, 428–30, 432
*Biochemical Cytology* (1967), J. Brachet,
  245
*Biochemistry and Morphogenesis* (1942), J.
  Needham, 47, 251, 253, 256
biogenetic law, 5, 10–12, 85
  dismissal of, 11, 52, 53, 60–1
  germ-layer doctrine, 27
  phenotype concept, 126
  *see also* recapitulation

*Biological Principles* (1929), J. Woodger, 314–15
blastopore formative powers, 215
*see also* dorsal blastopore lip
Bonnet, Charles
regeneration, 333, 334
Bordieu, Pierre
competition between research fields, 138
Born, G., xv
experimental approach, 170
heteroplastic grafting, 190
Bourne, Gilbert C., 36, 51
Boveri, Theodor, xiv, xv, 9, 181
on A. Weismann, 214
*Ascaris*, 185, 187, 214, 352
chromosomes, 187, 212ff, 400–1
cultural values of biology, 204–5
cytoplasmic and nuclear roles, 186–7, 365, 367
genes and embryo properties, 400–1
gradients and polarity, 349–50
and H. Spemann, 184–5, 186, 205, 213–14, 217
Kaiser–Wilhelm Institut für Biologie, 216
nationalistic views, 245–6
*Paracentrotus*, 187
privileged regions, 187, 188, 220
sex determination, 100
Bower, Frederick Orpen, 14
Brachet, A
in Brussels, 246, 247
and the organiser, 252
in U.S.A., 247
Brachet, J.
DNA in chromosomes, 248
DNA synthesis, 248–9
organiser's inducing factor, 253–4, 255
respiration sites in eggs, 250–1
RNA detection, 249–50, 300
*Biochemical Cytology* (1967), 245
*Chemical Embryology* (1950), 251, 253, 256
Braus, Hermann
double assurance, 191
protoplasmic bridges and nerve fibre development, 157–8
Bridges, Calvin B., xvi
genes and chromosomes, 121, 142
Britain
Cambridge, 41–50, 267
institutionalised biology *v.* private endeavour, 69
organicism, 69–70
Oxford, 50–63
response to organiser, 226
Brooks, William Keith
coelenterate phylogeny, 15
heredity, continuity theories, 29

at John Hopkins University, 109, 116
Browne, E. N., 353
Brucke, C.
cell organisation, xii, 280, 287
Bryn Mawr College, 117, 122, 129, 190
Burbank, Luther, 136
Bütschli, O, xiii, 181

*Caenorhabditis*, 336
Cajal, Ramon y, xiv
nerve fibre development, 151, 152, 153, 159
California Institute of Technology (Cal Tech)
Division of Biology, 281
establishment of, 123
*Campanularia gelatinosa*, 18
canalization, 321, 324
Carnegie Foundation
heredity research, 136–7
Carrel, Alexis
tissue culture, 164–5
Castle, W. E.
*Drosophila* breeding, 141
and *Journal of Experimental Zoology*, 168
*Cell, The* (1895), O. Hertwig, 6
*Cell, Development and Heredity, The* (1925), E. B. Wilson, xv, 4, 140, 144, 368–70
*Cell in Development and Inheritance, The* (1896), E. B. Wilson, 93, 95, 111
cell lineages, 87, 110
and homologies, 89
cell organelles
central role re-established, 300–1
genetic continuity, 301
cells
colloids in, 280, 292–3
crystallisation structures, 288
dynamic forces, 287–8, 289
microscopical structures, 275
properties and reductionism, 279–80
significance of cytologists' structures, 280f
sizes of, 282, 283, 286
structural organisation, 299–300
submicroscopic entities, 284–5
surface effects in, 299
Chamberlain, Houston Stewart, 205, 207
*Chemical Embryology* (1950), J. Brachet, 251, 253, 256
*Chemical Embryology* (1930), J. Needham, 245, 315
Child, Charles Manning, xvi, 59, 98, 181
gradient theory, 201, 227, 341, 343, 352, 355
and the organiser, 226, 227, 252
organism a colloidal system, 293

Child, Charles Manning, (*cont.*)
  *Planaria* regeneration, 351
  polarity and gradients, 351–2
  psychic factors, 218
  sex determination, 100
  *Individuality in Organisms* (1915), 352
*Chironomus*, 297
X-chromosome, 120, 374–5
chromosomes, 94, 95, 187, 214, 215–16
  combinations of, and embryogenesis,
    400–1
  and genes, 121, 128, 129ff
  lampbrush, and DNA, 248
  sex determination, 99–100, 129
Chuang, H,-H.
  inductive signals, 269
classical zoology
  aims and beginnings, 35
  embryology origins, 38
Claus, Carl, 14
  hydroid evolution, 17–18
*Clava Squamata*, 18
cleavage, 88, 89, 96–7
*Clepsine*, 87–8, 88–9, 110
  germ-layers and germ bands, 88–9
clonal restriction, 381, 382, 383
Coleman, William, 124
*Collecting Net, The* (1937), E. G. Conklin,
    109
colloid aggregate theory, 292
  and macromolecules, 294ff
colloids, 291ff
  in cells, 280, 292–3
  dimensions of particles, 283, 292
  dispersed systems, 278
  dynamical theories, 293–4
  properties, 292
  protoplasmic peculiarities, 280, 292
  widespread occurrence, 292–3
Colucci, lens formation, 333
Conklin, Edward Grant, 93, 95–8
  cleavage and development, 96–7
  *Cynthia*, 111
  cytoplasmic factors in development, 96, 97
  heredity and embryology, 125–6
  and *Journal of Experimental Zoology*, 168
  preformation-epigenesis, 97
  sex determination, 100
  on Spemann, 226
  *The Collecting Net* (1937), 109
Cope, Edward Drinker, 11
*Coronata*, 18
Correns, xv, xvi, 181, 216
Crick, Francis, 283, 325, 421
  fundamental dogma, 250
*Crystals, Fabrics and Fields* (1976),
    D. J. Haraway, 281
*Ctenophores*, 111

Cuvier, G. L., 38, 40, 41, 189
cybernetics, 310, 324–5
*Cynthia*, 111
cytoplasm
  environmental factors and development,
    125
cytoplasmic gradients
  privileged region, 187, 188, 220
cytoplasmic *v.* nuclear control
  development, 90, 95–6, 364–5, 367, 398,
    400
  heredity, 125, 131–2, 134, 369–70

Dalcq, Albert M., 246, 257
  cytoplasmic role *v.* classical genetics, 134
  gradients and threshold concept, 354
  and J. Brachet, 250, 252
*Daphnia*, 10, 13
Darwin, Charles R., xii, xiii, 7, 14
  gemmules, 4, 284, 285
  homologies, 39
  pangenesis, 4
Darwinism, 5
*Das Keimplasma* (1892), A. Weismann, xiv,
    9, 75, 185, 363
Davenport, C. B., 136, 148
de Beer, Gavin, 37, 38
  biogenetic law, 11, 27
  gradient-fields, 353–4
  heterochrony and evolution, 62
  larval adaptations, 44
  and the organiser, 226
  paedomorphosis, 61
  phylogeny, 62–3
  recapitulation, 58
  *Elements of Experimental Embryology*
    (1942), 60
  *Embryology and Evolution* (1930), 47, 60,
    62
de Vries, Hugo, xiv, xv
  hereditary mechanisms, 6
  mutation theory, xv, 120
  *Intracellular Pangenesis* (1889), 286
*Dentalium*, 111
Descartes, René, 3
determinants, 76ff, 187
developmental biology, 229–30
*Die physiologische Theorie der Vererbung*
    (1927), R. Goldschmidt, 368, 377
dimensions
  dispersed systems, 277, 278
  in physiological explanations, 276–84
  submicroscopic entities, 284–5
dimensions, molecular
  in cells, 284
  structural continuity, 281, 282
dimensions of organisms, 276–7
  minimum size, 277, 286

dimensions of organisms (*cont.*)
  and other objects, 283
*Diptera*, 10
DNA (deoxyribonucleic acid), 247–8
  synthesis or migration? 248–9
aDNA (thymonucleic acid), 247
Dohrn, Anton, 37
  pycnogonid derivation, 117
dorsal blastopore lip, 188, 220, 251
  heteroplastic grafting, 195, 199
  respiratory metabolism, 252, 253–4
  *see also* blastopore
Driesch, Hans, xiv, xv, 7, 9, 74–5, 117–18,
    181, 210
  cytoplasm and nucleus in ontogenesis,
    364–5
  position and harmonious equipotential
    system, 347ff
  regulation, regeneration and asexual
    reproduction, 340–1
  regulative development, 82–4, 186, 347
  and W. Roux, 82–4, 117–18, 186
  vitalism, 348, 365
*Drosophila*, 5, 100, 113, 141–2
  body segment genetic determination,
    383–4
  X-chromosome, 374–5
  clonal restriction in wing, 382, 383
  fate map model, 387
  genotype–phenotype distinction, 128, 129
  gynandromorphs, 375–6
  ontogenetic gene action, 371ff
  white-eyed, 119, 120, 126
  wing development, 319
*Drosophila* mutants
  *bicaudal*, 380
  *bithorax*, 380, 382, 383, 384
  *dicephalic*, 380, 385
  *gaunty*, 372
  *Notch* (*facet*), 374
  *vermilion*, 375–6
*Dugesia lugubris*, 339, 343
Durken, B., 200

ear formation, 185, 192
East, E. M., 137
*Elements of Experimental Embryology* (1942),
    J. S. Huxley and G. de Beer, 60
*Embryology and Evolution* (1930), G. de
    Beer, 60
*Embryonic Development and Induction*
    (1938), H. Spemann, 183, 195ff, 251
Emerson, R. A., 121, 137
entocodon phylogeny, 18, 19, 20
environmental factors, 125
*Epestia*
  *a/a*, 376, 377
epigenesis, 3, 124

  *see also* preformation–epigenesis debates
epigenetic landscape, 318, 320–1, 371
Errera, Leo
  minuteness limit of microorganisms,
    286
*Eudendrium* germ cells
  migration, 25
  origin sites, 15, 16
*Eudendrium racemosum*
  germ cell migration, 24
*Eudendrium ramosa*, 18
*Euphausiacea*
  recapitulation absence, 52
*Euscelis*, 356
evolution, 63, 310, 321, 325
  heredity and development, 118–20
  paedomorphosis and rate genetics, 62
  preformation–epigenesis debates, 75ff
  in small steps, 57–8
  sporosac, 17–21
*Evolution: The Modern Synthesis* (1942),
    J. S. Huxley, 63
*Evolution and Adaptation* (1903), T. H.
    Morgan, 118
experimental approach
  change to, at turn of century, 166–8
  non-acceptance by some scientists, 165–6,
    171–2
  suspicions about, 211
  validity of results, 172–3
*Experimental Embryology* (1909), W.
    Jenkinson, xv, 51, 52
eye regeneration, 332, 333
  *see also* lens induction

Farley, John, 13
  *The Spontaneous Generation Controversy*
    (1974), 5
Fell, Honor B., 313, 322
field concept, 143
  and gradients, 252–3, 352ff
  and the organiser, 196, 198
  regeneration, 342–4, 353–4
Fischer, Albert, 164, 165
  *Tissue Culture* (1925), 163f
Fisher, R. A., xvii
  evolution, 57–8
Fletcher, Walter, 313
Ford, E. B.
  experimental biology, 59
  rate genes, 36, 60
  recapitulation, 55–6
*Form and Function* (1916), E. S. Russell, 5,
    69–70, 179
Foster, Michael, 41
  on work of D'Arcy Thompson, 48
  homologies, 48
*Für Darwin* (1863), F. Müller, 35

*Galleria*, 356
Galton, Francis
    Law of Ancestral Inheritance 1889, 39
    *Natural Inheritance* (1889), 46
*Gammarus*
    rate genes and eye colour variants, 59, 60
Gardner, Stanley
    on F. M. Balfour, 41
Garstang, W., 56
    larval adaptations, 44
    paedomorphosis, 61
gastraea theory, xiii, 5, 52, 84–5
Gegenbaur, Carl, xii, 14, 29, 35, 181, 184
    cell theory, 212
    hydroid evolution, 17
Geinitz, Bruno, 225
    heteroplastic grafting of dorsal blastopore
        lip, 199
gemmules, 4, 284, 285
*Generelle Morphologie der Organismen*
        (1866), E. H. Haeckel, xii, 27, 29, 35–6
genes
    actin, 403, 404
    chromosome puffing patterns, 375
    and chromosomes, 121, 128, 129ff
    and competence, 319–20
    and differentiation, 368
    and embryogenesis, 367ff, 398, 400–1,
        402ff
    genetics–embryology split, 114, 128ff
    'gens', 127, 130
    histone, 403
    and lethal factors, 374–5
    and organisers, 319–20
    phenocopies, 372–4
    rate genes, 36, 59, 60, 61–2
    reaction chains and pigment development,
        375–7
    selector genes, 381ff, 387, 388
    sequence of gene effects, 371–5
    single trait expression, 375–7
    spatial pattern generation, 380ff, 386ff
    strategy of, 324–6
    synergetics, 385–9
    temperature-sensitive mutants, 372
    *see also* mutation theory
genetics
    classical school period, 121
    and embryology, 318–19
    and epigenetics, preformation and
        epigenesis, 101–2
    establishment as science, 132ff
    function of fertilisation, 398
    and recapitulation, 56
    rise of, 215–17
genetics–embryology split, 114–15, 131ff,
        364ff
    inevitability, 141–2

reunification attempts, 139–40, 143–4
    sociological factors, 137–9
genotype–phenotype distinction, 123, 124,
        126–8
Gerard, P., 246
germ bands, 88–9
germ cell origin, 15, 16–17, 21–4, 28
germ-layers, 10, 14
    biogenetic law, 27
    cell lineage and phylogeny, 89
    specificity of, 21–4, 27, 28
germ-plasm, 5–6, 9, 77
    continuity of, 9–10, 13
    *see also* hydroids and germ-plasm
        continuity
Germany
    chemical nature of induction, 267, 269
    Darwin's theory, 5
    embryologists and their American links, 180
    experimental approach, 179ff
    microscopical art, 69
    19th century embryology, 35–6
    organ-forming germ regions, 85, 86
    political and cultural effects, 205ff, 221–2
    predetermination–epigenesis debates,
        74–80
    humanities–sciences relationships,
        210–11
    Mendelism, 215–16
    mosaic or regulative development, 80–4
Goldschmidt, Richard B., xvi, 181, 205,
        216, 319
    genetics–embryology split, 139, 366
    nurse cells and polarity, 368, 385
    phenocopies, 372
    rate genes, 36, 60
    substrate-induced differential gene
        activation, 368, 369
    *Die physiologische Theorie der Vererbung*
        (1927), 368, 377
Golgi, Camillo, 151
    silver staining, xiii, 152
Goodrich, Edwin Stephen
    ancestral stages in ontogeny, 44
    biogenetic law, 54ff
    at Oxford, 51, 53ff
    recapitulation, 51, 54ff, 57
    visits to Stazione Zoologica, 37
Götte, A., 7
Gould, Stephen Jay
    on change to experimental approach, 52
    on F. M. Balfour, 42
    recapitulation, 58
    replacement of recapitulation by
        Mendelism, 56
    *Ontogeny and Phylogeny* (1977), 11–12
gradients
    emergence of concept, 349

gradients (*cont.*)
  and fields, 252–3, 352ff
  and pattern formation, 383–5
  and polarity, 349ff
  positional information, 359
  threshold concept, 252, 354
  *see also* Child, gradient theory;
    cytoplasmic gradients
Gray, James, 313, 319, 322
  on Creswell Shearer, 49
  organicism, 70
*Growth and Form, On* (1917), D'Arcy
    Thompson, xvi, 70, 276, 280, 281,
    283, 286, 293
gynandromorphs, 375–6

Haeckel, Ernst Heinrich, 14, 35, 181, 210
  biogenetic law, 11, 12, 85
  cell souls, 218
  Darwinism, 5
  gastraea theory, xiii, 84–5
  germ-layers, 27
  heredity and adaptation, 28–9
  recapitulation, 11, 12
  *Generelle Morphologie der Organismen*
    (1866), xii, 27, 29, 35–6
Haecker, Valentin
  timing in ontogenesis and phenogenetics,
    371
Haldane, J. B. S., 62
  *Animal Biology* (1927), 58–9
Haller, Albrecht, 73
Hamann, Otto, hydroid evolution, 18
Hamburger, Viktor, 134, 206, 209, 225, 228,
    229
Haraway, Donna J., 317
  *Crystals, Fabrics and Fields* (1976), 281
Hardy, Alister, 54, 59
  paedomorphosis, 61
Hardy, William Bate
  protoplasm structure, 280
Harrison, Ross, xv, 59, 181
  allometry, 63
  *Amblystoma*, 227–8
  approach to experimentation, 168–73
  early career, 147, 149, 153–4
  genetics–embryology split, 133
  innervation and muscle development,
    155–6
  later work of, 163
  lens induction, 227–8
  on non-acceptance of experimental
    methods, 165–6, 171–2
  and the organiser, 226, 227
  photograph, 223
  physicochemical approach, 170–1
  polarity and symmetry, 148, 163, 171
  on validity of experimental approach, 172–3

Harrison, Ross, nerve fibre development
  controversies, 152, 153
  fibre outgrowth *v.* protoplasmic bridges,
    156–9
  heteroplastic grafting, 154–5
  nerve outgrowth mechanism, 159–62
  peripheral, in vivo, 156–7
  stereotropism and fibre pathways, 162
  tissue culture, 150, 159–62, 172
Harvey, William, epigenesis, 3
Heape, Walter, 42
  physiology of reproduction, 46
  *Textbook of Embryology*, 46
Held, Hans
  nerve fibre development, 151, 152, 153
*Helix*, 297
Henking, H., xiv
  insect chromosomes, 99
Hensen, V., nerve fibre origin, 150, 151
Herbst, C., xiv, xv, 181, 189, 205
  lens development, 353
heredity
  and adaptation, 28–9
  blending theory, 39
  chromosome theory, 398
  germ structure, 285, 286
  inheritance of acquired characters, 39–40
  meanings of
    in 19th century, 5–6, 21, 27–8
    in early 20th century, 115, 124–7
  nuclear *v.* cytoplasmic control, 125,
    131–2, 134, 369–70
  recapitulation, 39–40
  submicroscopic entities, 284ff
  and variation, 29–30
heredity and development, 102–3
  and evolution, 118–20
  inseparable aspects, 363–4
  sex determination, 98, 100–1
heredity research
  agricultural breeding programmes, 135–7
  financial support, 136–7
  genetics–embryology split and
    sociological factors, 137–9
Hertwig, Oscar, xiii, xiv, 9, 14, 27, 181, 210,
    398
  on A. Weismann's predetermination
    theory, 77, 78, 79
  blastopore formative powers, 215
  cell theory, 213
  epigenesis and internal and external
    factors, 74, 78–9
  germ-layer theory, 21
  hydroid evolution, 18
  sea urchin eggs, 82
  on W. His' theories, 86
  *The Cell* (1895), 6
  *Preformation or Epigenesis* (1894), 77

Hertwig, Richard, 9, 14, 27, 181, 210
  germ-layer theory, 21
  hydroid evolution, 18
  nucleus and chromosomes, 212
  on T. Boveri, 204–5
heterochrony and evolution, 62
heteroplastic grafting, 153–5
  lens induction, 190
  nerve fibre development, 154–5
  organiser discovery, 262, 263
His, Wilhelm, xii, xiii, xv, 7, 35, 36, 75, 181
  biogenetic law rejection, 85
  blastopore formative powers, 215
  gastraea theory rejection, 84, 85
  nerve fibre origin, 150–1
  organ-forming germ regions of egg, 85, 86
  transmitted movement theory, 85
Höber, R., colloids, 293
Hofmeister, Wilhelm, xv, 14
holism, 289, 310, 312, 315
  emergent properties, 424, 426–7
  gene action, 319
  and physical determinationism, 424ff
  reductionism, 404
  value of, 430–1
Holtfreter, Johannes, xvii, 200, 228, 229
  'Die Institutskutsche', 224–5
  explantation technique, 251
  organiser's inducing factor, 201, 230, 253,
    255
  political and cultural effects, 206
  target cells of primary induction, 268
homeotic mutations, 381
homologies, 10, 38–9, 48
  and cell lineages, 89
  germ bands, 88–9
  germinal layers, 51–2
Hopkins, Frederick Gowland, xv, xvi, 249,
    252, 284, 293–4
  organicism, 70
Hörstadius, S., 348
  animal–vegetal gradients, 352
Hunt, John Wesley, 116
Husserl, 206, 417
Hutchinson, Evelyn, 49
Huxley, Julian S.
  allometry, xvi, 53, 63
  approaches to biology, 70
  evolutionary process, 63
  experimental biology, 58–9
  genetics and evolutionary theory, 70
  gradient fields, 353–4
  heterochrony, 62
  metamorphosis induction, 36
  on the organiser, 226
  at Oxford, 58–9
  rate genes, 36, 60
  recapitulation, 58–9

species and speciation, 63
visit to German embryologists, 36
visits to Stazione Zoologica, 37
*Animal Biology* (1927), 58
*Elements of Experimental Embryology*
    (1942), 60
*Evolution: The Modern Synthesis* (1942),
    63
*Problems of Relative Growth* (1932), xvii,
    53, 56, 59, 70
*The Science of Life* (1927–8), 59
Huxley, Thomas Henry, xii, 5, 7, 29
  homologies and comparative anatomy, 35,
    38–9
  hydroid evolution, 17
Hyatt, Alpheus, 11
  biogenetic law, 12
*Hydra*, 155, 357
  regeneration, 332–3, 339, 342
hydroids and germ-plasm continuity, 12ff
  dislocation of germ site, 15–17, 20, 23, 24, 25
  germ cell maturation sites, 15, 16
  germ cell migration, 15, 20–1, 24, 25
  germ cell origin sites, 15, 16–17
  and phylogeny, 17
  sporosac evolution, 17–21
*Hydromedusen* (1883), A. Weismann, 13,
    23, 25, 26, 28
*Hydrozoa*, 10
hypotheses
  'upward' and 'downward' linkage and
    progress, 414ff

immunology
  genetics–embryology reunification, 144
*Individuality in Organisms* (1915),
    C. M. Child, 352
induction
  autoneuralisation, 230, 255, 268
  competitive variability, 319–20
  embryonic field concept, 352–4
  organiser discovery, 262–3
  and pattern formation, 231
  slow development in field of, 261, 269–70
  target cells, 267–8, 271
  *see also* lens induction
inductors
  differential action theories, 230
  masked, 230, 255
  nature of, 266–7, 269–72
  regional specificity of, 263–6
  *see also* organiser's inducing factor
*Integrative Action of the Nervous System,
    The* (1906), C. S. Sherringham, 148
*Intracellular Pangenesis* (1889), H. de
    Vries, 286
*Introduction to Theoretical and Applied
    Colloid Science*, W. Ostwald, 281

Jenkinson, J. Wilfred, 36, 37, 58
  homologies of germinal layers, 51–2
  *Experimental Embryology* (1909), xv, 51,
    52
Jennings, Herbert Spencer, 134
  gene theory of inheritance, 134
Johannsen, Wilhelm
  genotype–phenotype distinction, xv, 123,
    124, 126ff
  'gens', 127, 130
  heredity, meaning of in early 20th century,
    126–7
Jones, Donald, 137
*Journal of Experimental Zoology*
  formation, xv, 168
Just, Ernst, 192, 219

Kaiser–Wilhelm Institut für Biologie, xvi,
    216–17
Keilin, D., 246, 252
Key, Francis Scott, 116
Kielmeyer, K. F., 3–4
Kleinenberg, Nicholaus, 7, 14, 15, 27
  recapitulation, 57, 58
Kölliker, R. A. von, xi, xii, xiv, 181
Kowalevsky, Alexander, 7, 9

Lakatos, Imre, 137–8
Lamarckism, 6, 11, 203, 217–19
*Lampeter*, 297
Lang, Arnold, 14
language as theoretical tool, 409, 410f
  appropriateness to chosen area, 413–14
  categories of terms, 411–13
  network model, 420–1
Lankester, Edwin Ray, 5, 9, 27, 29, 50
  biogenetic law, 12
  paedomorphosis, 61
  recapitulation, 35, 40–1
  visit to German embryologists, 36
  visits to A. Dohrn, 37
larval embryonic development, 43–4, 57
Lehmann, F. E., 199
lens induction, 188–91, 192, 227–8, 353
  double assurance, 191, 198
  double origin, 189
  Wolffian regeneration, 189, 333, 339
*Leptomedusae*, 14
Leuckart, Rudolf, xii, 181
  insect parthogenesis, 13
Levene, P. A., 247–8
Lewis, Margaret, xv, 165, 166, 171
Lewis, W. H., 190, 353
Lewis, Warren, 165, 166, 171
Lillie, Frank R., xvi, 98
  gene theory and heredity, 133
  and *Journal of Experimental Zoology*, 168
  sex determination, 100

Lillie, R. S.
  and the organiser, 226, 227
  limb regeneration, 340, 341
  limb transplantation
  nerve fibre development, 157–8
Liu, C. K., 26, 27
  continuity of germ-plasm, 13
Loeb, Jacques, xiv, xv, 92, 98, 111, 117, 181
  atomistic gene theory, 134
  and *Journal of Experimental Zoology*, 168
  mechanistic conception of life, xvi, 122–3
  psychic factors, 218
  sex determination, 100
Loeb, Leo, 159, 164
  stereotropism, 162
Lorenz, Konrad, 207
*Lumbricus*, 297
  germ bands, 89
*Lymnea peregra*, 367

MacBride, E. W., 37, 47, 61
McClintock, Barbara, 121, 137
McClung, Clarence, xv
  chromosomes and sex determination, 99
Malebranche, Nicholas, 3
Mangelsdorf, Paul, 137
Mangold, Otto, 190, 193, 195, 206, 207, 225,
    228
  Einsteckung technique, 251, 264
  and the organiser, xvi, 253
  regional specificity of inductor, 263–4
Marine Biological Laboratory (MBL),
    Woods Hole, xiv, 87–93, 109–11, 133
  E. B. Wilson and T. H. Morgan at, 129
  experimental approach, 147
Marinesco, G., 151, 153
Marshall, A. Milnes
  ancestral stages in ontogeny, 44
  recapitulation, 46–7, 57
  *Vertebrate Embryology* (1893), 46
Martin, H. Newall, 109
Marx, 200
  archenteric induction, 199
*Materials for the Study of Variation* (1894),
    W. Bateson, xv, 45
maternal effect, 377
Maupertius, P. L. M. de, 3
Maxwell, Clark, 277
  germ structure and heredity, 285–6
  minuteness of submicroscopic structures,
    284ff
  size limits, lower, of organisms, 277
MBL, *see* Marine Biological Laboratory,
    Woods Hole
*Mechanism of Mendelian Heredity, The*
    (1915), T. H. Morgan, 130, 131
mechanistic materialism, 121–3
Meckel, J. F., 3–4

Medawar, J. S., 101–2
Medawar, P. B., 48, 101–2
 on experimental biology, 59
Mencl, E. lens development, 19
Mendelism, 45, 128, 215–16, 398
 and chromosome theory, 113–14, 120ff,
  215–16
 recapitulation, 56
metaplasia and regeneration, 338, 339ff
Metchnikoff, Iliya, 7
Meyer, E., 37
*Micrococcus*, 286
*Micromonas progrediens*, 286
Miescher, F. J.
 Nuclein, 280
molecular biology, 230
 dimensions of molecules, 284
 dispersed systems, 277, 278
 dynamic forces, 288
 early instrumentation, 277, 279, 283–4,
  303, 316
 emergence of, 247, 316
 from T. Boveri's insights, 401ff
 fundamental dogma, 250
 growth of, 256
 reductionism, 421
 reductionist *v.* holistic approach, 404–5
 structural continuity, 281, 282, 283
 X-ray analysis, xvii, 148, 279, 292, 297
molecular genetics, 230
 genetics–embryology reunification, 144
Morgan, John Hunt, 116
Morgan, Thomas Hunt, xv, xvi, xvii, 5, 94,
  98, 181
 *Ctenophores*, 111
 cytoplasm *v.* chromosomes in
  embryogenesis, 400
 *Drosophila*, 100, 113, 119, 120, 126
 epigenesis and predetermination, 74
 on experimental approach, 118, 147
 genes
  and chromosomes, 121, 128, 129ff
  and differentiation, 367–8
  functional and biochemical nature, 131
 genetics–embryology split, 114–15, 131ff,
  139ff, 365–6
 genotype–phenotype distinction, 126–8
 gradients and polarity, 350–1
 gynandromorphs, 375
 heredity, 113ff
  development and evolution, 118–20
  genotype–phenotype distinction, 127–8
  nucleus, not cytoplasm, seat of, 131–2
 mechanistic materialism, 118, 121–3
 Mendelism and chromosome theories,
  113–14, 120ff
 and the organiser, 226, 228
 *Planaria*, 350

pycnogonid derivation, 116–17
 regeneration, 118, 331, 332, 350–1
 sociological factors forwarding research,
  138–9
 *Tubularia*, 350–1
 *Evolution and Adaptation* (1903), 118
 *The Mechanism of Mendelian Heredity*
  (1915), 130, 131
 *The Physical Basis of Heredity*, (1919),
  123, 131
 *Regeneration* (1901), 335
 *The Theory of the Gene* (1926), xvi, 132
Morgan, T. H., career
 Division of Biology establishment at
  CalTech, 123
 early background, 116
 at Johns Hopkins University, 116–17
 and *Journal of Experimental Zoology*, 168,
  169
 at MBL, 129
 at Stazione Zoologica, 117–18, 122–3
Mori, T., 247
mosaic development, 74–5, 80–2
 *v.* regulative totipotency, 83, 347
 *see also* preformation–epigenesis debates
Mosely, Henry Nottidge, 50
Müller, Fritz, 29, 181
 *Für Darwin* (1863), 35
Muller, H. J.
 genes and chromosomes, 121, 142
Müller, Johannes, P., 7, 211
 mutation theory, 120

Nägeli, Carl, xi, 181, 398
 hereditary mechanisms, 6, 29
Nanney, Donald, 138
National Animal Breeding and Genetics
  Research Organisation, 322
*Natural Inheritance* (1889), F. Galton, 46
natural selection, 38, 43, 57
 formalised theory, 410–11
Needham, D. M.
 organiser's inducing factor, 253, 255
 phosphorus metabolism, 248
Needham, Joseph, 49, 226, 249
 cellular biochemistry–morphology fusion,
  299–300
 masked inductors, 230
 organicism, 70
 organiser's inducing factors, 201, 253,
  254, 255, 267
 organisers and recapitulation, 47
 phosphorus metabolism, 248
 recapitulation, 47, 57, 58
 *Biochemistry and Morphogenesis* (1942),
  47, 251, 253, 256
 *Chemical Embryology* (1930), xvii, 245,
  315

Needham, Joseph (*cont.*)
  *Order and Life* (1936), 276, 301
neoblasts and regeneration, 339ff
*Nephrops*, 297
*Nereis*, 93
nerve fibre development, 172
  controversies, 151–3
  Golgi's silver staining, 152–3
  heteroplastic grafting, 154–5
  limb transplantation, 157–8
  nerve fibre outgrowth *v.* protoplasmic
    bridges, 156–9
  nerve outgrowth mechanisms, 159–62
  origination theories, 150–1
  peripheral, in vivo, 156–7
  stereotropism and fibre pathways, 162
*Neurospora*, xvii, 376
Newton, A. E., 41
*Non-Specificity of the Germ Layers, The*
  (1940), J. M. Oppenheimer, 27
nuclear *v.* cytoplasmic control
  development, 90, 95–6, 364–5, 367, 398,
    400
  heredity, 125, 131–2, 134, 369–70
Nussbaum, Moritz, 147, 153–4, 155, 158,
    181

*Octopus*, 297
*Oenothera*, mutation theory, 120
Oken, Lorenz, 3–4
Olby, Robert, 13
  on D'Arcy Thompson, 71
*Oncopeltus*, 357
*Ontogeny and Phylogeny* (1977), S. J.
    Gould, 11–12
Oppenheimer, J. M., 212, 221
  biogenetic law and germ layer doctrine, 27
  *The Non-Specificity of the Germ Layers*
    (1940), 27
*Order and Life* (1936), J. Needham, 276, 301
organicism, 69–70
organiser, 315–16
  discovery of, 193–5, 262–3
  field properties, 196, 198
  and genes, 319–20
  gradient theories, 353
  interpretations of mechanisms of effect,
    196ff
  reactions to, by H. Spemann's
    contemporaries, 223, 226ff
  and recapitulation, 47
  role of responding cells, 199, 200
  World War II and after, 228–31
  *see also* induction
organiser, conceptual origins, 195ff
  cell theory, 212–15
  genetics rise, 215–17
  leadership analogy, 211, 219–22

psychic analogies, 217–19
organiser's inducing factors (chemistry of
    organiser), 251ff
  carbohydrate metabolism, 254
  gradient theories, 252–3
  heterogeneous nature, 200–1, 253–6,
    315–16
  respiratory metabolism, 253–4
  and unmasking action, 255
*Organisms and Genes* (1940), C. H.
    Waddington, 319–20
Osterhout, W. J. V., 123
Ostwald, Wolfgang
  dispersed systems, 278
  *The World of the Neglected Dimensions*,
    291–2
  *Introduction to Theoretical and Applied
    Colloid Science*, 281
Owen, Richard
  homologies, 38, 39
  parthenogenesis, xii, 13

paedomorphosis, 11
  rate genes, 61–2
*Palinurus*, 297
Pallas, Peter S., regeneration, 333
Pander, Christian H., 7, 211
pangenesis, 4, 75–6, 285
  pangenes, 286
*Paracentrotus*, 187
parthenogenesis, 3, 13
Pasteels, J., 252, 354
*Patella*, 111
pattern formation, 356, 357ff
  clonal restriction, 382, 383
  French Flag problem, 357–8
  and gradients, 383–5
  mathematical models, 386ff
  selector genes, 381ff, 387, 388
  spatial, and genes, 380–5
Pauly, August, 181, 184, 205, 210
  psycho-Lamarckism, 191, 218–19
Pearson, K., 46
Pederson, K. O.
  *The Ultracentrifuge* (1940), 295, 296
*Peripatus*, 42
Petersen, Hans, 194–5
  blastopore influences, 194–5
Pflüger, E., xii, xiv, 181
phenocopies, 372–4
phenogenetics, 371
phenotype, 127
  *see also* genotype–phenotype distinction
*Physical Basis of Heredity, The* (1919),
    T. H. Morgan, 123, 131
physical determinationsm, 408
  Beth's theorem, 408, 429–30
  bridge laws, 420–1

determination of 'higher level' concepts by
'low level' facts, 418–20, 423–4
holism and emergent properties, 424ff
physical exhaustion, 423
and reductionism, 418–19, 422–4
physics
laws of, and laws of embryology,
418–19
and 'higher level' sciences, 419–20
pigment synthesis
gene action, 375–7
maternal effect, 377
Piper, John
epigenetic landscape, 320–1
planaria regeneration, 333, 334, 339, 342–3,
350, 351
*Planaria maculata*, 350
*Planaria simplicissima*, 350
*Podocoryne carnea*, 339
polarity
emergence of concept, 349
and gradients, 349ff
oocyte, and nurse cells, 368, 385
regeneration, 342
and symmetry, 148, 163, 171, 188
*see also* symmetry
positional information, 357ff
French Flag problem, 348, 357–8
and gradients, 344, 359
origin of concept, 347, 356–7
philosophy of embryology, 416–17
and prepatterns, 379
Poulson, Donald F., 374–5
Poulton, E. B., 39
preformation, 3
preformation–epigenesis debates, 73ff
evolution theory, 75ff
genetics and epigenetics, 101–2
Germany, 74–80
organ-forming germ regions, 85, 86
pre-embryonic organisation, 87ff
self-differentiating mosaic or regulative
totipotency, 80–4
sex determination, 98–9
theories and types of thinker, 73–4
transmitted movement theory, 85–6
USA, 90–2
*Preformation or Epigenesis* (1894),
O. Hertwig, xv, 77
primordial germ cells, 23, 24, 25
*Problems of Relative Growth* (1923), J. S.
Huxley, xvii, 53, 56, 59, 70
Pröscholt, Hilde,
organiser discovery, 193, 195, 225, 262,
263
protein synthesis
at fertilisation, 251
nucleic acids, 250

proteins
molecular morphology, 297
nature of, 295–6
Purkinje, J. E., xi, 181
pycnogonids, 116–17

Rabl, C., xiv, 181
*Rana esculenta*, lens formation, 191
*Rana fusca*, lens formation, 188
*Rana palustris*, nerve fibre growth, 161
*Rana pipiens*, nerve fibre growth, 161
Rathke, H., xi, 7
recapitulation, 4, 11, 12, 46–7, 57, 58–9
ancestral and non-ancestral developmental
stages, 43–4
and evolutionary regression, 20
and genetics, 56
and heredity, 39–40
introduction into embryology, 40–2
Oxford in early 20th century, 54ff
replacement by genetics, 56
*see also* biogenetic law
reductionism, 407–8, 421–2
application to embryology, 431–2
Beth's theorem, 408, 429–30
cell properties, 279–80
emergent properties, 427–8
holism, 404
and physical determinationism, 418–19
set theory, 422
regeneration, 118
axial gradient determination, 341ff
and determination, 338ff
and embryology, 336–7
and evolution, 118
field gradient systems, 342–4, 353–4
hydra, 332–3, 339, 342
meanings of, 331–2
and metaplasia, 338, 339ff
mirror-image, 342
and modulation, 338, 340
and neoblasts, 339ff
polarity, 342
and gradients, 350–1
widespread occurrence, 332–3
*see also* Wolffian regeneration
*Regeneration* (1901), T. H. Morgan,
335
regulative development, 82–4, 186,
347
Reichert, Karl B., xi, xii, 7, 181
Remak, R., xii, 181
reproduction,
interest in 19th century, 13–14
reptiles, 332
respiration
dorsal blastopore lip, 253–4
site in eggs, 250–1

respiratory pigments
  chemical taxonomy, 295
  molecular dimensions, 278
*Rhodnius*, 356
RNA (ribonucleic acid)
  detection of, 249–50
  Unna–Brachet test, 250
aRNA (zymonucleic acid), 247
mRNA, 251
Robinson, Gloria, 29
Roe, Shirley, 73
Rolleston, George, 40, 50
Romanes, G. J., homologies, 39
Röntgen, 184
Rose, S. M., gradients and patterns, 356
Rotmann, E., 201
Roux, Wilhelm, xiii, xiv, xv, 9, 35, 79, 181
  and H. Driesch, 82–4, 117–18, 186
  self-differentiating mosaic development,
    74–5, 8off, 347
Runnström, J., gradients and polarity, 352
Russell, E. S.
  biogenetic law, 5, 10–11, 27
  organicism, 69–70
  *Form and Function* (1916), 5, 69–70, 179

Sachs, J. von, 181, 184, 218
Sapp, Jan
  competition between nuclear and
    cytoplasmic genetics, 138
  heredity, meanings of, in early 20th
    century, 124, 126, 134
*Sarsia Tubulosa*, 18
Schotte, Oscar, 201, 228, 229
  embryologist and geneticist view of cell,
    364
Schultze, M., xii, 181
Schultze, O., xv, 181
Schwann, Theodor, xi, 181
  cell theory, 4, 280
  nerve fibre origin, 150, 151
Schwann cells, 156, 157, 172
*Science of Life, The* (1927–8), J. S. Huxley,
    59
sea spiders, 116–17
sea urchins, 82–3, 187, 248ff, 347ff, 357,
    400ff
Sedgwick, Adam, 41, 42, 44–5, 46
  ancestral stages in ontogeny, 44
  at Cambridge, 44–5
  on von Baer's Laws, 44–5
Seibold, Theodore von
  insect parthenogenesis, 13
Semon, Richard, 181, 217, 218–19
Semper, Karl, 7
sex determination,
  chromosomes, 99–100, 129
  heredity and development, 98, 100–1

predeterminism and epigenesis, 98–100
Sexton, E. W.
  *Gammarus* eye colour variants, 60
sexual organs
  primary and secondary, correlation
    between, 53
sexual reproduction, plants, 3, 14
Shearer, Cresswell, 49
Sherrington, C. S.
  *The Integrative Action of the Nervous
    System* (1906), 148
Shipley, Arthur E., 39
  on F. M. Balfour, 41, 42
*Siphonophora*, 10, 14
*situs inversus*, 192
Smith, Geoffrey, 37, 51
  internal generation of form, 53
  recapitulation rejection, 51–2
  visit to R. Hertwig, 36
  visits to Stazione Zoologica, 37
Smyth, H. F.
  on tissue culture applications, 165
snail regeneration, 333
*Some Difficulties in Darwinism* (1984),
    D'Arcy Thompson, 289
Sonneborn, Tracy, 138
  gene theory of inheritance, 134
Spallanzani, Lazaro, 335
  criticism of, 337
  regeneration, 333, 334
Spemann, Hans, xv, xvii, 9, 59, 181
  chemical nature of induction, 266
  correlative phenomena, 189
  dorsal blastopore lip, 188, 220
  double assurance, 191, 198
  double-headed embryos, 186, 187, 189
  field concept, 353
  genetic role, reluctance to stress, 366–7
  genetics–embryology split, 139–40
  gradient theory, 354
  and H. Driesch, 210
  hair loop experiments on blastulae, 186,
    187–8, 210
  lens induction, 188–9, 198, 353
  methodology, 209ff
  polarity and bilateral symmetry, 188
  psychic factors, 202, 203, 217ff
  *Strongylus paradoxus*, 184, 214
  and T. Boveri, 184–5, 186, 205, 213–14,
    217
  and W. Roux, 210
  *Embryonic Development and Induction*,
    (1938), 183, 195ff, 251
Spemann, Hans, career
  and cultural context, 202ff
  cultural value of biology, 204ff, 207, 210
  diversification of interests, 192–3
  early, 184–6, 202

Kaiser-Wilhelm Institut für Biologie, 216–17
    nationalistic views, 205–6
    neglect of, by historians, 183, 203, 213–20
    photograph, 223
    political background to science, 205ff, 207–9
    position of eminence, 198
    retirement, 201–2
    scientific output related to cultural context, 222–3
Spemann, Hans, and the organiser, 193ff, 251, 353
    cell theory, 212–15
    discovery of organiser, xvi, 193–5, 262–3
    field properties, 196, 198
    leadership analogy, 211, 220–1
    mechanisms of effect of, 196ff
    nature of stimulus, 200–1, 253
    psychic analogies, 217ff
    role of responding cells, 199, 200
Spencer, Herbert, 14
Spiegelman, S.
    principle of limited realisation, 355–6
spontaneous generation, 3, 5
*Spontaneous Generation Controversy* (1974), J. Farley, 5
sporasac evolution, 17–21
    endodermal lamella, 20
    entocodon development, 18, 19, 20
Stadler, J., 121, 137
Standinger, Hermann
    macromolecules and colloid aggregates, 294
Station for the Experimental Study of Evolution, Cold Spring Harbor, 136–7
Stazione Zoologica, Naples, xiii, 37, 117–18, 122
Steenstrup, Johannes Japetus
    *On the Alternation of Generations* (1845), xii, 13
stereotropism, and nerve fibre pathways, 162
Stern, Curt, prepattern concept, 377–9
Stevens, Nettie, 181
    sex determination, 100, 129, 142
Strangeways Research Laboratory, 49, 321
    tissue culture, 312–13
Strasburger, E. H., 398
*Strategy of the Genes, The* (1957), C. H. Waddington, 322, 324–6
*Strongylus paradoxus*, 184, 214
Sturtevant, Alfred H., xvi, 375–6
    genes and chromosomes, 121, 142
Svedburg, T., xvi
    macromolecules and colloid aggregates, 294ff
    respiratory pigment chemical taxonomy, 295

*The Ultracentrifuge* (1940), 295, 296
symmetry, 88, 89
    and polarity, 148, 163, 171, 188
    *see also* polarity
synergetics and gene action, 385–9
synthetic polymers, 294

technology transfer in biology, 277, 279
*Textbook of Embryology*, W. Heape, 46
Theoretical Biology Club, 228, 314
    gene structure, 317
    organicism, 70
    structural continuity, 281
*Theory of the Gene, The* (1926), T. H. Morgan, xvi, 132
thigmotropism, *see*, stereotropism
Thompson, D'Arcy, 37, 42, 44, 47–8, 319
    on C. Maxwell, 286, 287
    colloidal forces, 293
    on F. M. Balfour, 41
    function presupposition of structure, 275–6
    holistic physical analogies, 289
    impact of, on biology, 70–1
    mathematical description of form, 290
    opposition to adaptation and biogenetic law, 290
    structural and dynamic theories, 286ff
    theory of transformations, 48
    on W. T. Astbury's protein morphology, 298
    *On Growth and Form* (1917), xvi, 70, 276, 280, 281, 283, 286, 293
    *Some Difficulties in Darwinism* (1984), 289
Thompson, John V., 39
thymonucleic acid (aDNA), 247
tissue culture, 150
    embryological studies, 165
    increase in use of, 163–6
    nerve fibre development, 159–62
    pathological problems, 165
    validity as experimental method, 172
*Tissue Culture* (1925), A. Fischer, 163f
Toivonen, S., inductive signals, 269
*Towards a Theoretical Biology*, C. H. Waddington, 323
*Trachylina*, 15
*Treatise on Comparative Embryology, A* (1880–1), F. M. Balfour, xiii, 42
Trembley, Abraham, 335
    regeneration in hydra, 332–3, 339
*Triturus*, 262, 263
*Tubularia*, 18, 342
    regeneration, 350–1
Turing, Alan Mathison
    pattern formation, 359, 386

Uexkull, J. von, xvi, 207, 216, 219
*Ultracentrifuge, The* (1940), T. Svedberg
    and K. O. Pederson, 295, 296
United States of America
    embryology-genetics split, 113ff
    German influences, 109, 110–11
    nerve growth analysis and tissue culture
        origins, 149ff
    and the organiser, 226ff
    origins of embryological tradition, 109–11
    phenogenetics, 371
    sex determination and roles of heredity
        and development, 98, 100–1
    *see also* Conklin, E. G.; Harrison, R. G.;
        Morgan; T. H.; Whitman, C. O.;
        Wilson, E. B.

Van Beneden, Edouard, 14
    hydroid evolution, 17
    pronuclear fusion and genome formation,
        397, 399
*Vertebrate Embryology* (1893), A. M.
    Marshall, 46
Virchow, Rudolf, xii, 181
    cell theory, 7, 27, 203
vitalism, 348, 365, 385, 386
Vogt, W.
    cellular basis of differentiation, 199–200
von Baer, Karl Ernst, xi, 4, 7, 35
von Baer's Laws, 38, 42, 45
von Koch, Gottlieb, hydroid evolution,
    18
von Neuman, set theory, 422

Waddington, Conrad, 49
    biochemistry of induction, 315–16
    canalisation, 321, 324
    cybernetics, 310, 324–5
    *Drosophila* wing development, 319
    evolution, 310, 321, 325
    gene structure, 316–17
    genetics–embryology split, 139
    genetics and embryology, 318–19
    holism in biology, 310, 312, 315, 319
    landscape metaphor, 318, 320–1, 371
    and the organiser, 201, 226, 253, 255
    recapitulation, 47, 57, 58
    terminology and concepts, 314
    theoretical studies, 310
    *Organisms and Genes* (1940), 319–20
    *The Strategy of the Genes* (1957), 322,
        324–6
    *Towards a Theoretical Biology*, 323
Waddington, Conrad, career, 311ff
    Edinburgh University, 322–3
    and G. Bateson, 311–12
    in Germany, 313
    and J. Woodger, 313–14

National Animal Breeding and Genetics
    Research Organisation, 322–3
    1950s scientific work, 323ff
    Operational Research, World War II, 322
    at Strangeways Research Laboratory,
        312–13, 321
    in USA, 317–18, 319
    and W. Bateson, 312
Walsech, 228, 229
Waldeyer, Wilhelm, 14
Weaver, Warren
    molecular biology instrumentation, 277,
        279, 283, 303, 316
Weismann, August, xiv, 35, 181
    biogenetic law, 12, 26–7
    and caterpiller markings, 5, 13, 26
    biophores, 76, 284, 285
    *Daphnia*, 10, 13
    determinants, 76, 77, 79
    germ line separation from soma, 6, 12, 36
    germ-plasm, 5–6
        autonomy, 77
        continuity, 9–10
        reduction division, 9
        *see also* Weismann, A., hydromedusae
            and germ-plasm theory
    heredity, 21, 27–8
    hydromedusae
        collection and examination, 14–15
        recapitulation and evolutionary
            regression, 20
        sporosac evolution, 18ff
    neo-Darwinism, 9, 12–13, 20, 30
    on pangenesis, 75–6
    photograph, 8
    revised predeterminism, 74, 75–7, 79
    *Das Keimplasma* (1892), xiv, 9, 75, 185,
        363
    *Die Entstehung der Sexualzellen bei den
        Hydromedusen* (1883), 13, 14, 23, 25,
        26, 28
Weismann, August, hydromedusae and
    germ-plasm theory, 10, 12ff
    dislocation of germ-cell maturation site, 20
    dislocation of germ site, 15–17, 20, 23, 24,
        25
    germ-layer specificity and germ-cell
        origin, 24, 28
    primordial germ-cell migration, 23, 24, 25
Weiss, Paul
    field theory, 143, 342, 343
Weldon, W. F. R., 42, 44, 45–6, 50–1
Wheeler, William M., 87, 93, 181
    preformation and epigenesis theories and
        types of thinker, 73–4, 102–3
Whitehead, Alfred North, xvi, 312, 314–15,
    325–6
Whitman, Charles Otis, xiii, xiv, 84, 110, 181

cell lineages, 87, 89, 92, 110
*Clepsine*, 87–8, 88–9, 110
on cooperation among biologists, 92
evolution, 88
germ bands and germ layers, 88–9
germ layers, cell lineages and phylogeny,
   89
in Germany, 87, 110
at MBL, 87, 109–11
nuclear and cytoplasmic roles, 90
pre-embryonic organisation, 87ff
preformation and epigenesis, 87, 90–2
and W. His, 86
Wilson, Edmund Beecher, 89, 111, 117, 181
chromosomes and development, 94, 95,
   401
on entelechy, 405
genetics-embryology split, 140
genetics of fertilisation, 398
and *Journal of Experimental Zoology*, 168,
   169
at MBL, 128–9
nuclear and cytoplasmic roles, 368–70
organ-forming stuffs, 94–5
preformation and epigenesis, 93–4, 95
sex determination, 100, 129, 142
on T. Boveri, 186–7
*The Cell in Development and Inheritance*
   (1896), 93, 95

*The Cell, Development and Heredity*
   (1925), xv, 4, 140, 144, 368–70
Wilson, James
heredity studies and profitable agriculture,
   136, 137
Wolff, Caspar Friedrich, xiv, 210
epigenesis-preformation debate, 73
Wolff, Gustav, 184
eye development, 188–9
Wolffian regeneration, 189, 333, 339
Woodger, J. H., 414
conceptual and logical analysis, 409–10
language as theoretical tool, 410
*Biological Principles* (1929),
   314–15
Wright, Sewall, xvii, 321

*Xenopus*, 265

yolk
gradient, 252, 354
mechanical effect, 43, 44

Zeleny, Charles, 137
Zsigmondy, R. A.
colloid ubiquity, 292–3
zymonucleic acid (aRNA), 247